Power Systems

Electrical power has been the technological foundation of industrial societies for many years. Although the systems designed to provide and apply electrical energy have reached a high degree of maturity, unforeseen problems are constantly encountered, necessitating the design of more efficient and reliable systems based on novel technologies. The book series Power Systems is aimed at providing detailed, accurate and sound technical information about these new developments in electrical power engineering. It includes topics on power generation, storage and transmission as well as electrical machines. The monographs and advanced textbooks in this series address researchers, lecturers, industrial engineers and senior students in electrical engineering.

Power Systems is indexed in Scopus

More information about this series at http://www.springer.com/series/4622

Thongchart Kerdphol · Fathin Saifur Rahman ·
Masayuki Watanabe · Yasunori Mitani

Virtual Inertia Synthesis and Control

Springer

Thongchart Kerdphol
Department of Electrical
and Electronic Engineering
Kyushu Institute of Technology
Kitakyushu, Japan

Fathin Saifur Rahman
School of Electrical Engineering
and Informatics
Institut Teknologi Bandung
Bandung, West Java, Indonesia

Masayuki Watanabe
Department of Electrical
and Electronic Engineering
Kyushu Institute of Technology
Kitakyushu, Japan

Yasunori Mitani
Department of Electrical
and Electronic Engineering
Kyushu Institute of Technology
Kitakyushu, Japan

ISSN 1612-1287 ISSN 1860-4676 (electronic)
Power Systems
ISBN 978-3-030-57960-9 ISBN 978-3-030-57961-6 (eBook)
https://doi.org/10.1007/978-3-030-57961-6

This Springer imprint is published by the registered company Springer Nature Switzerland AG
The registered company address is: Gewerbestrasse 11, 6330 Cham, Switzerland

This book is dedicated to the
Engineering Profession
in recognition of the pivotal role played by
electricity in the design and analysis of power
system

Foreword

Over a decade, the increasing integration of inverter/converter interfaced power sources (e.g., wind power, solar power, and battery technologies) has posed new challenges to power systems. An important one is the reduction of system inertia, which is the significant ability to maintain the frequency at the nominal value. Subsequently, power system operation, stability, and resiliency will be critically affected, causing frequency oscillations, instability, and cascading failures. The concept of a virtual synchronous machine (VISMA) or virtual synchronous generator (VSG) has opened up new possibilities to monitor and control such a challenge.

The research depicted in this book is an excellent beginning toward understanding inertia control approaches. Single and multiple virtual inertia control systems need to be simple to design and establish, stable, scalable, and robust with the capability to skillfully diminish during major contingencies. The author's extensive familiarity with this problem has made this book a rich source of information both to academia and industry. The book emphasizes real-time simulations, design, control, and optimization under several operating conditions. It clearly reveals the damping inefficiency caused by renewable sources and presents new lessons learned with solutions. Moreover, the authors have collaborated with researchers from all over the world. Thus, this book will have a broad appeal.

It gives me great pleasure to write the foreword message for this timely book, which I am confident that it will be of great value to engineers, operators, researchers, and university students in the field of power engineering.

Dr.-Ing. Dirk Turschner
Pioneer of VISMA, Director of Power Mechatronics
and Electric Drives Engineering
Clausthal University of Technology (TU Clausthal)
Clausthal-Zellerfeld, Germany

Preface

Virtual inertia equipped with proper control provides an appealing solution for overcoming challenges caused by the penetration of renewable energy sources (RESs) and distributed generators (DGs) into today and future power systems. Compared to a conventional power systems dominated by synchronous generators, the modern power systems with a high share of RESs/DGs will have less inertia and damping properties, which are the main elements of power system stability. The inertia and damping properties from synchronous generators play an important role in limiting the frequency deviation during a disturbance and slowing down the oscillations, and thus maintain system stability.

The recent trend in power system development is the widespread integration of RESs to deal with the energy crisis and the environmental issues caused by conventional generating plants such as coal power plants. Hence, the capacity of RESs and DGs in power systems rapidly increases worldwide and the higher penetration is targeted for the next two decades. While the growing penetration of RESs and DGs is a good thing in terms of the utilization of the RESs, it is also detrimental to power system stability, particularly frequency stability. Due to the inertia-less nature of power electronics interface (i.e., converter/inverter), the increasing penetration of RESs and DGs will lead to the further reduction of system inertia and will also affect the damping properties of the system. Hence, the increasing penetration of RESs will result in negative impacts on power system stability and introduce new problems in regulating power system frequency stability.

As the share of power from RESs in a power system become higher, the power electronics converter/inverter would be massively utilized to connect the DGs, RESs, and DC loads into the grid. If the RESs' penetration keeps increasing, in the future, it would be possible for the power system to operate with close to 100% supply from the power-electronics-based RESs/DGs. In that condition, system operation, stability, and resiliency will be critically affected, leading to rapid frequency/voltage oscillations, system instability, undesirable load shedding, cascading outages, or even wide-area power blackout.

One of the solutions toward stabilizing such power systems with massive DG/RES penetration is by synthesizing additional inertia and damping properties virtually. Virtual inertia can be established by using a power electronics inverter/converter equipped with short-term energy storage and an appropriate control mechanism to emulate inertia and damping properties into the power system, maintaining safe and stable grid operation. Consequently, the concept of virtual inertia and its synthesis provide a key for maintaining a high share of RESs/DGs in future power systems without compromising system stability and resiliency. Ultimately, unlike an actual synchronous generator, the parameters of a virtual inertia system can be controlled and manipulated to enhance the dynamic response of the system. Therefore, understanding the dynamics of virtual inertia and how to control it using suitable control methods are important issues for today and future power system operation and control. The authors have taken this opportunity to compile the mentioned concepts and materials that have appeared and have been developed in very recent years.

Virtual Inertia Synthesis and Control provides a thorough understanding of the basic principles, synthesis, analysis, and control of virtual inertia systems using the latest technical tools to mitigate power system stability and control problems under the presence of high RES penetration. The material contained in this book is not specifically original since it is based on information that we have published in other forms, either an engineering journal or a conference proceeding. This book uses a simple virtual inertia control structure based on the frequency response model, complemented with various control methods and algorithms to achieve an adaptive virtual inertia control with respect to the frequency stability and control issues. The chapters are easy to understand and are sufficiently detailed to capture the important aspects in virtual inertia synthesis and control, with the objective of solving the stability and control problems regarding the changes of system inertia caused by the integration of DGs/RESs. Different topics on the synthesis and application of virtual inertia are thoroughly covered with the description and analysis of numerous conventional and modern control methods for enhancing the full spectrum of power system stability and control. Filled with illustrative examples, this book gives the necessary fundamentals and insights into practical aspects. The book covers the author's long-term research, teaching, and practical experiences on the virtual inertia synthesis and control. The materials contained in this book are mainly the research outcomes and original results from the research works conducted by two laboratories: the *Power System Engineering Laboratory* at Kyushu Institute of Technology (Japan) and the *Institute of Electrical Power Engineering and Energy Systems* at Clausthal University of Technology (Germany).

This book would be useful for engineers, operators, academic researchers, and university students interested in power system dynamics, analysis, stability, and control. This book describes the synthesis, dynamics, modeling, and control of virtual inertia from the introductory to the more advanced levels. This book could also be useful as a textbook or reference for university students in electrical engineering at both undergraduate and postgraduate levels in the standard course of modern power system control and micro/smart grids. The presented materials will be served to stimulate further research and to offer practical solutions to real-world power system stability and control problems with respect to the system inertia variation triggered by the integration of RESs/DGs.

Kitakyushu, Japan
Bandung, Indonesia
Kitakyushu, Japan
Kitakyushu, Japan
June 2020

Thongchart Kerdphol
Fathin Saifur Rahman
Masayuki Watanabe
Yasunori Mitani

Acknowledgments

Information, insights, and outcomes described in this book were achieved through long-term teaching and research with practical experiences performed by the authors and their research groups on power system analysis with respect to virtual inertia control over several years in Kyushu Institute of Technology, Japan (2013–2020), and Clausthal University of Technology, Germany (2017–2020). It is a pleasure to acknowledge the received awards and supports from all mentioned sources and sponsors.

The authors would like to thank Prof. Dr.-Ing. Hans-Peter Beck (Clausthal University of Technology, Germany), Dr.-Ing. Dirk Turschner (Clausthal University of Technology, Germany), Dr.-Ing. Ralf Benger (Clausthal University of Technology, Germany), Prof. Dr. Hassan Bevrani (University of Kurdistan, Iran), Prof. Dr. Mohammad Lutfi Othman (Universiti Putra Malaysia, Malaysia), Prof. Dr. Issarachai Ngamroo (King Mongkut's Institute of Technology Ladkrabang, Thailand), Asst. Prof. Dr. Komsan Hongesombut (Kasetsart University, Thailand), Asst. Prof. Dr. Sanchai Dechanupaprittha (Kasetsart University, Thailand), Asst. Prof. Dr. Dulpichet Rerkpreedapong (Kasetsart University, Thailand), Asst. Prof. Dr. Sinan Küfeoğlu (University of Cambridge, UK), Asst. Prof. Dr. Yaser Qudaih (Higher Colleges of Technology, United Arab Emirates), Asst. Prof. Dr. Ravi Nath Tripathi (Kyoto University of Advanced Sciences, Japan), Thanakorn Penthong (RWTH Aachen University, Germany), Dr. Nanang Hariyanto (Institut Teknologi Bandung, Indonesia), Dr. Muhammad Nurdin (Institut Teknologi Bandung, Indonesia), Dr. Kevin Marojahan (Institut Teknologi Bandung, Indonesia), Dr. Pradita Octoviandiningrum Hadi (Institut Teknologi Bandung, Indonesia), and Rizky Rahmani (Institut Teknologi Bandung, Indonesia) for their active role and continuous support. We are also indebted to Anthony Doyle (Executive Editor of Engineering, Springer), Bhagyalakkshme Sreenivasan, Rajan Muthu, and anonymous reviewers, who have contributed to the revision and production of this book. This book would not be possible without the encouragement and dedication of our friends and colleagues. Last but not least, the authors offer their deepest personal gratitude to their families and students for all the patience and support during the preparation of this book.

Contents

1 An Overview of Virtual Inertia and Its Control 1
1.1 Introduction 1
1.2 Overview on Virtual Inertia 4
1.3 Literature Review on Virtual Inertia 6
1.4 Summary 8
References 9

2 Fundamental Concepts of Inertia Power Compensation and Frequency Control 13
2.1 Fundamental Frequency Regulation 13
2.2 Inertia Power Compensation 16
2.2.1 Calculation of Inertia Constant 19
2.2.2 Minimum Inertia Levels 19
2.3 Primary and Secondary Control 21
2.4 Structure of Frequency Response Model 24
2.5 Frequency Regulation in a Single-Area Power System 28
2.6 Frequency Regulation in Interconnected Power Systems 32
2.7 Analysis of Steady-State Frequency Response 36
2.8 Participation Factor for Frequency Control 42
2.9 Physical Constraints for Frequency Control 43
2.9.1 Governor Dead Band and Generation Rate 43
2.9.2 Time Delay 44
2.10 Generation Droop Characteristics 46
2.11 Reserve Power 50
2.11.1 Frequency Operating Standards 53
2.12 Summary 54
References 57

3 Virtual Inertia Synthesis for a Single-Area Power System 61
3.1 Fundamental Virtual Inertia Synthesis and Control 61
3.2 Droop Characteristics of Virtual Inertia Control 66
3.3 Frequency Regulation for Virtual Inertia Synthesis 68
3.4 Frequency Response Model for Virtual Inertia Control 70
3.5 Frequency Analysis for Virtual Inertia Control 71
3.6 State-Space Modeling of a Single Area Power System 74
3.7 Simulation Results 76
3.7.1 Effect of Virtual Inertia Control Droop 81
3.7.2 Effect of Virtual Inertia Constant 82
3.7.3 Effect of Virtual Damping 83
3.7.4 Effect of Time Delay 84
3.8 Summary 85
References 88

4 Multiple-Virtual Inertia Synthesis for Interconnected Systems 91
4.1 Introduction to Interconnected Systems 91
4.2 Modeling of Multiple-Virtual Inertia Control 93
4.3 State-Space Modeling of Interconnected Systems 97
4.4 Multiple Virtual Inertia Control Droops 99
4.4.1 Sensitivity Analysis for Multiple Inertia Control Units 100
4.5 Simulation Results 102
4.5.1 Efficacy of Multiple-Virtual Inertia Control 102
4.5.2 Stability Analysis Under Continuous Disturbances 103
4.6 Summary 108
References 109

5 Application of PI/PID Control for Virtual Inertia Synthesis 111
5.1 Introduction to PI/PID Control 111
5.2 Fundamental Feedback Control 113
5.3 Actions of PI/PID Control 114
5.3.1 Proportional Action 114
5.3.2 Integral Action 115
5.3.3 Derivative Action 116
5.4 Structures of PI/PID Control 117
5.4.1 Modeling of PI Controller 117
5.4.2 Modeling of PID Controller 117
5.5 Tuning Rules for PI/PID Control 120
5.5.1 Classical Tuning 120
5.5.2 Modern Tuning 122
5.6 Modeling of PI/PID-Based Virtual Inertia Control 127

5.7 MATLAB-Based PI/PID Tuning Approach . . . 129
5.7.1 Optimal PI Control Gains . . . 130
5.7.2 Optimal PID Control Gains . . . 132
5.8 Simulation Results . . . 135
5.9 Summary . . . 138
References . . . 139

6 Model Predictive Control for Virtual Inertia Synthesis . . . 141
6.1 Introduction to Model Predictive Control . . . 141
6.2 Fundamental MPC Strategy . . . 144
6.3 MPC Disturbances . . . 147
6.4 MPC Constraints . . . 148
6.5 MPC-Based Virtual Inertia Control . . . 148
6.6 MATLAB-Based MPC . . . 150
6.7 Simulation Results . . . 155
6.7.1 Efficacy of MPC-Based Virtual Inertia Control . . . 155
6.7.2 Robustness Against Inertia and Damping Reduction . . . 157
6.7.3 Robustness Against Time Delay . . . 159
6.7.4 Robustness Against High Penetration of Renewables . . . 160
6.8 Summary . . . 164
References . . . 165

7 Fuzzy Logic Control for Virtual Inertia Synthesis . . . 167
7.1 Introduction to Fuzzy Logic Control . . . 168
7.2 Fundamental Fuzzy Logic . . . 170
7.2.1 Fuzzy Set . . . 170
7.2.2 Shapes of Fuzzy Set . . . 172
7.2.3 Fuzzy Rule Base . . . 176
7.2.4 Fuzzification . . . 177
7.2.5 Fuzzy Inference System . . . 178
7.2.6 Defuzzification . . . 180
7.3 Fuzzy-Based Virtual Inertia Synthesis . . . 181
7.4 MATLAB-Based Fuzzy Logic Control . . . 186
7.5 Simulation Results . . . 187
7.5.1 Effect of Low RESs Penetration . . . 188
7.5.2 Effect of High RESs Penetration . . . 192
7.5.3 Mismatch Parameters of Primary/Secondary Control . . . 194
7.6 Summary . . . 198
References . . . 198

8 Synthesis of Robust Virtual Inertia Control 203
8.1 Introduction to Robust Virtual Inertia Control 203
8.2 $H\infty$ Robust Control Theory 206
8.3 Design of $H\infty$ Robust Virtual Inertia Control 209
8.4 Modeling of Uncertainty and Disturbance 209
8.4.1 $H\infty$ Controller Design 211
8.5 Closed-Loop Nominal Stability and Performance 212
8.5.1 Closed-Loop Robust Stability and Performance 212
8.6 Order Reduction of $H\infty$ Controller 214
8.7 Simulation Results 215
8.7.1 Effect of Abrupt Change 216
8.7.2 High Penetration of RESs and Loads 218
8.8 Summary 224
References 224

9 Optimization of Virtual Inertia Control Considering System Frequency Protection Scheme 227
9.1 Introduction 227
9.2 Particle Swarm Optimization 229
9.3 Underfrequency Load Shedding (UFLS) 231
9.4 Design of Virtual Inertia Control Optimization Considering System Frequency Protection 233
9.5 System Modeling 235
9.5.1 Test System 235
9.5.2 Virtual Inertia Control Model 236
9.6 Simulation Results 238
9.6.1 Default High Inertia Condition and the Result of Optimization 238
9.6.2 Low Inertia Condition 240
9.6.3 Impact on the Existing Underfrequency Load Shedding (UFLS) Scheme 244
9.7 Summary 245
References 245

10 Technical Challenges and Further Research in Virtual Inertia Control 249
10.1 Introduction 249
10.2 Main Technical Aspects of Virtual Inertia Control 250
10.2.1 Improvement in Modeling, Aggregation, and Control of Virtual Inertia Control 250
10.2.2 Optimization of Virtual Inertia Control 251
10.2.3 System Inertia Estimation 252

10.3 Supporting Aspects for the Integration of Virtual Inertia Control Systems 253
10.3.1 Economic Valuation for Inertia Service 253
10.3.2 Standard and Regulation 254
10.4 Summary 254
References 255

Appendix 257

About the Authors

Thongchart Kerdphol obtained the B. Eng. and M. Eng. degrees in Electrical Engineering from the Kasetsart University, Bangkok, Thailand, in 2010 and 2012, respectively. He received his Ph.D. degree in Electrical and Electronic Engineering from the Kyushu Institute of Technology (Kyutech), Kitakyushu, Fukuoka, Japan, in 2016. In 2017, he was a Post-Doctoral Fellow with the Department of Electrical and Electronic Engineering, Kyutech. From 2018 to 2019, he was a Visiting Lecturer at the Institute of Electrical Power Engineering and Energy Systems, Clausthal University of Technology, Clausthal-Zellerfeld, Germany.

In 2020, he became a Lecturer/Senior Research Fellow at the Department of Electrical and Electronic Engineering, Kyutech, participating in a national R&D project about inertia estimation using phasor measurement units supported by the New Energy and Industrial Technology Development Organization (NEDO), Japan. His research interests include power system stability, robust power system control, intelligent optimization, and smart/micro-grid control.

Fathin Saifur Rahman obtained the B.Sc. degree in Electrical Power Engineering and M.Sc. degree in Electrical Engineering from Institut Teknologi Bandung, Indonesia, in 2012 and 2013, respectively. In 2019, he received a Ph.D. degree in Electrical and Electronic Engineering from Kyushu Institute of Technology (Kyutech), Fukuoka, Japan.

Currently, he is a Lecturer in the School of Electrical Engineering and Informatics, Institut Teknologi Bandung, West Java, Indonesia. His research interest includes power system stability, smart grid and clean energy, optimization in power system, and application of synchrophasor in power system.

Masayuki Watanabe received the B.Sc., M.Sc., and Dr. Eng. degrees in Electrical Engineering from the Osaka University, Japan, in 2001, 2002, and 2004, respectively. Later, he received a PMU licensed patent. Until present, he has authored numerous books and over 100 journals and conference papers.

Currently, he is an Associate Professor in the Department of Electrical and Electronic Engineering, Kyushu Institute of Technology (Kyutech), Fukuoka,

Japan. His research interest is in the area of the analysis and control of power systems. He has authored books/book chapters and over 100 journal/conference papers.

Yasunori Mitani received the B.Sc., M.Sc. and D. Eng. degrees in Electrical Engineering from the Osaka University, Japan, in 1981, 1983, and 1986, respectively. From 1994 to 1995, he was a Visiting Research Associate at the University of California, Berkeley, USA. Then, he received a PMU licensed patent. From 2016 to 2018, he was the President of the Institute of Electrical Engineers of Japan (IEEJ), Power and Energy Society. So far, he has authored numerous books and book chapters and over 300 journals and conference papers.

Presently, he is the Vice President and Professor at the Kyushu Institute of Technology (Kyutech), Fukuoka, Japan. He is one of the Distinguished Lecturers of the IEEJ, Power and Energy Society. His research interests are in the areas of power system stability, dynamics, and control.

Chapter 1
An Overview of Virtual Inertia and Its Control

Abstract Today, due to the widespread penetration of renewable energy sources (RESs) and distributed generators (DGs), a new power system stability issue has emerged. This issue is the reduction and variation of inertia in the power system and is triggered by the utilization of power electronics interfaces to connect the RESs and DGs into the system, leading to a higher system uncertainty that needs more complex system operation and control. To maintain system reliability and providing efficient use of RESs and DGs, the synthesis and control of virtual inertia should be a key technology to achieve a flexible operation in today and future power systems. This chapter provides an introduction to the fundamental aspects of synthesis and control of virtual inertia for the purpose of the power system controls. An overview of the low inertia issue in the system with a high share of RESs and the role of virtual inertia are highlighted. The concept of virtual inertia emulation is briefly presented. Finally, the past achievements in the synthesis of virtual inertia respect to power system stability and control are briefly reviewed.

Keywords Distributed generation · Frequency stability · Low inertia · Renewable energy · Virtual inertia control · Virtual inertia synthesis

1.1 Introduction

In recent years, due to increasing concern in the issues related to the long-term adequacy of non-renewable energy sources such as petroleum, coal, and natural gas and the environmental problem caused by the utilization of those resources, including for electricity generation, the penetration of renewable energy sources (RESs) in the power system rapidly increases and becoming a necessity. The increasing concern in the aforementioned issues, followed by the changes in the energy regulation, makes the increasing penetration of RESs-based generation such as photovoltaic and wind-turbine generation in the power system inevitable. As an example, in Japan, up to 64 GW of photovoltaic is expected to be connected to the grid in 2030 [1]. Meanwhile, in countries such as Denmark, Ireland, and Germany, the annual penetration level of RESs of more than 20% has been achieved at the national level [2]. At a global level, in 2018, 103 GW of photovoltaic (PV) generation units have been installed globally.

T. Kerdphol et al., *Virtual Inertia Synthesis and Control*, Power Systems,
https://doi.org/10.1007/978-3-030-57961-6_1

With these additional installations, the total installed capacity of more than 512 GW has been achieved in 2018 [3].

To enable the appropriate transfer of electrical energy from RESs to the power system, the inverter is normally required to integrate the RESs-based generation units into the power system. However, the inverter (and another power-electronics interface in general) is inertia-less, due to the absence of rotating mass as the source of inertia. Thus, the increasing penetration of inverter-based generation units implies a reduction in system inertia. In the power system dominated by the inverter-based renewable generation units, the overall system inertia would be significantly lower compared to the traditional power system dominated by the traditional synchronous generators (SGs).

As a result, even though the rapid increase in the penetration level of inverter-based RESs generation units is a good thing from the environmental perspective and implies a better utilization of available source of renewables, it is detrimental to the stability of the power system, particularly to the frequency stability [2, 4, 5], since the frequency stability of the system is closely related with the amount of inertia in the system. The illustration of the correlation between inertia and frequency is shown in Fig. 1.1. It is clearly shown that with lower system inertia, frequency nadir subject to a frequency event would be lower. In addition, other than the direct impact to the overall system inertia, the increasing penetration of RESs-based generation

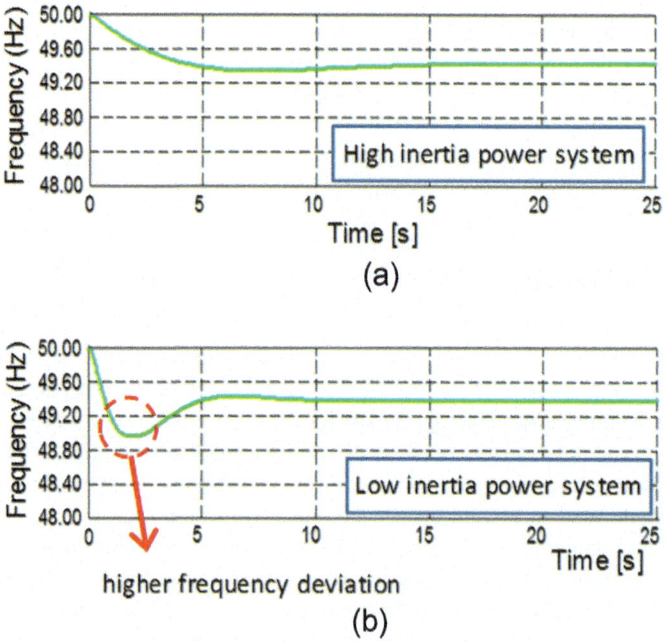

Fig. 1.1 Illustration of the correlation between inertia and frequency: frequency response to a particular frequency event in the high inertia power system (**a**) and low inertia power system (**b**)

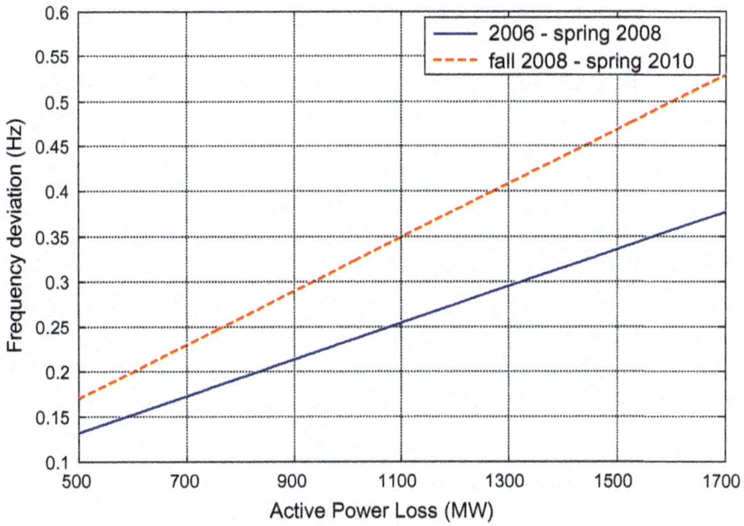

Fig. 1.2 Frequency deviation in the ERCOT system from 2006 to 2010 [8]

units could also lead to negative effects such as excessive electricity supply in the system in the case of maximum electricity generation by RESs-based generation units, power fluctuation caused by variable nature of RESs, and the deterioration of frequency regulation [6].

An example of the impact of increasing penetration of RESs in the system can be found in the Electricity Reliability Council of Texas (ERCOT) grid. In the ERCOT grid, the frequency deviation is higher in 2008–2010 than in the previous period for the same active power loss, as shown in Fig. 1.2. The continuous decline of system inertia in the system corresponds to the increased penetration of non-synchronous resources in the system [7, 8]. Hence, there is a close correlation between the amount of RESs penetration and the reduction in overall system inertia.

Therefore, to enable a high penetration of RESs in the power system, a new control strategy that could also provide the inertia support to the power system is developed. The control strategy is called a virtual inertia control. In general, the virtual inertia control is defined as the concept of providing virtual inertia to the power system by using an inverter, energy storage system (ESS), and proper control for virtual inertia emulation. This concept is also known as a virtual synchronous machine (VISMA) [9], virtual synchronous generator (VSG) [10], or synchronverter [11]. The aforementioned strategies have the same common objective, which is to provide additional inertia virtually by utilizing an inverter and an energy storage system (ESS), supported by a proper virtual inertia control mechanism. By using the aforementioned strategies, the kinetic energy reservoir in the rotating mass of a conventional synchronous generator could be imitated on the inverter-based generator, and hence, enable the emulation of virtual inertia by the inverter-based generator.

Due to its capability of providing additional inertia support in the low inertia power system, virtual inertia control units would be an integral part of the future power system dominated with RESs. Thus, the research on VSG is important to achieve a stable operation of the power system with high penetration of RESs-based generation units.

1.2 Overview on Virtual Inertia

The implementation of a virtual inertia control is based on the emulation of the typical swing equation of a synchronous generator (SG) in the control of inverter. The typical swing equation of an SG can be written as

$$\bar{P}_m - \bar{P}_e = \bar{P}_a = \frac{2H}{\omega_0}\frac{d^2\delta}{dt^2} = \frac{2H}{\omega_0}\frac{d\Delta\omega_r}{dt} \tag{1.1}$$

where $\bar{P}_m$ is mechanical power input [p.u.], $\bar{P}_e$ is electrical power output [p.u.], $\bar{P}_a$ is acceleration power [p.u.], H is inertia constant [MW.s/MVA], ω_0 is rated angular velocity of the rotor [rad/s], ω_r is angular velocity of the rotor [rad/s], δ is rotor angle [rad], and t is time [s]. $\bar{P}_m$ is related to the power supplied by the SG unit, while $\bar{P}_e$ is related to the power demand from the load. When the damping component is also included, the equation above becomes

$$\bar{P}_m - \bar{P}_e = \frac{2H}{\omega_0}\frac{d\Delta\omega_r}{dt} + K_D\frac{\Delta\omega_r}{\omega_0} \tag{1.2}$$

where K_D is the damping coefficient. Equation (1.2) could also be represented in frequency (Hz) as

$$\bar{P}_m - \bar{P}_e = \frac{2H}{f_0}\frac{d\Delta f}{dt} + K_D\frac{\Delta f}{f_0} \tag{1.3}$$

where f_0 is rated frequency of the power system [Hz] and f is the frequency of the power system [Hz]. The term $\frac{d\Delta f}{dt}$ is well known as the rate-of-change-of-frequency (RoCoF) of the power system.

The swing equation shows the relationship between the active power and the angular rotor velocity of an SG and is also correlated to the system frequency, as the term $\frac{d^2\delta}{dt^2}$ indicates the change in system frequency or the angular rotor velocity of an SG. Based on the swing equation, the system frequency can increase or decrease depending on the balance between the mechanical power input P_m and the electrical power output P_e. When $(\bar{P}_m - \bar{P}_e)$ is positive, the acceleration power $\bar{P}_a$ is positive.

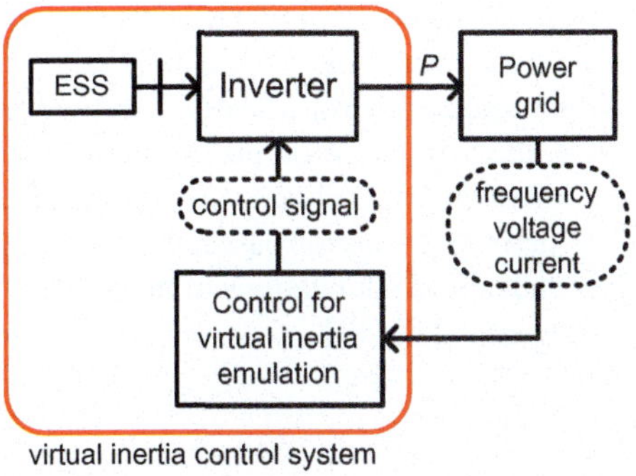

Fig. 1.3 The basic diagram of virtual inertia control system

In this condition, system frequency will increase, and vice versa. At a steady-state operating point, the system frequency is maintained by regulating the generation-load balance using a speed governing system in the SG units.

The development of the virtual inertia control is based on the swing equation described above. There are various proposed topologies for the emulation of virtual inertia, as summarized in [12, 13]. However, all of these topologies have the same main objective: to provide the additional inertia virtually into the power system using the power-electronics interface. The general concept of the virtual inertia control system is provided in Fig. 1.3. In general, the virtual inertia control system consists of an energy source, the inverter, and proper control for virtual inertia emulation. The energy source in the virtual inertia control system is usually in the form of an ESS. Another energy source, such as energy from the wind turbine could also be used. The emulation of virtual inertia by using wind turbines (i.e. doubly-fed induction generator (DFIG) wind turbines) is more commonly referred to as 'synthetic inertia'. However, for better operational flexibility, an ESS should be used as the energy source.

The idea of virtual inertia is based on the implementation of the swing equation of an SG into the inverter of the inverter-based RESs generation units so that the inverter (which is inertia-less) could be controlled to emulate the inertia characteristic of an SG. The term 'virtual inertia' refers to the fact that the inertia characteristic of an SG is emulated without utilizing any kind of rotating mass. The control for virtual inertia emulation is used to determine the required inertia power output from the virtual inertia control system.

To emulate the virtual inertia, there are various available approaches to virtually emulate the inertia characteristic of an SG, as summarized in [6, 12, 13]. Among the available approaches, the simplest and the most fundamental method for emulating virtual inertia is the virtual emulation based on the rate-of-change-of-frequency (RoCoF) of the system. The virtual inertia power in this virtual inertia emulation method is calculated by using (1.4), based on (1.3) as

$$P_{VI} = K_{VI}\frac{d\Delta f}{dt} + K_D\Delta f \tag{1.4}$$

where P_{VI} is the output virtual inertia power of virtual inertia control units and K_{VI} is the virtual inertia constant/gain. In this method, the virtual inertia power of virtual inertia control units is directly emulated by using a derivative term ($\frac{d\Delta f}{dt}$). Thus, the virtual inertia could be emulated by simply incorporating the derivative control in the controlled inverter based on the frequency measurement of the system. The derivative control-based virtual inertia control is utilized in [14–16] and the more advanced applications of the derivative control-based virtual inertia control are presented in [17–19].

1.3 Literature Review on Virtual Inertia

In this section, a brief literature review on the achievements in the synthesis and control of virtual inertia will be presented. This literature review would serve to give insight on various aspects related to virtual inertia and would be presented as brief as possible to avoid exhaustive explanation. Interested readers could refer to the provided references for a more detailed explanation.

In the future power system with high-share of RESs, the overall inertia of the system would be significantly lower due to the incorporation of inverter-based RESs generation units, which are inertia-less. The high-share of inverter-based generation units and the resulting reduction in overall system inertia, could have significant impacts on the stability and operation of the power system. Such impacts of high-share of inverter-based generation units and low inertia on the stability and operation are discussed in various references, such as [20–22].

To mitigate the impacts of low system inertia and improve the stability of low inertia power system, particularly the frequency stability, the emulation of additional inertia into the power system without using actual rotating mass in terms of virtual inertia becomes one of the promising solutions. There are several topologies to emulate the virtual inertia, as summarized in [12, 13]. All of these topologies are developed based on a similar basic concept. However, they differ in terms of the level of detail in their implementation. To give an insight on various topologies for virtual inertia emulation, several notable topologies would be briefly discussed and the difference between the topologies would be highlighted.

In general, the virtual inertia emulation topology could be divided into three main categories. These three main categories are [13]:

1. Synchronous generator model-based topology
2. Swing equation-based topology
3. Frequency-power response-based topology.

- **Synchronous Generator Model-Based Topology**

Synchronous generator model-based topology is the topology for virtual inertia emulation based on the full modeling of the dynamics of a synchronous generator (SG). One of the examples of topologies in this category is synchronverter [11]. In synchronverter, both the electrical part (e.g. the interaction between windings) and the mechanical part of an SG (i.e. the rotating mass and inertia) are modeled. Hence, the dynamics of an SG could be accurately replicated. Several research works on the analysis and the improvement of synchronverter are discussed in [23–25]. The other topologies in this category (i.e. synchronous generator model-based topology) are virtual synchronous machine (VISMA), which is also known as IEPE's topology [9, 26–28], and Kawasaki Heavy Industries (KHI)'s topology [29].

- **Swing Equation-Based Topology**

Swing equation-based topology is the topology for virtual inertia emulation based on the swing equation of an SG. Hence, rather than full modeling of an SG, only the swing equation is modeled to emulate the virtual inertia. One of the well-known topologies in this category is Ise Lab's topology introduced in [30]. The topology works based on the measurement of grid frequency and the active power output of the inverter. Several research works on the analysis and the improvement of Ise Lab's topology are discussed in [31–34]. The other topology in this category is the synchronous power controller (SPC) [35, 36].

- **Frequency-Power Response-Based Topology**

Frequency-power response-based topology is the topology for virtual inertia emulation based on the response to the frequency change. This topology uses the measurement of the derivative of the frequency change to emulate the virtual inertia. The virtual inertia emulation using the derivative of the frequency change has been briefly discussed in the previous section. One of the topologies in this category is the virtual synchronous generator (VSG). Several research works on the implementation of VSG are discussed in [14, 37]. The VSG is also known as the VSYNC's topology.

The approaches and topologies described above are based on the emulation of the inertia characteristic of an SG and differ only in the level of emulation detail. Other than these three categories of virtual inertia emulation topologies, several other topologies have been proposed to improve the frequency regulation in a low inertia power system without directly emulating the inertia characteristic of an SG. One of them is the approach based on droop control. This approach is based on the parallel operation of a synchronous generator and could be used for regulating the active and reactive power of RESs-based generation units. The development of frequency-droop control for autonomous operation of an inverter-based microgrid is discussed in [38, 39]. In the frequency-droop control, a low pass filter employed in the measurement of its output power could be used to approximate the behavior of virtual inertia control [40, 41]. The comparison between the dynamic performance of virtual inertia and droop control is described in [33]. The other research works on the design, implementation, and improvement of the droop-based approach are discussed in [42, 43].

In terms of the virtual inertia emulation, the virtual inertia could also be emulated from the control of doubly-fed induction generator (DFIG) wind turbines, which is more commonly referred to as synthetic inertia. Several research works on the emulation of synthetic inertia could be found in [44–46].

Regarding the application of virtual inertia control, various control techniques, such as the robust control technique, could be applied to further improve the virtual inertia control unit. The implementation of various control techniques to improve the performance of the virtual inertia control unit is presented in [17–19].

To improve the performance of the virtual inertia control units, several research works have been conducted. The adaptive control of virtual inertia control units by alternating inertia method has been discussed in [32, 47]. In this method, virtual inertia control parameters are controlled based on the state of the angular velocity oscillation (i.e. accelerating or decelerating). Related to the adaptive control of virtual inertia control units, a self-tuning virtual synchronous machine to minimize frequency deviation and the power flow through the energy storage system is proposed in [48].

The improvement of the virtual inertia control unit could also be achieved by optimizing its control parameter. The research works on this field are presented in [16, 49–51]. In [52], the optimization of virtual inertia constant/gain is performed by also considering a frequency protection scheme to achieve coordination between the virtual inertia control unit and frequency protection scheme.

Besides the optimization of virtual inertia control parameters, the improved performance of virtual inertia control could also be achieved by selecting optimal placement locations for virtual inertia control units. Some of the research works in this field are presented in [53–55]. Reference [53] suggests that the resilience of a power system is dictated by the placement of inertia and the location of disturbance, rather than the total inertia in the power system.

From the presented references, most of the research works are focused on the implementation of virtual inertia control in the microgrid system. However, the implementation of virtual inertia control would also be important in the interconnected system, due to the widespread utilization of RESs. The analysis of virtual inertia control applications in the interconnected system is presented in [16, 19, 54, 56].

1.4 Summary

This chapter presents a brief introduction to the low inertia problem in the power system dominated by inverter-based generation units. The importance of virtual inertia emulation in a low inertia power system and its concept have also been briefly explained. Finally, the literature review on the achievements related to the virtual inertia emulation is provided to give insights on the previous research works in virtual inertia emulation.

References

1. Ministry of Economy Trade and Industry, Long-term Energy Supply and Demand Outlook (2015). https://www.enecho.meti.go.jp/committee/council/basic_policy_subcommittee/mitoshi/pdf/report_02.pdf. Accessed 22 Jan 2019
2. B. Kroposki et al., Achieving a 100% renewable grid: operating electric power systems with extremely high levels of variable renewable energy. IEEE Power Energy Mag. **15**(2), 61–73 (2017)
3. G. Masson, I. Kaizuka, *Trends in Photovoltaic Applications*, Report, IEA PVPS T1-36 (2019)
4. R.F. Yan, T.K. Saha, N. Modi, N.A. Masood, M. Mosadeghy, The combined effects of high penetration of wind and PV on power system frequency response. Appl. Energy **145**, 320–330 (2015)
5. J.G. Slootweg, W.L. Kling, Impacts of distributed generation on power system transient stability, in Proc. *IEEE Power Engineering Society Summer Power Meeting,* 862-867 (2002)
6. H. Bevrani, *Robust Power System Frequency Control*, 2nd ed. (Springer, New York, USA, 2014)
7. J. Matevosyan et al., Proposed future Ancillary Services in Electric Reliability Council of Texas, in Proc. *2015 IEEE PowerTech*, 1-6 (2015)
8. Electric Reliability Council of Texas, Future Ancillary Services in ERCOT, Final Report (2013)
9. H.P. Beck, R. Hesse, Virtual synchronous machine, in Proc. *International Conference on Electrical Power Quality and Utilisation*, 1-6 (2007)
10. J. Driesen, K. Visscher, Virtual synchronous generators, in Proc. *IEEE Power Energy Society General Meeting (IEEE PES GM)*, 1-3 (2008)
11. Q.C. Zhong, G. Weiss, Synchronverters: inverters that mimic synchronous generators. IEEE Trans. Ind. Electron. **58**(4), 1259–1267 (2011)
12. H. Bevrani, T. Ise, Y. Miura, Virtual synchronous generators: a survey and new perspectives. Int. J. Electr. Power Energy Syst. **54**, 244–254 (2014)
13. U. Tamrakar, D. Shrestha, M. Maharjan, B. Bhattarai, T. Hansen, R. Tonkoski, Virtual inertia: current trends and future directions. Appl. Sci. **7**(7), 654 (2017)
14. M.P.N. Van Wesenbeeck, S.W.H. De Haan, P. Varela, K. Visscher, Grid tied converter with virtual kinetic storage, in Proc. *IEEE Bucharest PowerTech*, 1-7 (2009)
15. V. Karapanos, S. De Haan, K. Zwetsloot, Real time simulation of a power system with VSG hardware in the loop, in Proc. *Annual Conference of the IEEE Industrial Electronics Society*, 3748-3754 (2011)
16. P. Rodriguez, E. Rakhshani, A. Mir Cantarellas, D. Remon, Analysis of derivative control based virtual inertia in multi-area high-voltage direct current interconnected power systems. IET Gener. Transm. Distrib. **10**(6), 1458–1469 (2016)
17. T. Kerdphol, F.S. Rahman, Y. Mitani, K. Hongesombut, S. Küfeoğlu, Virtual inertia control-based model predictive control for microgrid frequency stabilization considering high renewable energy integration. Sustainability **9**(5), 773 (2017)
18. T. Kerdphol, F. S. Rahman, Y. Mitani, M. Watanabe, S. Kufeoglu, Robust virtual inertia control of an islanded microgrid considering high penetration of renewable energy. IEEE Access. **6**, 625-636 (2018)
19. T. Kerdphol, F. Rahman, Y. Mitani, Virtual inertia control application to enhance frequency stability of interconnected power systems with high renewable energy penetration. Energies. **11**(4), 981 (2018)
20. A. Ulbig, T.S. Borsche, G. Andersson, Impact of low rotational inertia on power system stability and operation. IFAC Proceedings Volumes. **47**(3), 7290-7297 (2014)
21. P. Tielens, D. Van Hertem, The relevance of inertia in power systems. Renew. Sustain. Energy Rev. **55**, 999-1009 (2016)
22. Y. Wang, V. Silva, M. Lopez-Botet-zulueta, Impact of high penetration of variable renewable generation on frequency dynamics in the continental Europe interconnected system. IET Renew. Power Gener. **10**(1), 10-16 (2016)

23. Q.C. Zhong, P.L. Nguyen, Z. Ma, W. Sheng, Self-synchronized synchronverters: inverters without a dedicated synchronization unit. IEEE Trans. Power Electron. **29**(2), 617-630 (2014)
24. Q.C. Zhong, Virtual synchronous machines: a unified interface for grid integration. IEEE Power Electron. Mag. **3**(4), 18-27 (2016)
25. Q.C. Zhong, G.C. Konstantopoulos, B. Ren, M. Krstic, Improved synchronverters with bounded frequency and voltage for smart grid integration. IEEE Trans. Smart Grid **9**(2), 786–796 (2018)
26. Y. Chen, R. Hesse, D. Turschner, H. P. Beck, Investigation of the virtual synchronous machine in the island mode, in Proc. *IEEE PES Innovative Smart Grid Technologies Europe*, 1-6 (2012)
27. R. Hesse, D. Turschner, H. P. Beck, Micro grid stabilization using the Virtual Synchronous Machine (VISMA), in Proc. *International Conference on Renewable Energies and Power Quality*, 676-681 (2009)
28. Y. Chen, R. Hesse, D. Turschner, H.P. Beck, Dynamic Properties of the Virtual Synchronous Machine (VISMA), in Proc. *International Conference on Renewable Energies and Power Quality*, 755-759 (2011)
29. Y. Hirase, K. Abe, K. Sugimoto, Y. Shindo, A grid-connected inverter with virtual synchronous generator model of algebraic type. Electr. Eng. Japan. **184**(4), 10-21 (2013)
30. K. Sakimoto, Y. Miura, T. Ise, Stabilization of a power system with a distributed generator by a Virtual Synchronous Generator function, in Proc. *International Conference on Power Electronics*, 1498–1505 (2011)
31. J. Liu, Y. Miura, H. Bevrani, T. Ise, Enhanced virtual synchronous generator control for parallel inverters in microgrids. IEEE Trans. Smart Grid **8**(5), 2268–2277 (2017)
32. J. Alipoor, Y. Miura, T. Ise, Power system stabilization using virtual synchronous generator with alternating moment of inertia. IEEE J. Emerg. Sel. Top. Power Electron. **3**(2), 451–458 (2015)
33. J. Liu, Y. Miura, T. Ise, Comparison of dynamic characteristics between virtual synchronous generator and droop control in inverter-based distributed generators. IEEE Trans. Power Electron. **31**(5), 3600–3611 (2016)
34. T. Shintai, Y. Miura, T. Ise, Oscillation damping of a distributed generator using a virtual synchronous generator. IEEE Trans. Power Deliv. **29**(2), 668-676 (2014)
35. P. Rodriguez, I. Candela, A. Luna, Control of PV generation systems using the synchronous power controller, in Proc. *IEEE Energy Conversion Congress and Exposition*, 993-998 (2013)
36. P. Rodriguez, I. Candela, J. Rocabert, R. Teodorescu, *Virtual Controller of Electromechanical Characteristics for Static Power Converters*, European Patent Office, EP2683075A1 (2012)
37. M. Torres, L.A.C. Lopes, Virtual synchronous generator control in autonomous wind-diesel power systems, in Proc. *IEEE Electrical Power and Energy Conference,* 1-6 (2009)
38. F. Katiraei, M.R. Iravani, Power management strategies for a microgrid with multiple distributed generation units. IEEE Trans. Power Syst. **21**(4), 1821-1831 (2006)
39. N. Pogaku, M. Prodanović, T.C. Green, Modeling, analysis and testing of autonomous operation of an inverter-based microgrid. IEEE Trans. Power Electron. **22**(2), 613-625 (2007)
40. S. D'Arco, J.A. Suul, Equivalence of virtual synchronous machines and frequency-droops for converter-based microgrids. IEEE Trans. Smart Grid. **5**(1), 394-395 (2014)
41. S. D'Arco, J.A. Suul, Virtual synchronous machines — Classification of implementations and analysis of equivalence to droop controllers for microgrids, in Proc. *IEEE Grenoble Conference* (2013)
42. N. Soni, S. Doolla, M.C. Chandorkar, Improvement of transient response in microgrids using virtual inertia. IEEE Trans. Power Deliv. **28**(3), 1830-1838 (2013)
43. J. Van De Vyver, J.D.M. De Kooning, B. Meersman, L. Vandevelde, T.L. Vandoorn, Droop control as an alternative inertial response strategy for the synthetic inertia on wind turbines. IEEE Trans. Power Syst. **31**(2), 1129-1138 (2016)
44. A. Bonfiglio, M. Invernizzi, A. Labella, R. Procopio, Design and implementation of a variable synthetic inertia controller for wind turbine generators. IEEE Trans. Power Syst. **34**(1), 754-764 (2019)
45. M.F.M. Arani, E.F. El-Saadany, Implementing virtual inertia in DFIG-based wind power generation. IEEE Trans. Power Syst. **28**(2), 1373-1384 (2013)

46. A. Gloe, C. Jauch, B. Craciun, J. Winkelmann, Continuous provision of synthetic inertia with wind turbines: implications for the wind turbine and for the grid. IET Renewable Power Generation. **13**(5), 668-675 (2019)
47. D. Li, Q. Zhu, S. Lin, X.Y. Bian, A self-adaptive inertia and damping combination control of VSG to support frequency stability. IEEE Trans. Energy Convers. **32**(1), 397-398 (2017)
48. M.A. Torres, L.A.C. Lopes, L.A.T. Morán, J.R.C. Espinoza, Self-tuning virtual synchronous machine: a control strategy for energy storage systems to support dynamic frequency control. IEEE Trans. Energy Convers. **29**(4), 833-840 (2014)
49. I. Serban, C.P. Ion, Microgrid control based on a grid-forming inverter operating as virtual synchronous generator with enhanced dynamic response capability. Int. J. Electr. Power Energy Syst. **89**, 94–105 (2017)
50. M. Hajiakbari Fini, M.E. Hamedani Golshan, Determining optimal virtual inertia and frequency control parameters to preserve the frequency stability in islanded microgrids with high penetration of renewables. Electr. Power Syst. Res. **154**, 13–22 (2018)
51. J. Alipoor, Y. Miura, T. Ise, Stability assessment and optimization methods for microgrid with multiple VSG units. IEEE Trans. Smart Grid **9**(2), 1462–1471 (2018)
52. F.S. Rahman, T. Kerdphol, M. Watanabe, Y. Mitani, Optimization of virtual inertia considering system frequency protection scheme. Electr. Power Syst. Res. **170**, 294–302 (2019)
53. B.K. Poolla, S. Bolognani, F. Dorfler, Optimal placement of virtual inertia in power grids. IEEE Trans. Automat. Contr. **62**(12), 6209-6220 (2017)
54. F.S. Rahman, T. Kerdphol, M. Watanabe, Y. Mitani, A study on the placement of virtual synchronous generator in a two-area system, in Proc. *IEEE Innovative Smart Grid Technologies Asia*, 782–786 (2018)
55. Israjuddin, N. Hariyanto, L. Chao-Yuan, L. Chih-Wen, Optimal placement of energy storage with synthetic inertia control on a grid with high penetration of renewables using mean-variance mapping optimization, in Proc. *IEEE PES Asia-Pacific Power and Energy Engineering Conference*, 1-6 (2019)
56. E. Rakhshani, P. Rodriguez, Inertia emulation in AC/DC interconnected power systems using derivative technique considering frequency measurement effects. IEEE Trans. Power Syst. **32**(5), 3338-3351 (2017)

Chapter 2
Fundamental Concepts of Inertia Power Compensation and Frequency Control

Abstract The inertia power generated by the rotating mass (rotor) of a synchronous generator plays a significant role in slowing down the frequency oscillation and has an active role in the system frequency stability during a disturbance. Lower system inertia could lead to a significantly faster change in the system frequency, resulting in the degradation of system frequency stability. A rapid frequency deviation can lead to system instability, collapse, and power blackout. Thus, by understanding the concept of inertia and its role in the power system, it would give a better insight on how to deal with the frequency problem caused by low system inertia. To give a clear understanding of the inertia compensation and frequency control, this chapter elaborates on the subject of active power-based inertia compensation regarding power system frequency control, including its basic concept and definition. Afterward, the primary, secondary, tertiary, and emergency control loops for power system frequency control are discussed in detail. A frequency response model is provided and its utilization for the sake of dynamic analysis and simulation regarding virtual inertia is elaborated. Finally, the past achievements regarding the inertia power compensation for system frequency control are discussed.

Keywords Area control error · Frequency control · Frequency response model · Inertia power compensation · Primary control · Secondary control · Tertiary control · Emergency control · Swing equation · Synchronous generator · Virtual inertia control

2.1 Fundamental Frequency Regulation

Frequency regulation is related to the energy balance of load demand and generation, which is of great significance and recognized as a high priority area by most operators [1]. Any disturbance that leads to the unbalance between generation and load can cause an abrupt change in system frequency, resulting in frequency oscillations. Frequency oscillations may affect system stability, operation, and resiliency. Large frequency oscillations can damage equipment, deteriorate load performance, overload transmission lines, trip protection relays, and in the worst-case, lead to system collapses and wide-area power blackouts.

T. Kerdphol et al., *Virtual Inertia Synthesis and Control*, Power Systems,
https://doi.org/10.1007/978-3-030-57961-6_2

The frequency of the system is proportional to the rotating speed of the generator. Thus, the frequency control issue may be directly transformed into a speed control issue of the generator-turbine units. This problem is solved by applying a governing system, which can track the generator speed and adjust the input value to change the mechanical power output to follow the load variation and reducing the frequency deviation. After that, the secondary control action will restore the frequency back to its nominal value.

Based on the frequency oscillation (deviation) territory (see Fig. 2.1), the natural response called inertia power compensation, along with primary control, secondary control, tertiary control, and emergency control may be needed to regulate system frequency. From Fig. 2.1, f_0 is the nominal frequency (e.g., 50 Hz or 60 Hz), and $\Delta f_1, \Delta f_2, \Delta f_3,$ and Δf_4 reveal frequency oscillation territories with respect to various operating conditions based on permittable frequency operating standards [1].

During normal operation, the small frequency oscillations could be diminished by the inertia compensation and primary control. For larger frequency deviations (abnormal operation), based on the available reserve power, the secondary control could restore the system frequency back to the steady-state or nominal value. For a critical load-generation mismatch with fast frequency variations during a significant disturbance/fault, the operation of secondary control may be insufficient to restore the system frequency back to the steady-state. In such circumstances, it is required

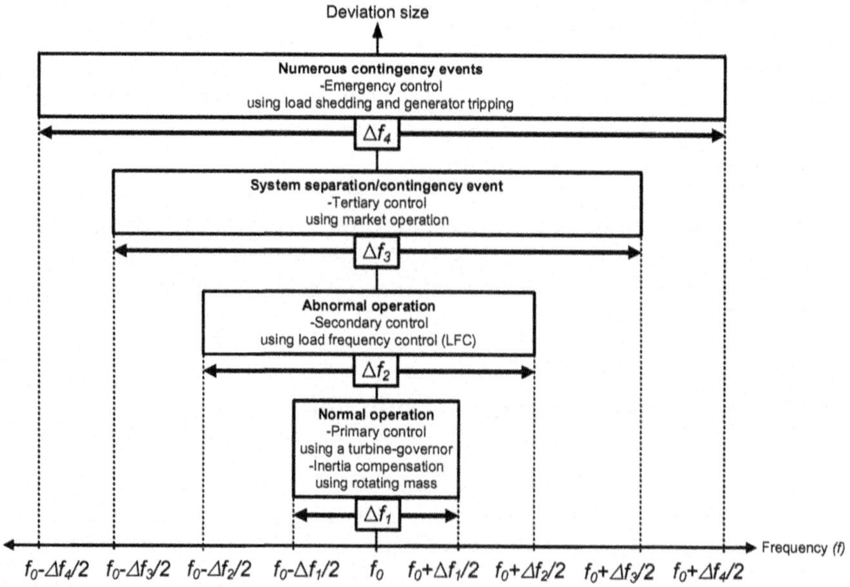

Fig. 2.1 Operating control schemes with regards to frequency deviation size

to activate the unusual operations of tertiary control, emergency control (e.g., under-frequency load shedding) and protection schemes, reducing the risk of cascading failures, load, or grid/network separation events, and additional generation events [2].

Following a disturbance or event (see Fig. 2.2), the inertia power compensation in the rotating units respond within less than 10 s to arrest the initial frequency deviation. Then, the primary control loop in a governor-turbine unit is activated within 3 s and fully released to the system within 10 s. This service will be maintained, where necessary, up to 20–40 s [1]. As soon as the system is stable, the system frequency recovers back to a fixed value, but it might be distinct from the nominal value as the generator droop generates a proportional type of action. Then, the secondary control is activated following the inertia compensation and primary control timescales and can be activated up to 30 min after a disturbance. This control can re-establish the nominal frequency and the interchanged power by a distribution of controlling power. In some critical disturbances or events, if the system frequency rapidly drops and reaches a decisive value, the tertiary and emergency controls are significantly needed to recover the frequency. If such actions are not taken, it may lead to critical under-speed, causing the tripping of generators, cascading failures, and wide-area power blackouts. The tertiary control is applied to manage eventual congestions, to recover the secondary control reserve, and to restore the frequency and tie-line power to their fixed values when the secondary control reserve is insufficient. This service is frequently called as the manual frequency control by connecting and disconnecting of power, relocating the output from frequency control participating units, and load/demand-side regulation.

In a conventional power system dominated by synchronous generators, the conceptual frequency response model, including four frequency control loops (i.e., primary, secondary, tertiary, and emergency controls) in a simplified structure, is depicted in Fig. 2.3. All frequency control loops are usually available. The market operator is capable of balancing the load-generation of the system considering economy, reliability, and resiliency. The operator can change participation factors,

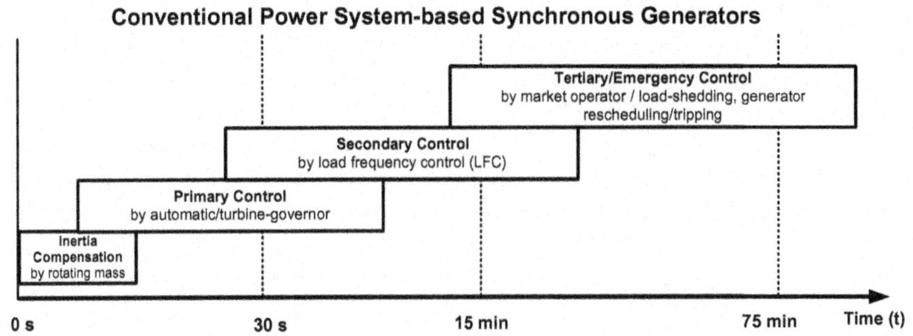

Fig. 2.2 Timescale of frequency dynamic control for conventional power systems dominated by synchronous generators

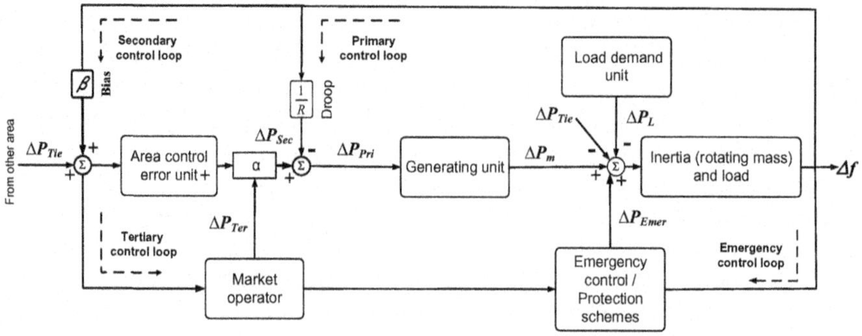

Fig. 2.3 Conceptual frequency response structure with frequency control loops for a conventional synchronous generators-based power system

the setpoint of all generators, and power dispatch via the secondary control and tertiary control. The operator also can perform a generator tripping or load shedding in an emergency situation.

In Fig. 2.3, ΔP_m is the generated power change from a generating unit, ΔP_{Tie} is the change of interchanged power among areas, ΔP_L is the power change from load demand (as a disturbance), β is the area bias factor, R is the primary droop constant, α is the participation factor of a generating unit in frequency control, Δf is frequency deviation of the system, and ΔP_{Pri}, ΔP_{Sec}, ΔP_{Ter}, and ΔP_{Emer} are the control action signals for primary, secondary, tertiary, and emergency controls, respectively.

2.2 Inertia Power Compensation

In traditional power systems dominated by synchronous machines, the synchronous generators generate active power and kinetic energy, regulating the system frequency. The rotating mass in the rotor of a synchronous machine generates inertia power with the unit of Joule-second ($J \cdot s$) or Watt-square second ($W \cdot s^2$) for compensating the disturbances. Inertia power performs an important function in regulating the stable frequency stability of the system.

Inertia is generated when the rotating/spinning mass or rotor or prime mover of a synchronous generator continues to spin unless braked to slow down or stop. Subsequently, the majority of inertia is contributed by the physical rotating mass related to the power outputs of synchronous generators, enhancing the inertial response of the system. Thus, a minimum level of total system inertia is indispensable for solving two main dynamic issues. The first is to decrease the initial rate of change of frequency (RoCoF) after the large disturbance (i.e., generation/load disconnection), preventing a cascading disconnection of generators. The second is to arrest the frequency decay and limit the frequency nadir following a generation trip or load increment. Also, the inertia is used to arrest the frequency raise and reduce the frequency zenith following a load trip or generation increment. The associated issues have been investigated and reported in Europe [3–7], North America [8–10], and Asia [11, 12], including Japan [13, 14].

A schematic block diagram of a synchronous machine considering the dynamic of inertia power response is depicted in Fig. 2.4. Figure 2.5 displays an overview of a typical 36 kVA synchronous generator equipped with an alternator manufactured by ABB company at the Institute of Electrical Power Engineering and Energy Systems (IEE), Clausthal University of Technology (TU Clausthal), Germany. The alternator unit generates the frequency which is matched to the prime mover speed from the generator unit. Thus, the generator speed must be accurately and responsively controlled by the inertia power and speed governor (i.e., primary control) inside the generator unit to ensure the constant/stable frequency. A sudden increase in a load will cause the generator and alternator units to slowing down momentarily

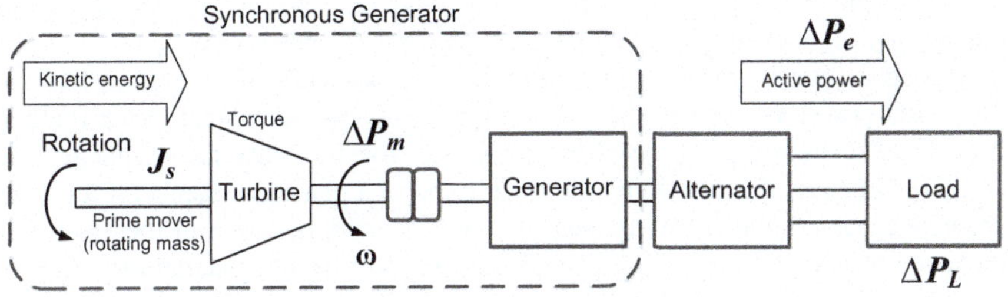

Fig. 2.4 Schematic block diagram of a synchronous machine with respect to inertia power response

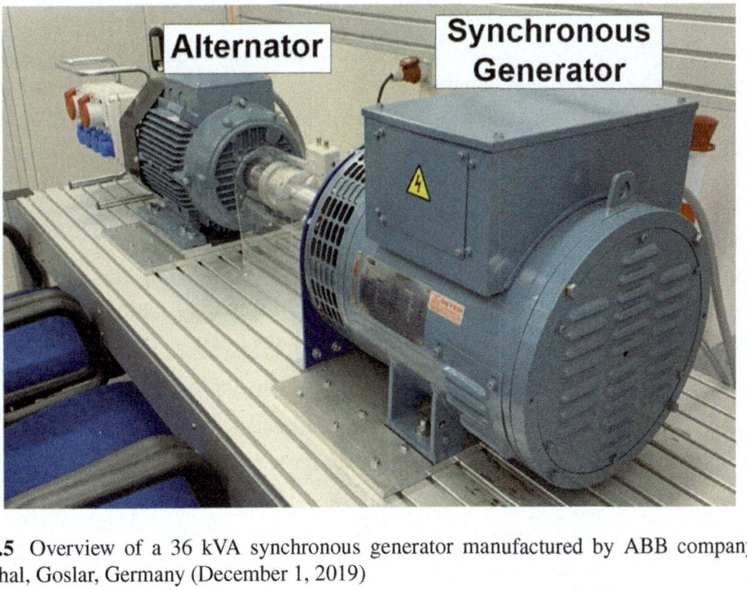

Fig. 2.5 Overview of a 36 kVA synchronous generator manufactured by ABB company, TU Clausthal, Goslar, Germany (December 1, 2019)

until the governor can adjust its speed. This will momentarily reduce both frequency and voltage of the system. If the load suddenly reduces, the generator and alternator will speed-up momentarily before the governor can adjust its speed, causing both the frequency and voltage of the system to momentarily increase.

To figure out what inertia compensation and control is, this section explains the dynamic model of frequency control, regarding inertial response. In traditional power systems, the dynamic behavior of a synchronous machine is based on the swing equation, defined as [1, 2]:

$$J_S \frac{d\omega}{dt} = T_m - T_e = \frac{P_m}{\omega} - \frac{P_e}{\omega} \tag{2.1}$$

where J_s is the moment of inertia with the unit of kg/m^2, ω is the angular velocity of the synchronous rotor (rad/s), and $\omega = 2\pi f_0, f_0$ is the nominal frequency (Hz), T_m and T_e are the mechanical and electrical torque for the generator, P_m and P_e means the mechanical and electrical/generated power for the generator (W).

During the power unbalance between the load demand and generation (ΔP) in traditional power systems, the kinetic energy ($E_{kinetic}$) accumulated in the rotating mass of a generating unit is used to compensate for the initial speed change. The system frequency is evaluated by the rotor speed of a generator. Hence, if the rotor speed changes, the system frequency will also change, resulting in the frequency deviation from a nominal value. The generated power from a generator in terms of the kinetic energy is expressed as:

$$\Delta P_{m\_inertia} = \frac{d}{dt} E_{kinetic} \tag{2.2}$$

Table 2.1 shows the relationship between the rotor speed of a generator with respect to inertia property. When the rotor speed is decelerating or accelerating due to the power mismatch, the acceleration or deceleration is inversely proportional to the moment of inertia. With a higher moment of inertia, the deceleration or acceleration of rotor speed will be lower. Similarly, with a lower moment of inertia, the deceleration or acceleration of rotor speed will be higher. It is noted that the rate of change of rotor speed is also known as the rate of change of frequency (RoCoF).

The inertia power is directly produced by the accumulated kinetic energy in a synchronous machine. The kinetic energy ($E_{kinetic}$) from the rotating mass, including spinning loads, is formed in the unit of Watt (W) as [1, 2]:

Table. 2.1 Relationship between the rotor speed and moment of inertia

Power mismatch	Moment of inertia (J_s)	Rate of change of rotor speed (df/dt or $d\omega/dt$)
$\neq 0$	High	Low
	Low	High
0	High or Low	Zero

Table 2.2 Typical inertia constant for generating units

Generation type	H (MW · s per MVA)
Hydraulic unit	2–4
Thermal unit, 3600 r/min (2 pole)	2.5–6
Thermal unit, 1800 r/min (4 pole)	4–10

$$E_{kinetic} = \frac{1}{2} J_S \omega^2 \tag{2.3}$$

2.2.1 *Calculation of Inertia Constant*

Usually, the inertia of a synchronous generator can be represented by inertia constant *H*, presented in second (s). The inertia is expressed as a proportion of kinetic energy and generator power rating as [1, 2]:

$$H_i = \frac{E_{kinetic}}{S_{SGi}} = \frac{J_S \omega^2}{2 S_{SGi}} \tag{2.4}$$

where i is the generator number, S_{SG} is the rated power of a synchronous generator (VA).

The dynamics of system frequency consist of the aggregated rotating/spinning dynamics. When numerous generators are lumped together as a group or system, their inertia constants (system inertia) are calculated by the weight of the rated power of each generator as follows [15]:

$$H = \frac{\sum_i (H_i S_{SGi})}{S_{PS}} \tag{2.5}$$

where S_{PS} is the rated power of the system.

The practical values of the inertia constant (H) for thermal and hydraulic generating units are given in Table 2.2 [2].

2.2.2 *Minimum Inertia Levels*

A minimum inertia level or a critical inertia floor is usually used to determine the minimum inertia of the system for assuring the stable and safe system operation [11]. As shown in Fig. 2.6, the minimum inertia level may be determined based on two constraints: (1) the maximum RoCoF and (2) the frequency nadir, based

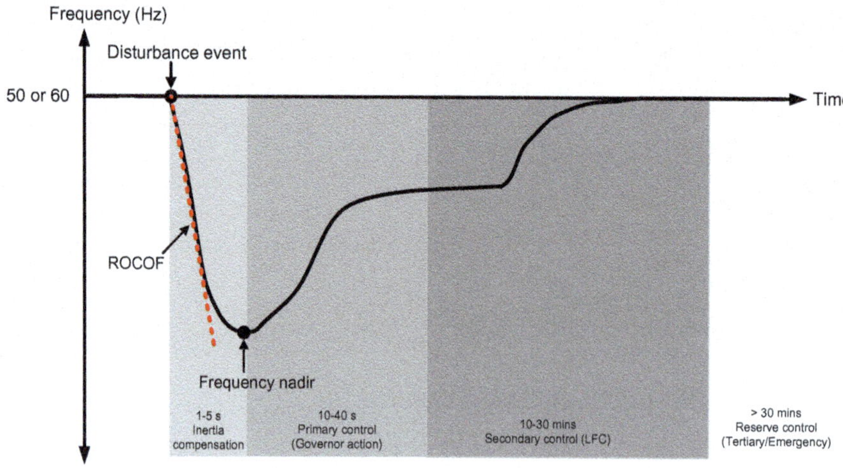

Fig. 2.6 Frequency unbalance response

Table 2.3 Criteria for determining minimum inertia levels

System operator	RoCoF	Frequency nadir
EirGrid (Ireland)	✓	
National Grid (UK)	✓	
AEMO (Australia)	✓	✓
ERCOT (Texas, USA)		✓
NORDIC (Scandinavia)		✓

on grid code requirements. However, different regions or system operators have different constraints, as shown in Table 2.3. Some system operators (e.g., EirGrid [16], National Grid [17], AEMO [18]) have concerned that high RoCoF is only mitigated by inertia. Other system operators (e.g., ERCOT [19] and NORDIC [4]) have concerned that the frequency nadir can also be mitigated by other means, such as fast frequency response.

Each system operator will have requirements based on their own needs, such as inertia, fast frequency response availability, size of contingency, the setpoint for frequency-based load shedding, and RoCoF setpoints. Practically, system operators in each region are those who are setting inertia requirements, as shown in Table 2.4. UFLS is the abbreviation for under-frequency load shedding, one of the protection schemes in the power system to maintain the frequency stability of the system.

Exercise 2.1 The turbine generating unit has the rating parameters of a 550 MVA, 24 kV, 0.9 p.f., 60 Hz, and 3600 RPM. The moment of inertia is given as 26,300 kg m^2. Please determine the following:

Table 2.4 Practical requirements for determining minimum inertia [18–20]

Constraint	ERCOT	Great Britain	Ireland	NORDIC	South Australia
UFLS stage 1	59.3 Hz	48.8 Hz	48.85 Hz	48.85 Hz	47.6 Hz
RoCoF	1 Hz/s	0.5 Hz/s	1 Hz/s	0.5 Hz/s	1.5 or 3 Hz/s
Contingency	2.75 GW	1.25 GW	0.5 GW	1.65 GW	0.35 GW
Inertia floor	100 GW	135 GW	23 GW	125 GW	6.2 GW
Peak demand	73 GW	60 GW	6.5 GW	72 GW	36 GW

(1) Inertia constant, MW · s/MVA rating
(2) Stored power, MW · s at the rated speed.

Solution 2.1

(1) Inertia constant is calculated from Eq. (2.4) as:

$$H = \frac{J_S \omega^2}{2S_{SG}} = \frac{J_S (2\pi f_0)^2}{2S_{SG}} = \frac{26{,}300(2\pi 60)^2}{2(550)} = 3.398\,\text{MW} \cdot \text{s/MVA}.$$

(2) The stored power at the rated speed can be determined by multiplying the inertia constant and rated capacity of the generator as:

$$P_{inertia} = H(MVA) = 3.398(550) = 1868.9\,\text{MW} \cdot \text{s}.$$

2.3 Primary and Secondary Control

Power system frequency stability relies on the active power balance between load and generation. The active power change at one point of a system would affect the whole network by the deviation of frequency. Thus, system frequency offers a useful index to signify an unbalance between system generation and load demand. Any short-term unbalance leads to an immediate variation in system frequency since the disturbance has been originally arrested by the inertia (kinetic) power of the rotor from 1 to 5 s. Significant loss of the generation without a suitable system control action creates severe frequency excursions outside the operating range of the power plant. Consequently, the primary and secondary control schemes are required to avoid such stability issues. These control schemes are the basic frequency control loops in power systems.

Based on the generation type, active power produced by a generator is governed by the mechanical power output of a rotor or prime mover (e.g., diesel engine, steam-turbine, hydro-turbine, gas-turbine). For the hydro/steam turbine, the mechanical power is governed by the closing or opening of valves, maintaining the input of water or steam flow into such a turbine. The water or steam inputted to generators has to be maintained continuously to pair actual power demand. The falling machine

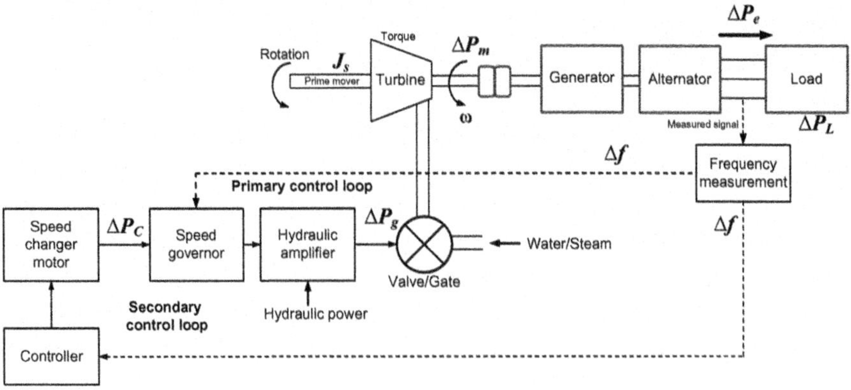

Fig. 2.7 A schematic diagram of a synchronous generator with primary and secondary controls

speed could result in the consequent change in frequency. For reliable operation of the power system, system frequency should be maintained almost constant at nominal frequency of 50 or 60 Hz depends on the system.

In terms of multiple operating generators, the primary control is equipped as a basic frequency control in all synchronous generators, while several large synchronous generators are provided with the secondary control. A schematic diagram of a synchronous generator equipped with the primary and secondary controls is displayed in Fig. 2.7.

From Fig. 2.7, when the frequency experiences the transient deviation (Δf) after a load change (ΔP_L), the feedback mechanism automatically activates and provides a suitable control signal (ΔP_g) for the turbine unit to decrease or increase the mechanical power (ΔP_m) in a generation unit, tracking the load change and restoring the system frequency. Table 2.5 shows the relation between power mismatch and frequency deviation. Generally, the frequency deviation indicates the power unbalance of the system and it should be regulated by using frequency control. Frequency control aims to balance the generated power and electrical demand to achieve a power system operation at the nominal system frequency. When the system frequency is higher than the nominal frequency due to over-generation ($P_m > P_e$ or P_L), to recover frequency back to its nominal value, the P_m must be reduced to achieve the balance with the P_e or P_L. On the contrary, when the frequency of the system is lower than the nominal frequency due to lack of generated power ($P_m < P_e$ or P_L), the P_m must

Table. 2.5 Relationship between power mismatch and frequency deviation

Power mismatch	Rotor speed (df/dt or $d\omega/dt$)	Frequency deviation (Δf)
$P_m > P_L$	Plus (acceleration)	Increasing
$P_m = P_L$	Zero (no acceleration)	Stable
$P_m < P_L$	Minus (deceleration)	Decreasing

be increased to achieve the balance with the P_e or P_L and restoring frequency to its nominal value.

A relation between the mechanical power and electrical power can be represented by the swing-equation as:

$$P_m - P_e = M\frac{df}{dt} + D\Delta f \quad (2.6)$$

where

$$M = 2H(\text{in per-unit})\text{or}M = 2H/f_0 \quad (2.7)$$

M is the moment of inertia with the unit of $\mathrm{J} \cdot \mathrm{s}$ or $\mathrm{W} \cdot \mathrm{s}^2$, H is the inertia constant, presented in a unit of second, and f_0 is the nominal frequency.

In addition to the primary control (see Fig. 2.7), the speed governor measures the variation in speed (i.e., frequency) through the loops of primary and secondary control. The hydraulic amplifier offers the required mechanical forces to adjust the primary valve against the high pressure from water or steam. The speed changer delivers a steady-state power output setting for the turbine.

At each generating unit, the speed governor delivers the primary speed control function, with all these generating units contribute to the frequency regulation without considering the locations of load variation. However, the primary control is not usually efficient to recover the frequency of the system, particularly in an interconnected system. Thus, the secondary control loop equipped in a large synchronous generator is applied to correct the load reference setpoint via the speed changer motor. The secondary control loop feedbacks the measurement of frequency deviation via a dynamic controller. In real practice, a simple integral (I) or proportional-integral (PI) controller is operated as the dynamic controller. Then, the output signal (P_c) is used to maintain system frequency.

Exercise 2.2 Considering the power system in Fig. 2.7, the generator delivers the active power of 450 MW (P_e) to the load at the nominal system frequency of 60 Hz. At the time $t = 5$ s, the load demand power is suddenly increased to 500 MW. Please determine a graph of the frequency change against time after the sudden load is increased. Assuming the mechanical power (P_m) is constant, before and after the sudden load change. The moment of inertia is given as 70 $\mathrm{MW} \cdot \mathrm{s}^2$.

Solution 2.2 At $t = 0$, before the load is suddenly increased, it is known that $P_m = P_e = 450$ MW (i.e., the balance between mechanical and electrical power). Thus, the frequency deviation is zero (i.e., $df/dt = 0$). It indicates that the system frequency is equal to the generator frequency (i.e., $f = f_0 = 60$ Hz).

At $t > 5$ s, after the load is suddenly increased, the electrical power increases from 450 to 500 MW, while the mechanical power is constant at 450 MW. Thus, considering Eq. (2.6), power parameters can be substituted as:

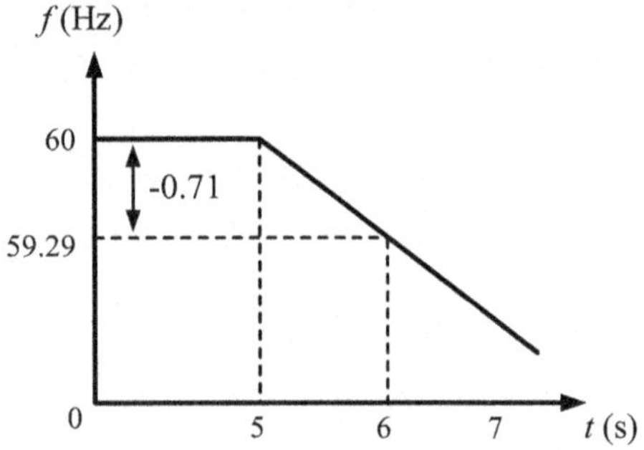

Fig. 2.8 Frequency deviation of Exercise 2.2

$$70\frac{df}{dt} = 450 - 500 = -50 \tag{2.8}$$

$$f(t) = f(5) + \int_0^t \left(\frac{50}{70}\right) dt = 60 - 0.71t \tag{2.9}$$

From the calculated Eq. (2.9), the graph of *df/dt* is plotted in Fig. 2.8.

2.4 Structure of Frequency Response Model

The dynamic characteristic of the power system is usually time-varying and nonlinear. However, to perform frequency control synthesis with respect to variation in load or output power of RESs, a linearized low-order structure is used. The dynamic effects of frequency response are quite slow compared with rotor angle and voltage dynamics, in the timescale of seconds to minutes.

Moreover, to perform the analysis of slow and fast power system dynamics by analyzing the dynamics of generation and load in detail, complex numerical techniques are required to allow changing of the simulation time step with the amount of fluctuation of system parameters [4, 21]. By ignoring the fast dynamics of rotor angle and voltage, the complexity of computation, data requirements, and modeling could be reduced. Thus, the results and analysis can be simplified.

A simplified structure of frequency response for the explained schematic diagram in Fig. 2.7 considering one generating unit is explained in this section.

The whole generator-load dynamic relation between the incremental mismatch power ($\Delta P_m - \Delta P_L$) and frequency deviation (Δf) can be described using the swing equation as [1, 2]:

$$\Delta P_m(t) - \Delta P_L(t) = 2H\frac{d\Delta f(t)}{dt} + D\Delta f(t) \tag{2.10}$$

where H is the inertia constant and D is the load damping coefficient.

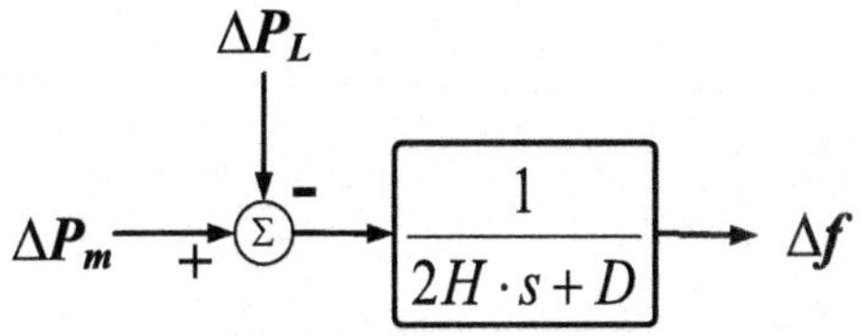

Fig. 2.9 Block diagram of the load-generator model for frequency control study

The load damping coefficient is directly determined as a percent variation in load for 1% variation in frequency. For example, a specified value of 2 for D indicates that a 1% variation in frequency could result in a 2% variation in load.

Then, Eq. (2.10) is transformed into the term of the Laplace as:

$$\Delta P_m(s) - \Delta P_L(s) = 2Hs\Delta f(s) + D\Delta f(s) \tag{2.11}$$

Equation (2.11) could be represented as a block diagram, as depicted in Fig. 2.9. The generator-load model significantly reduces the complexity of the schematic block diagram of a closed-loop synchronous generator model in Fig. 2.7. The reduced block diagram can be simply represented as Fig. 2.10.

Where ΔP_C is the change (signal) of secondary control action and ΔP_g is the change (signal) of governor control action.

The frequency response of a system relies on the integrated effects of control droops of speed governors (in generators) and loads. For such a system in Fig. 2.10, the steady-state frequency deviation can be expressed as [2]:

$$\Delta f_{ss} = \frac{-\Delta P_L}{\left(\frac{1}{R_1} + \frac{1}{R_2} + \cdots + \frac{1}{R_i}\right) + D} = -\frac{\Delta P_L}{\frac{1}{R_T} + D} \tag{2.12}$$

where

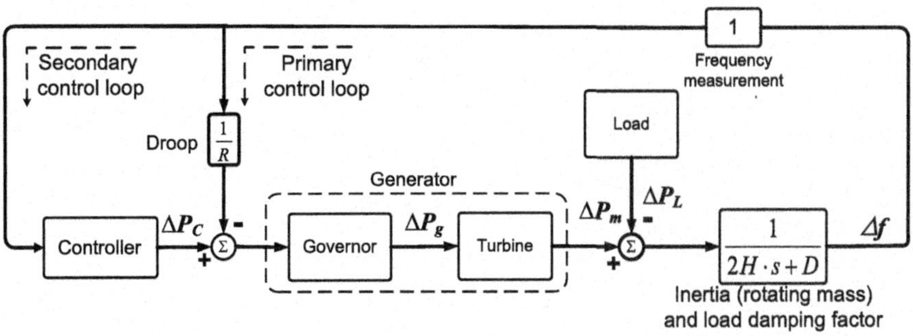

Fig. 2.10 Reduced block diagram of Fig. 2.7

$$R_T = \frac{1}{\frac{1}{R_1} + \frac{1}{R_2} + \cdots + \frac{1}{R_i}} \tag{2.13}$$

where R_i is the droop characteristic at the generator i, and R_T is the equivalent droop characteristic.

Exercise 2.3 A power system comprises of 3 identical 200 MVA generating units feeding a total load of 300 MW at 60 Hz. The inertia constant (H) for each unit is four on the 200 MVA base. The load changes 2.5% for a 1% change in system frequency. Then, the load suddenly dropped by 30 MW. Please determine the following:

(1) The system block diagram contains H and D expressed on 600 MVA base.
(2) The system frequency deviation, if there is no speed governor control.

Solution 2.3

(1) For three units on 600 MVA base, the inertia constant can be determined as:

$$H = 4(200/600)(3) = 4\ \text{s}.$$

Expressing D for the remaining load (300–30 = 270 MW) on 600 MVA base,

$$D = 2.5(270)/600 = 1.125\%.$$

(2) When there is no speed governor control ($\Delta P_m = 0$), the system block diagram with parameters expressed in p.u. on 600 MVA is shown in Fig. 2.11.

Referring to the standard form of the gain and time constant, K, and T can be represented as Fig. 2.12.

where,

$$K = \frac{1}{D} = \frac{1}{1.125} = 0.88 \tag{2.14}$$

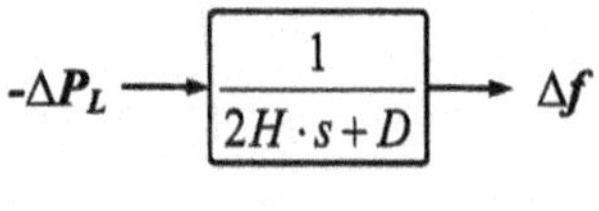

Fig. 2.11 Block diagram of the generator-load model for Exercise 2.3

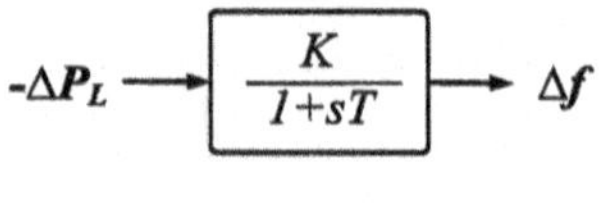

Fig. 2.12 Block diagram of the standard form of the gain and time constant

$$T = \frac{2H}{D} = \frac{2(4)}{1.125} = 7.11\text{s} \tag{2.15}$$

The sudden load change from MW to p.u. is calculated as:

$$\Delta P_L = -30 = \frac{-30}{600} = -0.05\text{p.u.} \tag{2.16}$$

For a step decrease in load by 0.05 p.u., the Laplace transform of the load change can be expressed as:

$$\Delta P_L(s) = \frac{-0.05}{s} \tag{2.17}$$

Thus, by considering Eq. (2.17) and Fig. 2.12, the frequency deviation can be obtained as:

$$\Delta f(s) = -\left(\frac{-0.05}{s}\right)\left(\frac{K}{1+sT}\right) \tag{2.18}$$

Applying the inverse transform to Eq. (2.18), it can be represented in terms of the time domain as:

$$\Delta f(t) = -0.05Ke^{-\left(\frac{t}{T}\right)} + 0.05\text{K} \tag{2.19}$$

Substituting K and T into Eq. (2.19), it can be represented as:

$$\Delta f(t) = -0.044e^{-(0.14t)} + 0.044 \tag{2.20}$$

Thus, the time constant T is 7.11 s, and the steady-state deviation can be computed as (see Fig. 2.13):

$$\Delta f_{ss} = 0.044 f_0 = 0.044(60) = 2.64\text{Hz} \tag{2.21}$$

Numerous low-order models for generator and turbine dynamic representation have been presented for the use in system frequency stability analysis and control design [1, 2, 22–27]. The slow system boiler dynamics and fast generator dynamics are neglected in these structures. The useful block diagrams of the turbine and speed governor for the steam and hydraulic types are proposed in Fig. 2.14 for frequency control study. R and R_h are the speed droop characteristics, which can provide the speed regulation according to the governor action. T_t, T_g, T_{tr}, T_r, T_{th}, and T_{gh} represent the turbine-generator time constants.

The integrated structure of the control blocks in Figs. 2.10 and 2.14 is shown in Fig. 2.15. It represents the control block for a non-reheat steam generator with regards to frequency control loops, including generator, turbine, governor, rotating mass (inertia compensation), primary control, secondary control, and load.

Fig. 2.13 System frequency deviation of Exercise 2.3

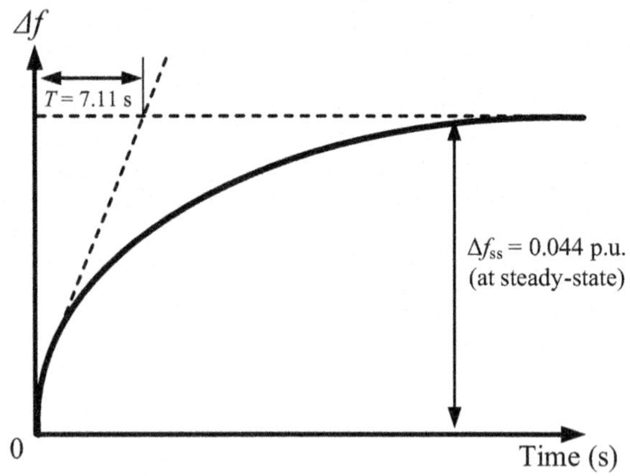

Where ΔP_P is the change (signal) in primary control action, ΔP_C is the change (signal) in secondary control action, and β is the frequency bias factor.

2.5 Frequency Regulation in a Single-Area Power System

With the approximation, an equivalent Eq. (2.11) can be used for the study of a single-area (isolated) power system. To perform frequency stability study and analysis for a single-area power system, it is common to model a multi-generator dynamic behavior using an equivalent generator model, as displayed in Fig. 2.15. Subsequently, the combined model in Fig. 2.15 can be used as an equivalent frequency analysis model for the whole generators of one area power system.

In addition to the system loads and generators, the equivalent model lumps their damping effects into a single damping factor/constant. The equivalent system inertia constant is supposed to be equal to the sum of the inertia constant of all generating units. Nevertheless, it is noted that the generators-turbines and individual control loops have similar control parameters and response behaviors. It should also be noted that the equivalent structure is useful only to simplify the frequency stability study and analysis of an isolated system.

The dynamic response of the single-area power system due to sudden (step) changes of the load disturbance (0.02 p.u. at 3 s and 0.01 p.u. at 15 s) is plotted in Fig. 2.16. This figure displays the importance of the primary and secondary control deployments during the disturbances. By only applying the primary control, it is obvious that the frequency of the system could not restore to its nominal value. By integrating the secondary control into the system, the system frequency could properly restore to its nominal value within a few seconds. Evidently, the frequency nadir and overshoot of the system significantly reduce. The system parameters for

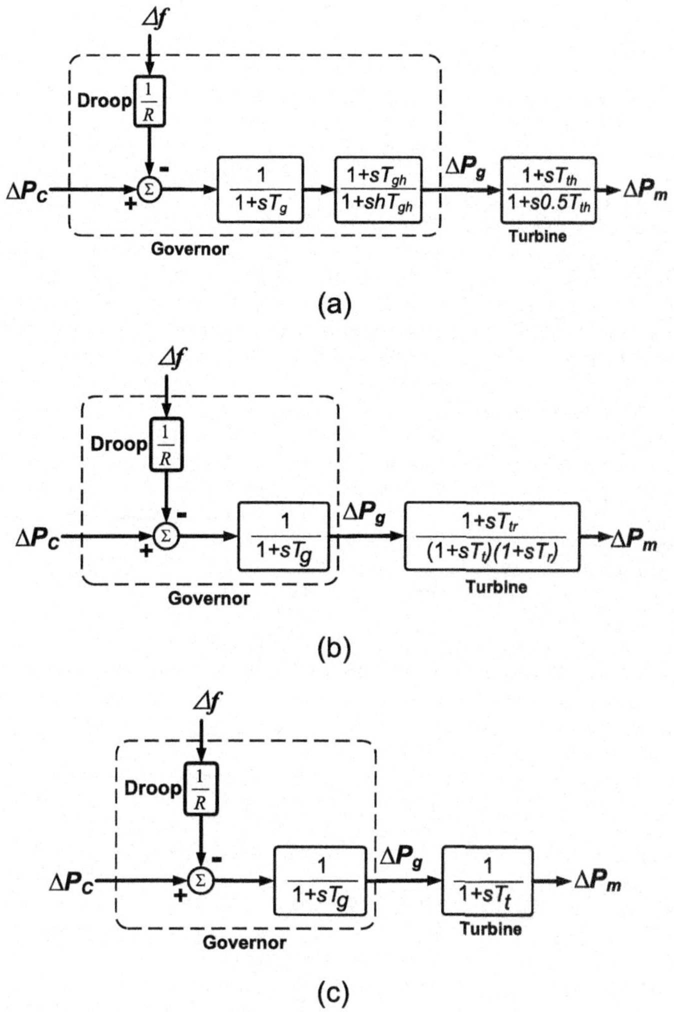

Fig. 2.14 Control block of the governor-turbine system; **a** hydraulic unit, **b** reheat steam unit, **c** non-reheat steam unit

the conducted simulation are shown in Table 2.6. The simulation is performed using the MATLAB/Simulink environment.

To investigate the effect of inertia power compensation, the system inertia constant is reduced by 50% from its nominal value and the dynamic response of the system is plotted in Fig. 2.17. Due to the reduction of inertia power, it is obvious that the inertial response of the system consequently reduces, resulting in higher RoCoF. As a result, it yields the larger frequency overshoot and nadir, which requires longer stabilizing time after the disturbance. Moreover, following the disturbance, the generated power is more fluctuating, leading to the stress in the generating unit.

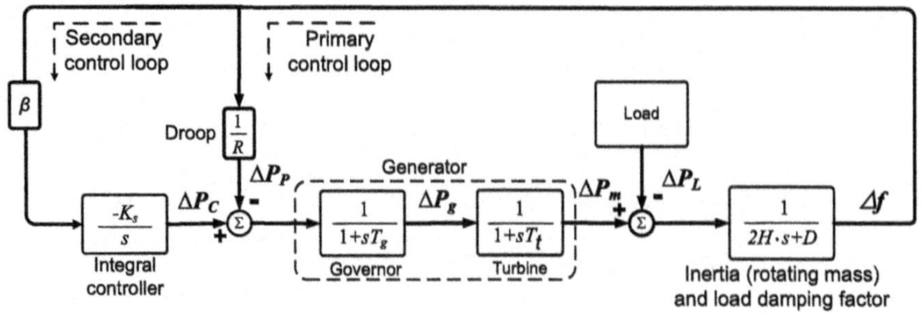

Fig. 2.15 Combined dynamic model of a non-reheat steam generator with inertia compensation, primary and secondary controls for frequency analysis

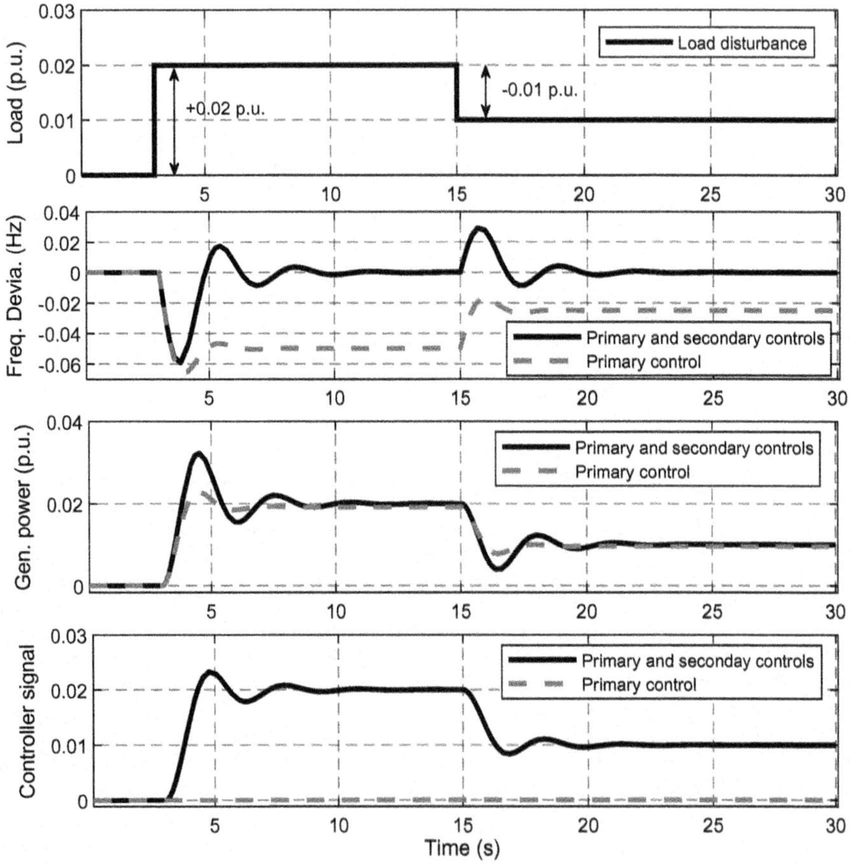

Fig. 2.16 Dynamic response of the single-area power system

Table 2.6 Simulated parameters for the isolated system

Parameter	Value
Gain of Integral controller, K_s (s)	0.35
Governor time constant, T_g (s)	0.07
Turbine time constant, T_t (s)	0.37
Governor droop constant, R (Hz/p.u.)	2.60
Bias factor, β (p.u./Hz)	0.98
System inertia constant, H (p.u. s)	0.083
System load damping coefficient, D (p.u./Hz)	0.016

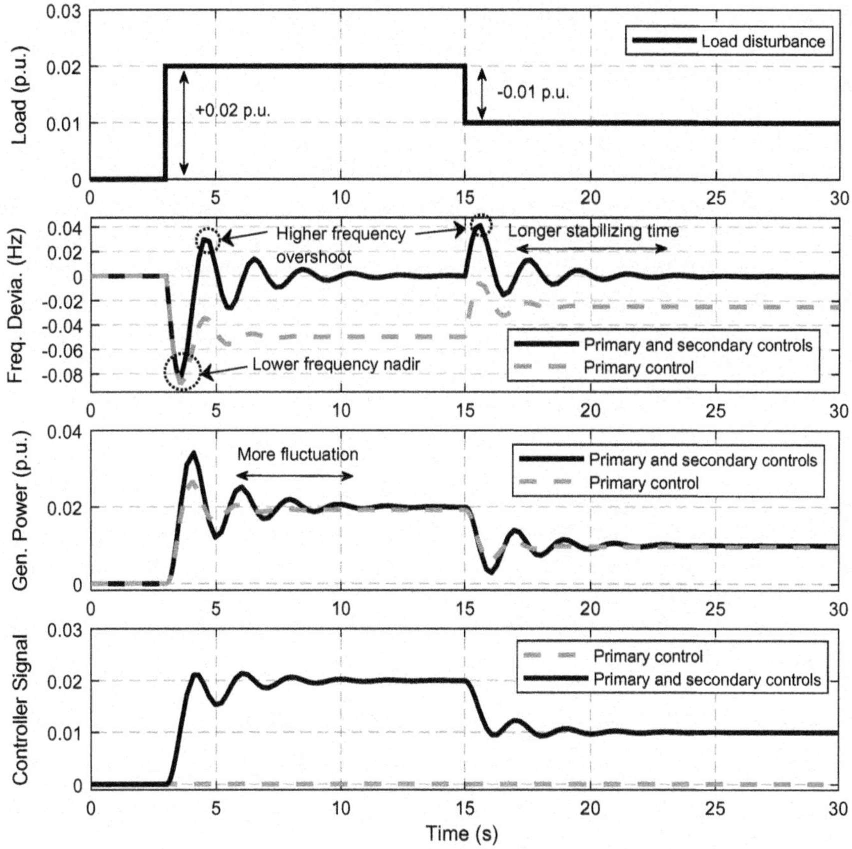

Fig. 2.17 Dynamic response of the single-area power system during the lack of inertia power compensation

2.6 Frequency Regulation in Interconnected Power Systems

In real practice, most power systems are interconnected via tie-lines or transmission lines to help each other in regulating and exchanging interchange power, improving system stability, reliability, and resiliency [28]. In the single-area power system, the interchange power regulation is not a control problem, and the operation of secondary frequency control is restricted to recover the frequency of the system to the fixed or nominal value. To configure the explained structure in Fig. 2.15 for a multi-area (interconnected) power system, the concept of the control area is required to be applied as it is a coherent area consisting of a group of loads and generators, where the whole generators respond to variations in speed changer setting or load in unison. It should be noted that the frequency is presumed to be similar in all nodes of a control area.

A multi-area power system usually contains the regions or areas, which are connected by the tie-lines or high-voltage transmission lines. The measured frequency trend in each control area is a key indicator of the mismatch power trend in the interconnected system. Additionally, the secondary frequency control in each area of the interconnected system could control the interchange power with the other control areas together with its local frequency. Hence, the explained dynamic structure in Fig. 2.15 must be re-established by considering the tie-line power signal. Figure 2.18 shows the concept of an interconnected system connected with *j*-control areas. The tie-line power flow from Area 1 to Area 2 can be represented as [1, 2]:

$$P_{Tie,12} = \frac{V_1 V_2}{X_{12}} sin(\delta_1 - \delta_2) \tag{2.22}$$

where V_1, V_2 are the voltages at the equivalent generator's terminals of Area 1 and 2. X_{12} is the tie-line reactance between Area 1 and 2, and δ_1, δ_2 are the power angles of the equivalent generators of Area 1 and 2.

Equation (2.22) is linearized at an equilibrium point (δ_1^0, δ_2^0) and represented as [2]:

$$\Delta P_{Tie,12} = T_{12} cos(\Delta\delta_1 - \Delta\delta_2) \tag{2.23}$$

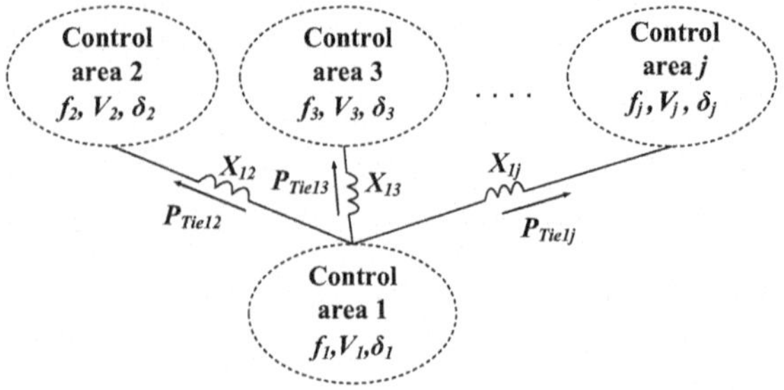

Fig. 2.18 The concept of an interconnected system connected with *j*-control areas

where T_{12} is the synchronizing torque coefficient expressed as:

$$T_{12} = \frac{|V_1||V_2|}{X_{12}} cos\left(\delta_1^0 - \delta_2^0\right) \tag{2.24}$$

Applying the relationship between frequency and area power angle, Eq. (2.23) can be rewritten as:

$$\Delta P_{Tie,12} = 2\pi T_{12}\left(\int \Delta f_1 - \int \Delta f_2\right) \tag{2.25}$$

where Δf_1 and Δf_2 are the frequency deviations of Area 1 and 2, respectively.

Applying the Laplace transform to Eq. (2.25), it can be rewritten as:

$$\Delta P_{Tie,12}(s) = \frac{2\pi}{s} T_{12}(\Delta f_1(s) - \Delta f_2(s)) \tag{2.26}$$

Likewise, the tie-line power change between Area 1 and Area 3 is obtained as:

$$\Delta P_{Tie,13}(s) = \frac{2\pi}{s} T_{13}(\Delta f_1(s) - \Delta f_3(s)) \tag{2.27}$$

Finally, the total tie-line power change between Area 1 and the other two areas can be determined as:

$$\Delta P_{Tie,1}(s) = \Delta P_{Tie,12}(s) + \Delta P_{Tie,13}(s)$$

$$= \frac{2\pi}{s}\left[(T_{12}\Delta f_1(s) + T_{13}\Delta f_1(s)) - \left(T_{12}\Delta f_2(s) + T_{13}\Delta f_3(s)\right)\right] \tag{2.28}$$

Similarly, for *j*-control areas, the total tie-line power change between Area 1 and the other areas can be obtained as [1]:

$$\Delta P_{Tie,j}(s) = \sum_{\substack{j=1 \\ j \neq k}}^{y} \Delta P_{Tie,jk}(s) = \frac{2\pi}{s}\left[\sum_{\substack{j=1 \\ j \neq k}}^{y} T_{jk}\Delta f_j - \sum_{\substack{j=1 \\ j \neq k}}^{y} T_{jk}\Delta f_k\right] \tag{2.29}$$

where y is the total number of control areas.

Then, Eq. (2.29) can be written in terms of a block diagram as depicted in Fig. 2.19. It is noted that the influence of tie-line power change for an area is equivalent to the

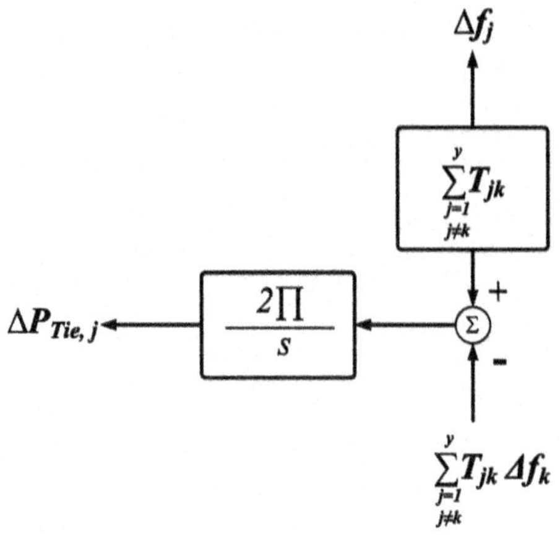

Fig. 2.19 Control block for the tie-line power change of the control area j

variation of load in that area. Thus, $\Delta P_{Tie,j}$ has to be included in the mechanical (generated) power change and area load variation.

Combining the control blocks between Figs. 2.15 and 2.19, the frequency response model for the j-control area (interconnected) system is shown in Fig. 2.20.

Later, the secondary control loop of the interconnected system must be modified due to the tie-power configuration. In the single-area power system, the secondary control provides a suitable control action/signal to suppress the steady-state frequency deviation to zero. In the case of the interconnected power system, the secondary control should regulate the area frequency and total interchange power with the other interconnected areas at scheduled values. This task can be done by

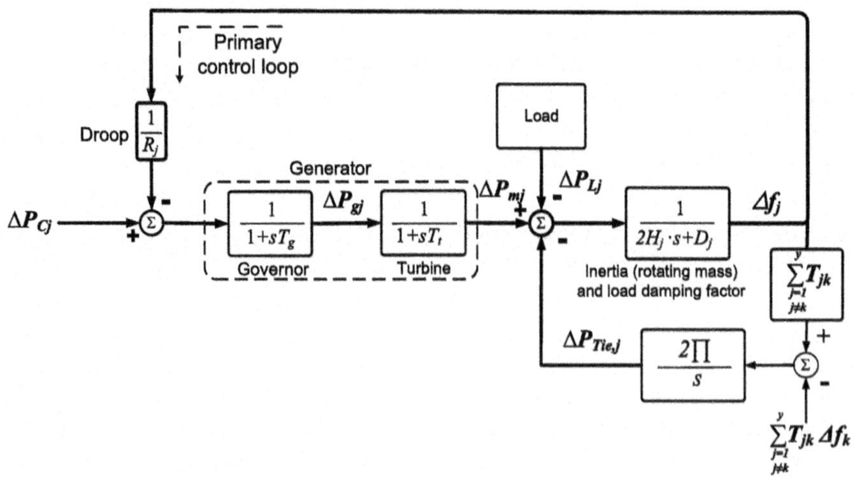

Fig. 2.20 Dynamic model for the control area j without secondary control

including the tie-line deviation to the frequency deviation in the secondary control loop, as shown in Fig. 2.21. An appropriate linear combination of tie-line power and frequency variations for the area j is called the area control error (ACE) [2]. The ACE equation is expressed as:

$$\Delta P_{Tie,j}(s) = ACE_j = \Delta P_{Tie,j} + \beta_j \Delta f_j \tag{2.30}$$

where β_j is the bias factor, with its optimal value can be determined as:

$$\beta_j = D_j + \frac{1}{R_j} \tag{2.31}$$

Figure 2.22 shows the schematic diagram of the frequency response of the three-area interconnected system. Each area monitors its own frequency and tie-line power flow at the area control center. The size of system inertia (H), which is the ability

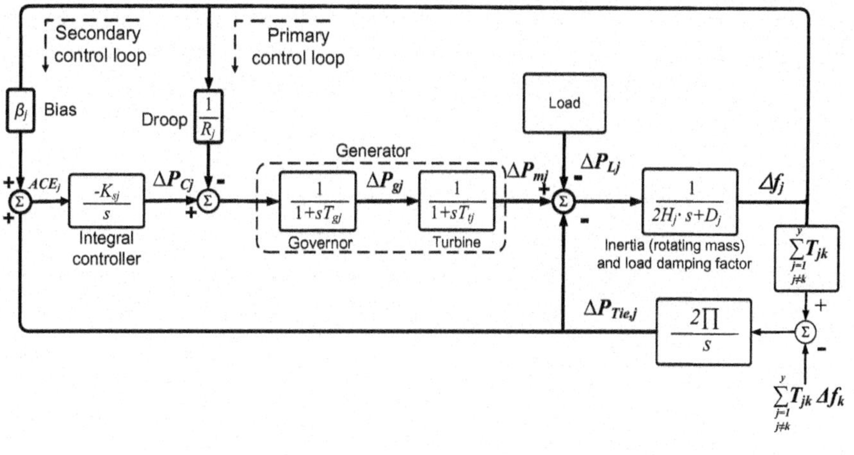

Fig. 2.21 Dynamic model for the control area j with secondary control

Fig. 2.22 The schematic diagram of the three-area interconnected system

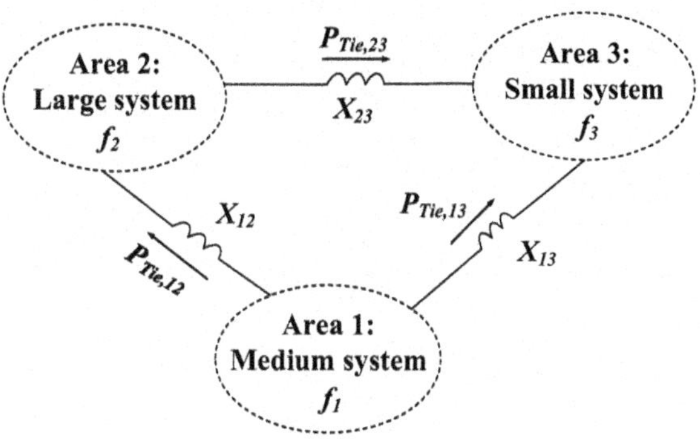

to resist a disturbance, indicates the strength of each area. Figure 2.23 shows the small-signal/dynamic model of the three-area interconnected system. The system parameters for the conducted simulation are shown in Table 2.7.

Figure 2.24 shows the dynamic response of the interconnected system due to multiple load disturbances in all areas. Following the first disturbances at 2 s (i.e., 0.025 p.u. in Area 1 and 0.015 p.u. in Area 3), the frequency drops, which are sensed by speed governors from all generators. After a few seconds, the additional power for responding to the disturbances is only generated from the disturbance areas (i.e., Areas 1 and 3) to compensate for local load changes. Similarly, at 15 s, the disturbance of 0.02 p.u. occurs in Area 2 and the additional power for responding to the disturbance is only generated from Area 2. Due to the interconnection in all areas, it is obvious that the disturbance in one area could affect frequency stability in all areas.

To examine the effect of inertia power compensation, the system inertia constants for all areas are reduced by 50% from its nominal values and the dynamic response of the interconnected system is plotted in Fig. 2.25. Due to the reduction of inertia power, it is obvious that the inertial response of all areas consequently reduces, resulting in higher RoCoF and larger frequency overshoot and nadir, with longer stabilizing time after the disturbances. Moreover, following the disturbance, the generated power in all areas is more fluctuating, leading to the stress in all generating units.

2.7 Analysis of Steady-State Frequency Response

To perform frequency response analysis, considering Fig. 2.26, it is supposed that all generator units i in a control area j are non-reheat steam systems and represented as:

$$M_{ij}(s) = \left[\frac{1}{1 + sT_{gij}}\right] \cdot \left[\frac{1}{1 + sT_{tij}}\right] \tag{2.32}$$

Then, the frequency deviation in an area j can be obtained as [1, 2]:

$$\Delta f_j(s) = \frac{1}{2H_j + D_j}\left[\sum_{i=1}^{n} \Delta P_{mij}(s) - \Delta P_{Lj}(s) - \Delta P_{Tie,j}(s)\right] \tag{2.33}$$

where

$$\Delta P_{mij}(s) = M_{ij}(s) \cdot \left[\Delta P_{Cij}(s) - \Delta P_{Pij}(s)\right] \tag{2.34}$$

and

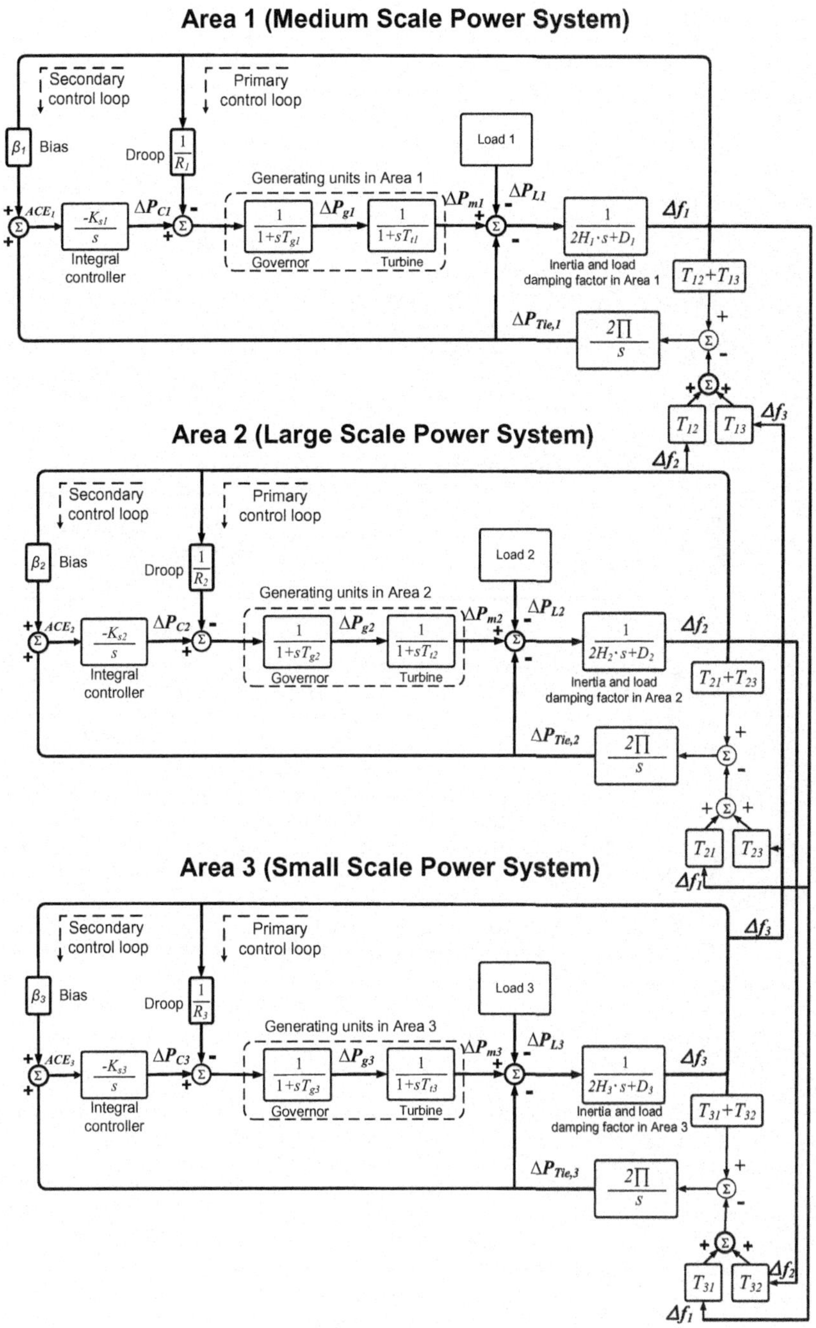

Fig. 2.23 Dynamic model of the three-area interconnected system

Table 2.7 Simulated parameters for the interconnected system

Parameter	Area 1	Area 2	Area 3
Gain of Integral controller, K_s (s)	0.35	0.25	0.40
Governor time constant, T_g (s)	0.07	0.05	0.06
Turbine time constant, T_t (s)	0.38	0.45	0.35
Governor droop constant, R (Hz/p.u.)	3.0	2.73	2.81
System inertia constant, H (p.u. s)	0.083	0.100	0.0623
System load damping coefficient, D (p.u./Hz)	0.015	0.017	0.014
Frequency bias factor, β (p.u./Hz)	0.34	0.38	0.36
Synchronizing torque coefficient, T (p.u./Hz)	$T_{12} = 0.2$	$T_{21} = 0.2$	$T_{31} = 0.25$
	$T_{13} = 0.25$	$T_{23} = 0.12$	$T_{32} = 0.25$

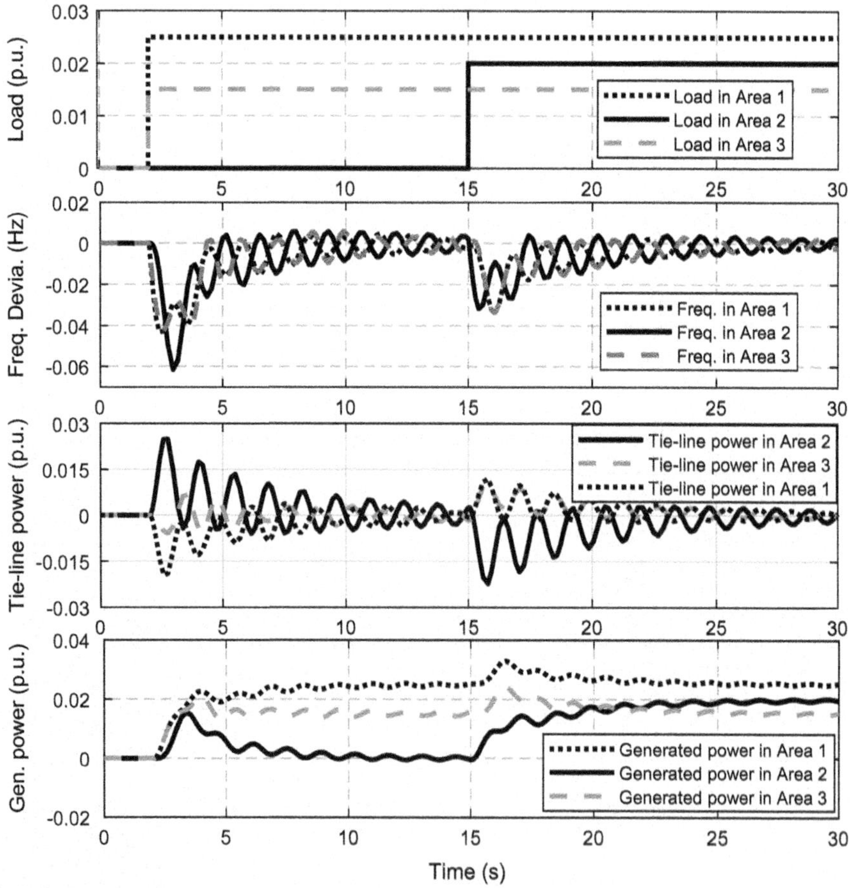

Fig. 2.24 Dynamic response of the three-area interconnected system

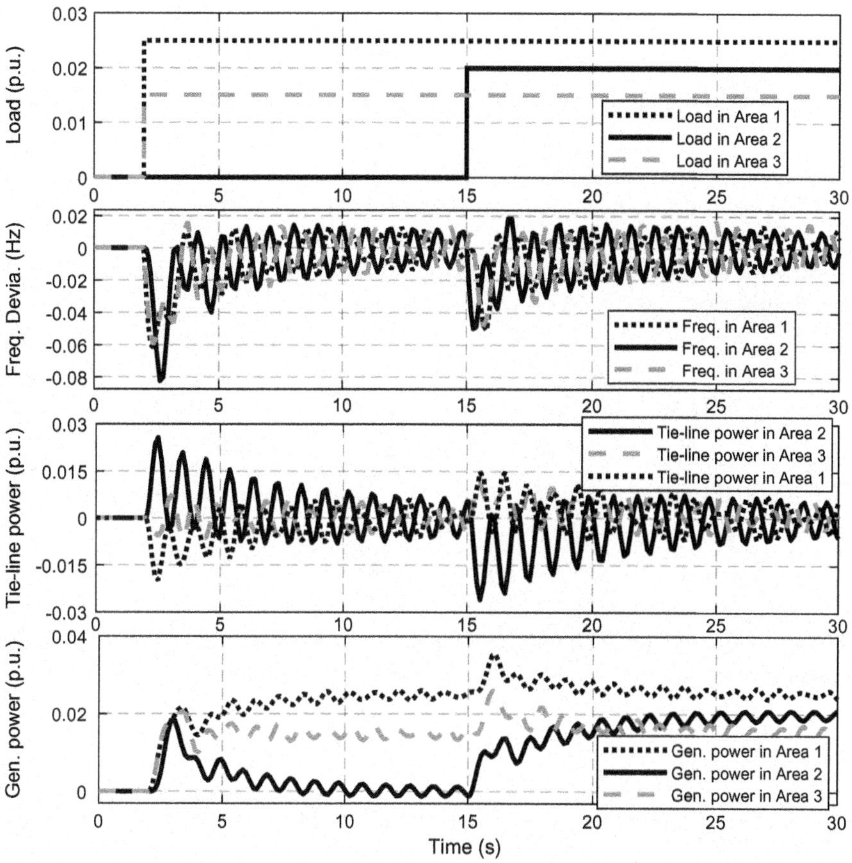

Fig. 2.25 Dynamic response of three-area interconnected system during the lack of inertia power compensation

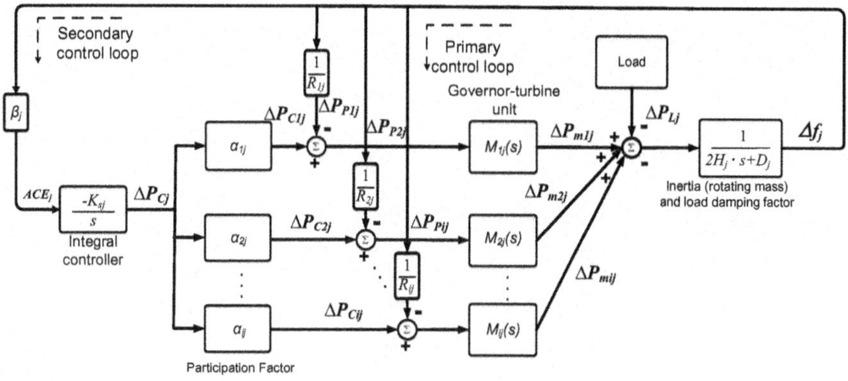

Fig. 2.26 Frequency response model of a single-area system with multiple generators

$$\Delta P_{Pij}(s) = \frac{\Delta f_j(s)}{R_{ij}} \tag{2.35}$$

In the equations above, ΔP_P is the change (signal) in primary control action, ΔP_C is the change (signal) in secondary control action, and ΔP_{Tie} is the tie-line power change for an interconnected power system. In a single-area power system, $\Delta P_{Tie} = 0$. The α_{ij} is the frequency control participation factor in the control area j for the generator unit i, which will be explained in the following section.

Substituting Eqs. (2.34) and (2.35) into (2.33), the resulting equation can be represented as:

$$\Delta f_j(s) = \frac{1}{2H_j + D_j}\left[M_{ij}(s) \cdot \left[\Delta P_{Cij}(s) - \frac{\Delta f_j(s)}{R_{ij}}\right] - \Delta P_{Lj}(s) - \Delta P_{Tie,j}(s)\right] \tag{2.36}$$

For load disturbance analysis, ΔP_L is considered as a step function as:

$$\Delta P_{Lj}(s) = \frac{\Delta P_{Lj}}{s} \tag{2.37}$$

Substituting Eq. (2.37) into (2.36), the resulting equation can be summarized and represented as:

$$\Delta f_j(s) = \frac{1}{L_j(s)}\left[M_{ij}(s) \cdot \Delta P_{Cij}(s) - \Delta P_{Tie,j}(s)\right] - \frac{\Delta P_{Lj}}{sL_j(s)} \tag{2.38}$$

where

$$L_j(s) = 2H_j + D_j + \frac{M_{ij}(s)}{R_{ij}} \tag{2.39}$$

Substituting Eq. (2.32) into Eqs. (2.38) and (2.39) and applying the final value theory, the steady-state frequency deviation of the system can be obtained as:

$$\Delta f_{ss,j} = \lim_{s\to 0} s\Delta f_j(s) = \frac{1}{L_j(0)}\left(\Delta P_{Cij} - \Delta P_{Lj}\right) \tag{2.40}$$

Assuming that ΔP_{Tie} is equal to zero at the steady-state, thus:

$$\Delta P_{Cj} = \lim_{s\to 0} s\left(\sum_{i=1}^{n} M_{ij}(s) \cdot \Delta P_j(s)\right) \tag{2.41}$$

$$L_j(0) = \sum_{i=1}^{n} \frac{1}{R_{ij}} + D_j = \frac{1}{R_{Tj}} + D_j \tag{2.42}$$

where R_{Tj} is the equivalent droop characteristic for an area j and represented as:

$$\frac{1}{R_{Tj}} = \sum_{i=1}^{n} \frac{1}{R_{ij}} \tag{2.43}$$

Based on Eq. (2.31), the $L_j(0)$ is equivalent to the frequency response characteristic of the system (β_j) as:

$$\beta_j = \frac{1}{R_{Tj}} + D_j \tag{2.44}$$

By applying Eq. (2.42), Eq. (2.40) can be rewritten as:

$$\Delta f_{ss,j} = \frac{\Delta P_{Cj} - \Delta P_{Li}}{1/R_{Tj} + D_j} \tag{2.45}$$

From Eq. (2.45), it can be described that if the magnitude of the disturbance matches with the available power reserve via secondary control ($\Delta P_{Cj} = \Delta P_{Lj}$), the system frequency deviation becomes zero at the steady-state condition. For large generating units, a suitable R_{ij} is between 0.05 and 0.1.

In the case of a small value of D_j and R_{Tj}, Eq. (2.45) can be reduced as:

$$\Delta f_{ss,j} = \frac{(\Delta P_{Cj} - \Delta P_{Li})R_{Tj}}{R_{Tj}D_j + 1} \cong (\Delta P_{Cj} - \Delta P_{Li})R_{Tj} \tag{2.46}$$

In the case of no secondary control ($\Delta P_{Cj} = 0$), the frequency deviation in steady-state depends on the magnitude of the disturbance as:

$$\Delta f_{ss,j} = \frac{(-\Delta P_{Li})R_{Tj}}{R_{Tj}D_j + 1} \tag{2.47}$$

To simplify the dynamic frequency analysis, the governor-turbine time constants are considered as a smaller value compared with the time constant of a power system (rotating mass and load). Thus, it is acceptable to assume that T_{gj} and T_{tj} are zero [2]. With this condition, Eq. (2.38) can be reduced as:

$$\Delta f_j(s) \cong -\frac{\Delta P_{Lj}}{s}\left(\frac{1}{2H_j + D_j + \frac{1}{R_{Tj}}}\right) \tag{2.48}$$

By simplifying Eq. (2.48) and arranging into partial fractions, the resulting equation can be represented as:

$$\Delta f_j(s) \cong \frac{(-\Delta P_{Lj})R_{Tj}}{R_{Tj}D_j + 1}\left(\frac{1}{s} + \frac{1}{s + \frac{R_{Tj}D_j+1}{2H_jR_{Tj}}}\right) \tag{2.49}$$

By substituting Eqs. (2.47) into (2.49), the resulting equation can be represented as:

$$\Delta f_j(s) \cong \Delta f_{ss,j}\left(\frac{1}{s} - \frac{1}{s + \tau_j}\right) \tag{2.50}$$

where τ_j is the time constant of the closed-loop system represented as:

$$\tau_j = \frac{1 + R_{Tj}D_j}{2H_jR_{Tj}} \tag{2.51}$$

By applying the inverse Laplace transformation into Eq. (2.50), the equation can be rewritten in terms of the time domain as:

$$\Delta f_j(t) \cong \Delta f_{ss,j}\left(1 - e^{-\tau_j(t)}\right) \tag{2.52}$$

2.8 Participation Factor for Frequency Control

In each control area, there are several generators/machines with different generator types and governor-turbine parameters. Nowadays, more distributed generators (DGs), usually based on renewable energy, are installed in the power systems to solve energy crisis and environmental issues. The generating units are more decentralized and may or may not engage in the frequency control task [1]. Thus, the generator participation rates are not the same for all generators. To consider the variety of generating dynamics and their shared rates in the secondary control action, the dynamic model of the control area j in Fig. 2.21 can be reconfigured as Fig. 2.27.

Where α_{ij} is the frequency control participation factor for the generator unit i in a control area j, and M_{ij} is the turbine-governor model for the generator unit i.

After a disturbance in the control area, the suitable secondary control signal is allocated among generators in proportion to their participation factor, forcing the generating units to follow the load change. For a control area, the sum of participation factors must be equal to one as [1, 2]:

$$\sum_{j=1}^{n} \alpha_{ji} = 1, 0 \leq \alpha_{ji} \leq 1 \tag{2.53}$$

where n is the total number of generators in the control area.

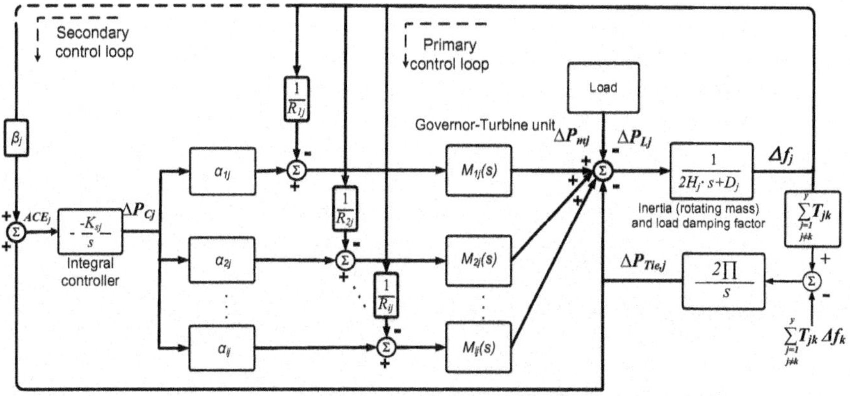

Fig. 2.27 Dynamic structure for the control area j with the participation factor

In real practice, participation factors are time-dependent parameters. Thus, system operators may dynamically calculate these factors based on availability, bid prices, costs, congestion problems, and other related problems.

2.9 Physical Constraints for Frequency Control

Previously, the simulated studies on frequency control performance have been modeled based on a linearized analysis. The explained models did not consider physical constraints. However, analyzing all dynamics in frequency response modeling may be difficult and not useful [1]. Thus, to obtain a precise perception of power system frequency study, it is essential to analyze the basic constraints and significant inherent requirements related to the physical dynamics of the system, modeling them for the goal of performance evaluation.

2.9.1 Governor Dead Band and Generation Rate

One of the significant physical constraints is the rate of change of power generation according to the restriction in mechanical and thermal aspects. The dynamic frequency model that does not consider the delays triggered by the crossover devices in the thermal generating unit or the characteristic of the penstocks in a hydraulic system, leads to a circumstance where the tie-line power and system frequency could be restored to their scheduled value within a second.

In the real practice of a frequency control system, fast-changing devices of system signals are nearly unobservable due to several filters related to the process. Thus, the designed control system performance is dependent on how generation units react to

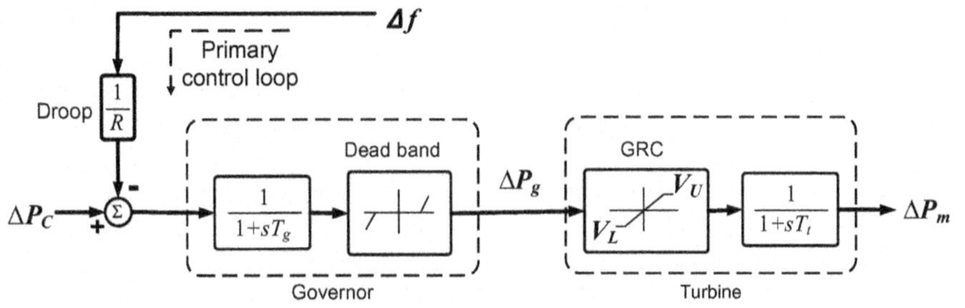

Fig. 2.28 Configuration of the dead band and GRC for a non-reheat generating unit

the control signals. A very rapid response from secondary control is neither desirable nor possible. Hence, the suitable control technique must regulate adequate levels of reserved control rate and control range.

Usually, the generation rate of the non-reheat thermal generating units is higher than the generation rate of the reheat thermal generating units [1, 29, 30]. The reheat units have a rate of about 3–10% p.u. MW per minute. For the hydro units, the rate is about 100% maximum continuous rating per minute [31]. The investigations and results of the impacts of the generation rate constraint (GRC) on the performance of frequency control systems have been fully explained in [32–34].

Another important constraint is the speed governor dead band. By varying an input signal, the speed governor might not suddenly respond until the input achieves a certain value. All governors should consider the dead band component, which is significant for frequency control under the disturbances. The dead band is set as the whole magnitude of a sustained speed change, in which there is no resulting variation in valve position. The governor's dead band effect is to improve the apparent steady-state speed control [35]. The maximum value of the governor dead band of traditional large-scale steam turbine is set as 0.06% (0.036 Hz) [36]. The investigations and results of the impacts of the governor's dead band on the performance of frequency control systems have been fully reported in [1, 35, 37–39].

The GRC and dead band can be studied by integrating the hysteresis pattern and limiter to the turbine-governor structure as depicted for a non-reheat steam turbine in Fig. 2.28. The V_L and V_U are the lower and upper constraints that limit the rate of valve/gate opening and closing speed.

2.9.2 *Time Delay*

In the real practice of a frequency control system, fast response and changing devices of frequency are almost unobservable due to several delays and filters related to the process. Any signal filtering and processing create delays that must be analyzed. Practical filters on ACE and tie-line measurements use about 2 s or more for the decision cycles and data acquisition of the frequency control system [1].

Currently, the communication delays in frequency control analysis have received a more important challenge according to the expanding and restructuring of functionality, complexity, and physical setups of power systems. During the last decades, research works on frequency control analysis have ignored issues related to the communication network. In a system using conventional communication links, it was considered as a valid assumption. Hence, the employment of an open communication infrastructure to assist ancillary services (e.g., secondary control) under the deregulated conditions is increasing concerns on the issues that may arise in the communication system.

Focusing on the time delays in the secondary control, the delays occur on the communication channels between the operating stations and the control center. Especially, the delays exist on the measurement of tie-line power flow and frequency from the remote terminal units (RTUs) to the control center and the control signal from the control center to individual generating units to the control center. The structure of such delays is described in Fig. 2.29. The delay is represented by an exponential function block $e^{-s\tau}$, where τ or t is the communication delay time, t_s and t_t are the communication delay time for secondary control and tie-line control units, respectively.

The investigations of the impacts of the time delay on the performance of frequency control systems have been fully reported in [14, 39–45]. The introduction of time delays in secondary control decreases the efficacy and performance of the controlled system. The frequency control performance significantly reduces due to the increase in the delay time. A compensation technique for communication time delay in the frequency control system is proposed in [40, 42, 43, 45, 46].

Fig. 2.29 Configuration of time delays in the secondary control

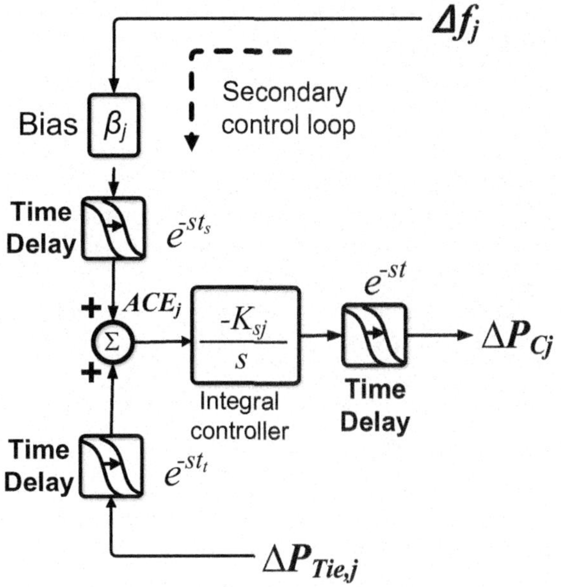

2.10 Generation Droop Characteristics

The droop characteristics of generating units are designed to preserve the frequency stability of the power system following a disturbance. To understand a concept of droop characteristics, a single-machine/generator infinite bus model is introduced by the swing equation of a synchronous machine as [1, 2]:

$$P_m - P_e = M\frac{df}{dt} + D\Delta f \tag{2.54}$$

where,

$$P_e = \frac{V_\infty V_G sin\delta}{x_d' + x_l} \tag{2.55}$$

In the equations above, M is the moment of inertia, D is the load damping coefficient, P_e is the electrical output, P_m is the mechanical input to the generator, V_∞ is the voltage of the infinite bus, V_G is the generator voltage, δ is the generator rotor angle, x_l is the line reactance, and x_d' is the transient reactance of the generator.

For simplicity, the resistances of the transmission line and generator are not considered. It is presumed that the V_G and V_∞ are constant, and $sin\Delta\delta \cong \Delta\delta$, $cos\Delta\delta \cong 1$. Thus, the output power deviation of the generator can be represented as:

$$\Delta P_e \cong \frac{V_\infty V_G cos\delta_0}{x_d' + x_l}(\Delta\delta) \tag{2.56}$$

where

$$\Delta\delta = \delta_G - \delta_\infty \tag{2.57}$$

in which δ_G and δ_∞ are the angles of V_G and V_∞, respectively,

For small $\Delta\delta$, the ΔP mainly relies on the $\Delta\delta$. This indicates that the generator determines the transferred active power (P_e) flows from itself to the network depends on the phase of its output voltage.

To enable a feedback loop to control the active power of the generator and the frequency, the following relationship is used [1]:

$$-R(P_G - P_{G0}) = f - f_0 \tag{2.58}$$

where f_0 is the nominal frequency, P_G is the generated active power, and P_{G0} is the nominal active power. Using this relationship, if the frequency increases above its nominal value, the generator would decrease its power output to counteract the frequency increase and vice versa. The change in the generator output power will depend on the droop characteristic R, which can be represented as:

$$\frac{\Delta f}{\Delta P_G} = -R \tag{2.59}$$

For example, a 10% droop indicates that a 10% change in frequency (i.e., from 60 to 54 Hz or from 50 to 45 Hz) results in a 100% change in output power. The droop characteristics for a generating unit is shown in Fig. 2.30, which is related to Eq. (2.58). It is noted that the unit of R is Hz/p.u.MW.

The generating units with different droop characteristics can simultaneously participate in tracking a load change and recovering the frequency, as shown in Fig. 2.30. From this figure, two generating units are operated at the same nominal frequency with different output powers. A load increase in the network will result in the reduction of system frequency (and the speed of generating units). To prevent the frequency from falling, the governors raise the output power of the generating units until the generation-load balance is restored at a new common operating frequency. The amount of generated power by each generating unit i to respond to the load change relies on the droop characteristics of each unit as:

$$\Delta P_{Gi} = -\frac{\Delta f}{R_i} \tag{2.60}$$

Thus, based on Fig. 2.30, the relationship between the droop and output power change can be expressed as:

$$-\frac{R_2}{R_1} = \frac{\Delta P_{G1}}{\Delta P_{G2}} \tag{2.61}$$

The droop characteristic in Eq. (2.58) is derived for a power system with inductive impedance ($X \gg R$) and a high amount of inertia [2]. In such a system, after a sudden disturbance, the power is generated to stabilize the frequency using inertia power from

Fig. 2.30 The change of active power output in generators with different droops

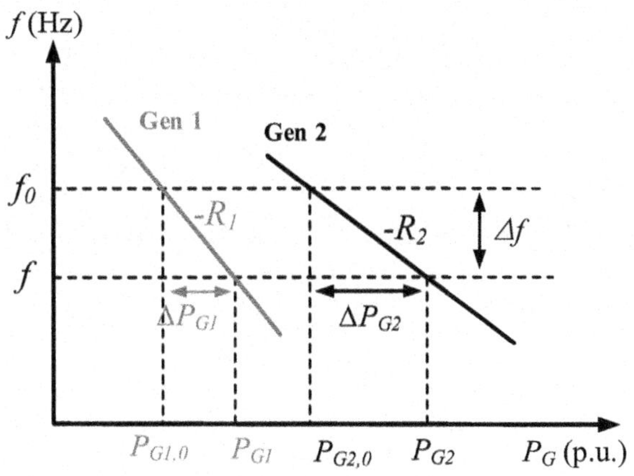

Table. 2.8 The control parameter for Exercise 2.4

Parameter	Generator 1	Generator 2
Droop (Hz/MW)	1/300	1/150
Rated power (MW)	600	300
Moment of inertia ($MW \cdot s^2$)	60	30

rotating mass. In the case of a low inertia power system, it requires a different type of the droop characteristic, which is based on the operation of inverter-based storage systems. This droop characteristic will be explained in the next chapter.

Exercise 2.4 A single-area power system consists of two generating units, which are generating power to a load with the power of P_L at the nominal frequency of 60 Hz. The load damping coefficient is not considered in this study. The control parameters for each generator are listed in Table 2.8.

Assuming the load is suddenly increased to 45 MW ($\Delta P_L = 45$ MW).

(1) Please determine the graph of frequency deviation against the time of each generator, while at the steady-state, generator frequency will not restore to the nominal system frequency.
(2) Please demonstrate that the mechanical power of generators is changed following the rated power.

Solution 2.4 If two generators are synchronized, both of generator frequencies are equal, representing as f. Any change in loads results in the same frequency in both generators. The characteristics of two generators can be represented by Eq. (2.54) without considering the load damping coefficient (D) effect as:

$$P_{m1} - P_{e1} = M_1 \frac{df}{dt} \tag{2.62}$$

$$P_{m2} - P_{e2} = M_2 \frac{df}{dt} \tag{2.63}$$

Ignoring the power loss in a transmission line, the sum of generated power from the units 1 and 2 is equal the electrical load power as:

$$P_{e1} + P_{e2} = P_L \tag{2.64}$$

The droop unit from each generator can be calculated using Eq. (2.60) as:

$$\Delta P_{G1} = -300 \Delta f \tag{2.65}$$

$$\Delta P_{G2} = -150 \Delta f \tag{2.66}$$

The dynamic equations of mechanical power and frequency for each generator after the load change can be expressed as:

$$P_{m1} = P_{m10} + \Delta P_{m1} \quad (2.67)$$

$$P_{m2} = P_{m20} + \Delta P_{m2} \quad (2.68)$$

$$P_L = P_{L0} + \Delta P_L \quad (2.69)$$

$$f = f_0 + \Delta f \quad (2.70)$$

where f_0 is the nominal frequency of both generators before the load change. P_{L0} is the initial load power before the load change. ΔP_L is the load change. P_{m10} and P_{m20} are the mechanical power of each generator before the load change, respectively. ΔP_{m1} and ΔP_{m2} is the mechanical power change of each generator, respectively.

Combining Eqs. (2.62) and (2.63) and substituting by Eqs. (2.67)–(2.70), the resulted equation can be expressed as:

$$(P_{m10} + P_{m20}) + (\Delta P_{m1} + \Delta P_{m2}) - (P_{L0} + \Delta P_L) = (M_1 + M_2)\frac{d\Delta f}{dt} \quad (2.71)$$

Before the load change, $P_{L0} = P_{m10} + \Delta P_{m20}$. Then, substituting in Eq. (2.71), it can be expressed as:

$$-\left(\frac{1}{R_1} + \frac{1}{R_2}\right)\Delta f - \Delta P_L = (M_1 + M_2)\frac{d\Delta f}{dt} \quad (2.72)$$

Thus,

$$\frac{d\Delta f}{dt} = -5\Delta f - 0.5 \quad (2.73)$$

(1) Before the load change at $t = 0$, Δf is equal to zero, which is the initial constraint for calculating the differential equation. The calculated equation can be obtained as:

$$\Delta f(t) = -0.1\left(e^{(-t/0.2)}\right) \quad (2.74)$$

Thus, the graph of the frequency deviation of each generator can be plotted in Fig. 2.31. It is obvious that when the time constant is equal to 0.2 s, the system frequency drops to 59.94 Hz (−0.063 Hz) from its nominal value. After that, the system frequency reaches 59.9 Hz during the steady-state, and it does not restore to its nominal value.

Fig. 2.31 Frequency deviation of the system

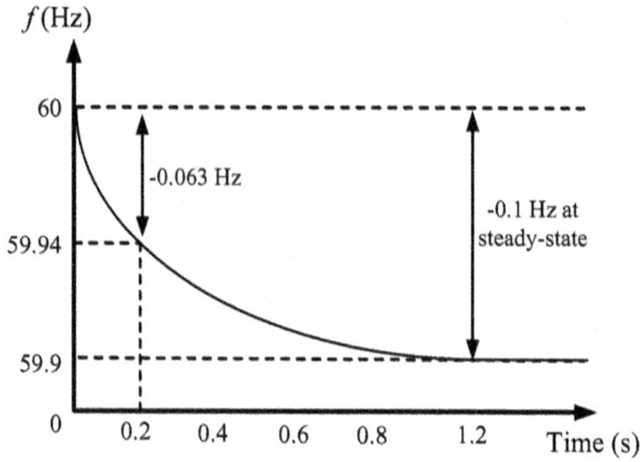

(2) Substituting Δf from Eq. (2.74) into Eqs. (2.65) and (2.66), the mechanical power deviation of each generator can be obtained as:

$$\Delta P_{G1}(t) = 30\left(1 - e^{(-t/0.2)}\right)\text{MW} \tag{2.75}$$

$$\Delta P_{G2}(t) = 15\left(1 - e^{(-t/0.2)}\right)\text{MW} \tag{2.76}$$

It is obvious that the ratio of mechanical power deviation for each generator is $\Delta P_{G1} : \Delta P_{G2} = 2 : 1$, which is equal to the rated power of both generators; $\Delta P_{N1} : \Delta P_{N2} = 600 : 300$. Therefore, it is confirmed the generated power from each generator to the load corresponds with the power capacity of the generators.

2.11 Reserve Power

To obtain a secure/reliable operation of the power system, a sufficient power reserve must be available. The reserve power capacity must be ready to be used when it is required to preserve stable power system operations after a large disturbance. Power management and control are considered as a serious issue in today and future power systems due to the employment of DGs and RESs. Applying such management needs the information of the entire power demand change of the grid or network.

In a power system, standby power reserves should be carefully prepared and purchased to achieve proper power management from a system/market operator. The system operator will trigger these power reserves to achieve the standard performance indices in an economically and timely manner. The spinning reserve or regulating power is generated by energy storage systems, pumped-storage systems, thermal stations, and gas turbines, operating at less than their maximum power output. This

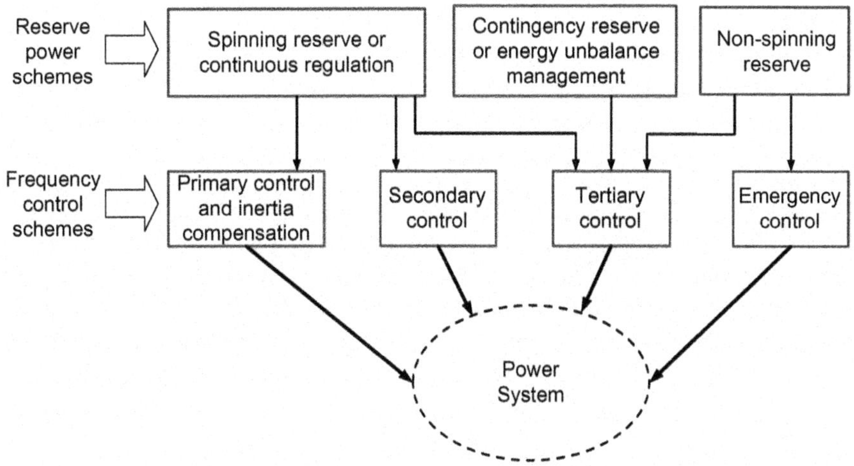

Fig. 2.32 Power reserves related to frequency control schemes

spinning reserve also needs to spare its power capacity for secondary control and tertiary control actions. The size of the needed reserve is depended on the size of load change, generating units, and schedule variations. Figure 2.32 displays the frequency control loops with power reserves. To evaluate a suitable amount of power reserve, it is essential to refer to the recent reliability standards and desirable performance issued by the technical committees. The amount of required power reserve depends on numerous factors, especially the size and type of the unbalance between load and generation.

In traditional synchronous generator-based power systems, the primary control is reserved for 30 s at maximum [1]. In modern RESs/DGs-based power systems, the reserve of primary control is significantly reduced. Virtual inertia synthesis and control can be applied as a powerful solution to assist the primary control and compensate for the rapid frequency variation. Such a control scheme is fully discussed in the next chapter.

The spinning reserve is needed during normal conditions and used for continuous frequency control and energy unbalance management. The main operation of this reserve is to follow minute to minute deviation with regards to load-generation change. It is supported by online systems with automatic controls. The spinning reserve is simply determined as the difference between the actual generation and the total generation capacity. The spare power capacity is required to deliver the essential regulation power for the required of primary and secondary control. The operating time of the primary spinning reserve is about 30 s, which is faster than the operating time of secondary spinning reserve (i.e., within 15 min) [1]. The regulation power is the needed power to restore the frequency of the system to its nominal value.

The energy unbalance management is another type of spinning reserve use, which functions as a bond between the half-hourly or hourly bid-in energy schedules and the regulation service. This type of spinning reserve is slower than the continuous regulation. It should be available within 30 min at a scheduled minimum rate, practically 2 MW per minute [1].

An instantaneous contingency or replacement reserve which is operated during system contingency is known as non-spinning reserves. A contingency is defined as a trip of a generator or transmission line, a loss of load, or the combination of both events. The occurrence of contingency might lead to other issues; such as the overload of a transmission line and significant deviation in voltage/frequency or voltage/frequency instability. Contingency reserves are a special percentage of generation capacity resources, which is reserved for use in emergency conditions. Related to frequency control problems, the non-spinning reserve can be divided into two types; that is replacement reserves and instantaneous contingency reserves.

Instantaneous contingency reserves are given by online generating units (e.g., pumped storage generation). This reserve is able to quickly increase its output after obtaining a control command in response to a significant disturbance. This reserve is known as quick-start reserve, with the response time about 10–15 min [1].

The replacement reserve is provided by generating units (e.g., combined-cycle gas turbine generation) with a slower response time up to 30 min. It will be activated when severe events (e.g., power unbalances or generator outages) occur.

Exercise 2.5 A power system has a total load of 1550 MW at 60 Hz. The system has 260 MW of spinning reserve of generation capacity with 5% regulation based on this capacity. All other generators are operating with valves wide open. The governor dead band is at 80% of the governors respond to the reduction in system load. Please determine the following.

(1) The total spinning generation capacity
(2) The contributing generation for frequency regulation due to the dead band
(3) If a 2.5% change in frequency occurs, how many % changes in power generation.

Solution 2.5

(1) The total spinning generation capacity is equal to:

$$\text{Load power} + \text{Reserve power} = 1550 + 260 = 1{,}810\ \text{MW}.$$

(2) Generation contribution to frequency regulation in respect to the governor dead band is calculated as:

$$(\text{Dead band}) \times (\text{total spinning capacity}) = (80/100) \times (1,810) = 1,448\ \text{MW}.$$

(3) Regulation of 5% means that a 5% change in frequency will cause a 100% change in power generation. Thus, if the 2.5% change in frequency, it will result in a 50% change in power generation.

Table 2.9 Control actions with regards to frequency operation [1]

Frequency deviation range	Situation	Control action
Δf_1	No contingency or load event	Normal operation
Δf_2	Load/generation event	Load frequency control operation
Δf_3	Separation/contingency event	Tertiary/emergency operation
Δf_4	Several contingency event	Emergency operation

2.11.1 Frequency Operating Standards

System frequency is a direct indicator of the generation-load balance in the system. The changes in frequency in large-scale power systems are an actual outcome of the unbalance between generated power and electrical load. After the large disturbance (e.g., loss of generation, sudden load change), the system frequency may rapidly fall if the remaining generation cannot match the existing loads. A significant loss of a generating unit without a proper system response leads to severe frequency excursion with large transient outside the operating range of the system. Large frequency excursion could damage equipment, attenuate load performance, overload transmission lines, trigger system protection schemes, and in the worst case, lead to wide-area power blackouts.

Based on the size of experienced frequency excursion, primary control (i.e., inertia power compensation and governor response), secondary control (load frequency control: LFC), tertiary control, and emergency control may be needed to maintain the frequency of the power system. Table. 2.9 shows relevant control actions respect to the experienced frequency deviations. At the normal operation, frequency is maintained close to nominal frequency by balancing the load and generation. That is for small frequency deviations up to the change of f_1. These deviations can be arrested by the primary control (i.e., inertia power compensation and governor response). At the load frequency control operation, the secondary control unit must be designed to regulate frequency of the system including time deviations within the constraints of specified frequency operating standards. This means the change of f_2 should be evaluted by the available amount of operating reserved power in the system. At the large frequency deviation events under the complex condition (i.e., the changes of f_3 and f_4), the emergency control and protection schemes should be activated to restore the system frequency. In addition to these severe events, it is noted that the secondary control unit is unable to control the system frequency because it is designed to operate during small and slow changes [1].

In all available standards, the accepted frequency deviation for nominal operation is about 1%. Higher frequency deviations may trigger the protection relays to trip the generators and suspend the power supply [39]. A practical generation trip range due to frequency deviation constraints is shown in Fig. 2.33.

Among different power systems, frequency operating standards could be different depending on the operating conditions. As expressed by the international grid codes,

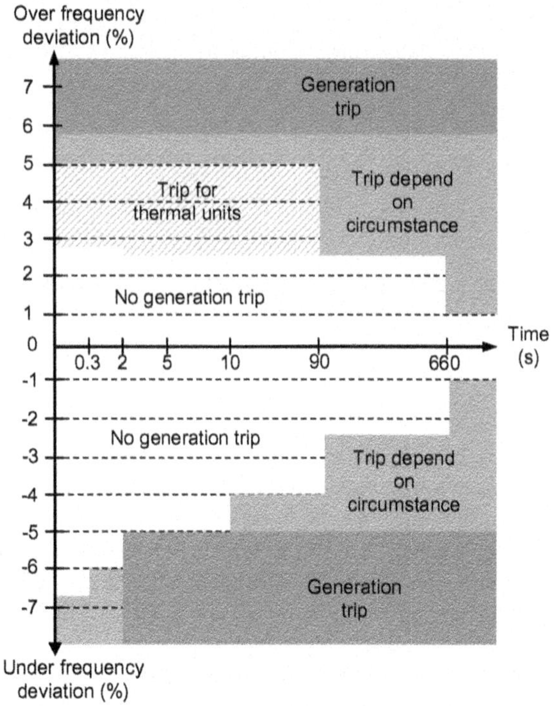

Fig. 2.33 Tripping conditions of generating units respect to frequency deviation

the active power balance between load and generation following a frequency event must be restored in a rapid-enough way, as shown in Table. 2.10.

2.12 Summary

The subject of inertia power compensation with regards to frequency control, its basic concepts, and definitions are presented. The frequency response mechanism of a single-area power system is firstly explained and then expanded to a multi-area (interconnected) power system. Participation factors, generation droop characteristics, and reserve power are fully discussed in detail. The effects of physical constraints (i.e., dead band, generation rate, and time delays) are emphasized. The important frequency operating standards in different power grids are briefly reviewed.

Problems for Chapter 2

2.1 The turbine generating unit has the rating parameters of a 350 MVA, 24 kV, 0.91 p.f., 50 Hz, and 3600 RPM. The moment of inertia is given as 21,500 kg m^2. Please determine the following:

(1) Inertia constant, MW · s/MVA rating
(2) Stored power, MW · s at the rated speed

Table 2.10 Frequency control standards provided by international grids [47]

Country/region	Japan	North America	Australia	Europe	Great Britain	China
Nominal frequency	Eastern region: 50 Hz or Western region: 60 Hz	60 Hz	50 Hz	50 Hz	50 Hz	50 Hz
Normal operating frequency range	±0.2 Hz or ±0.3 Hz	Target range: Eastern interconnection: ±0.018 Hz Western interconnection: ±0.022 Hz Texas interconnection: ±0.03 Hz Quebec interconnection: ±0.021 Hz	Interconnected system: ±0.15 Hz Islanded system: ±0.5 Hz	±0.2 Hz	±0.5 Hz Target range: ±0.05 Hz	System ≥ 3GW: ±0.2 Hz System < 3GW: ±0.5 Hz
Operational frequency tolerance range	N/A	Under-frequency load shedding: Eastern interconnection: 59.5 Hz Western interconnection: 59.5 Hz Texas interconnection: 59.3 Hz Quebec interconnection: 58.5 Hz	49–51 Hz Extreme frequency tolerance range: 47–52 Hz	49.2–50.8 Hz	49–51 Hz Under-frequency load shedding: 48.8 Hz	49–51 Hz

2.2 Considering the power system in Fig. 2.7, the generator delivers the active power of 610 MW (P_e) to the load at the nominal system frequency of 50 Hz. At the time $t = 10$ s, the load demand power is suddenly increased to 660 MW. Please determine a graph of the frequency change against time after the sudden load is increased. Assuming the mechanical power (P_m) is constant, before and after the sudden load change. The moment of inertia is given as 40 MW $\cdot$ s^2.

2.3 Please construct the single-area power system based on Fig. 2.15 and Table 2.6 using MATLAB/Simulink software. The system base is 200 MVA. The nominal system frequency is 60 Hz. The gain of an integral controller is set at −0.5. If the step load change (ΔP_L) of 10 MW is added at the time $t = 3$ s, please determine the following.

(1) System frequency nadir in Hz
(2) Steady-state system frequency in Hz
(3) Generated power of the generator at steady-state in MW

2.4 A multi-area (interconnected) power system consists of generating units and loads, as depicted in Fig. 2.34. The control parameters are based on Table 2.7. The governor dead band for each area is 0.03% (0.018 Hz). The generation rate constraint (GRC) for each area is ±0.15 p.u. MW. The system base is 1800 MVA. The nominal system frequency is 50 Hz. Please construct the system using MATLAB/Simulink software. If loads of 120 MW in Area 2, and 70 MW in Area 3 are suddenly increased at the time $t = 5$ s, please determine the following.

(1) System frequency nadir in each area in Hz
(2) Generated power of the each generating unit at steady-state in MW
(3) Maximum tie-lines power change in each area in MW.

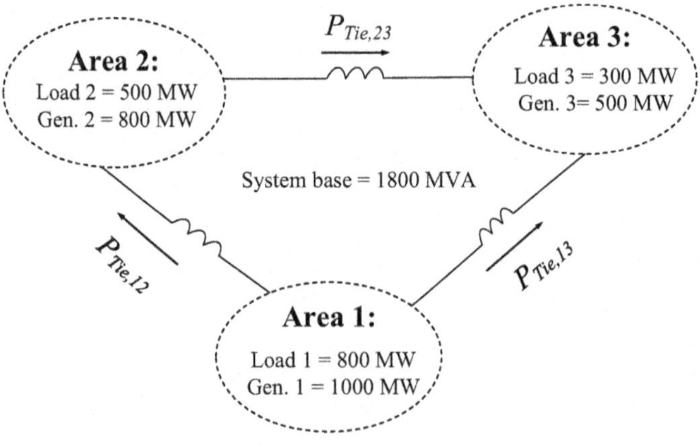

Fig. 2.34 Three-area power systems for Problem 2.4

2.5 From Problem 2.4, the inertia constant and load damping coefficient of each area is reduced by 30% and 40% from its nominal values, respectively, if loads of 150 MW in Area 1, and 90 MW in Area 3 are suddenly increased at the time $t = 5$ s, please determine the following.

(1) System frequency nadir in each area in Hz
(2) Generated power of the each generating unit at steady-state in MW
(3) Maximum tie-lines power change in each area in MW

2.6 A power system has a total load of 2750 MW at 50 Hz. The system has 750 MW of spinning reserve of generation capacity with a 10% regulation based on this capacity. All other generators are operating with valves wide open. The governor dead band is at 65% of the governors respond to the reduction in system load. Please determine the following.

(1) The total spinning generation capacity
(2) The contributing generation for frequency regulation due to the dead band

References

1. H. Bevrani, *Robust Power System Frequency Control*, 2nd ed. (Springer, New York, USA, 2014)
2. P. Kundur, *Power System Stability and Control* (McGraw-Hill, New York, USA, 1994)
3. ENTSO-E, Frequency Stability Evaluation Criteria for the Synchronous Zone of Continental Europe: Requirements and Impacting Factors, Final Report, Brussels, Belgium (2016)
4. ENTSO-E Nordic Analysis Group, Future system inertia, Final Report, Brussels, Belgium (2015)
5. I. Dudurych, M. Burke, L. Fisher, M. Eager, K. Kelly, Operational security challenges and tools for a synchronous power system with high penetration of non-conventional sources. CIGRE Sess. **46** (2016)
6. P.F. Frack, P.E. Mercado, M.G. Molina, Extending the VISMA concept to improve the frequency stability in micrsogrids, in Proc. *International Conference on Intelligent Systems Application to Power Systems (ISAP 2015)*, 1-6 (2015)
7. T. Kerdphol, F.S. Rahman, M. Watanabe, Y. Mitani, D. Turschner, H.P. Beck, Extended virtual inertia control design for power system frequency regulation, in Proc. *IEEE PES GTD Grand International Conference and Exposition Asia*, 97-101 (2019)
8. V. Gevorgian, Y. Zhang, E. Ela, Investigating the impacts of wind generation participation in interconnection frequency response. IEEE Trans. Sustain. Energy **6**(3), 1004–1012 (2015)
9. S. Sharma, S.H. Huang, N.D.R. Sarma, System inertial frequency response estimation and impact of renewable resources in ERCOT interconnection, in Proc. *IEEE Power and Energy Society General Meeting (IEEE PES GM)*, 1-6 (2011)
10. J. Conto, Grid challenges on high penetration levels of wind power, in Proc. *IEEE Power and Energy Society General Meeting (IEEE PES GM)*, 1-3 (2012)
11. H. Gu, R. Yan, T.K. Saha, Minimum synchronous inertia requirement of renewable power systems. IEEE Trans. Power Syst. **33**(2), 1533–1543 (2018)
12. R. Yan, T.K. Saha, N. Modi, N. Al Masood, M. Mosadeghy, The combined effects of high penetration of wind and PV on power system frequency response. Appl. Energy **145**(1), 320–330 (2015)

13. J. Alipoor, Y. Miura, T. Ise, Stability assessment and optimization methods for microgrid with multiple VSG units. IEEE Trans. Smart Grid **9**(2), 1462–1471 (2018)
14. T. Kerdphol, F.S. Rahman, M. Watanabe, Y. Mitani, D. Turschner, H.P. Beck, Enhanced virtual inertia control based on derivative technique to emulate simultaneous inertia and damping properties for microgrid frequency regulations. IEEE Access **7**(1), 14422–14433 (2019)
15. H. Bevrani, I. Francois, T. Ise, *Microgrid Dynamics and Control* (Wiley, Hoboken, New Jersey, USA, 2017)
16. DNV KEMA, RoCoF, An independent analysis on the ability of generators to ride through rate of change of frequency values up to 2Hz/s., Final Report, London, UK (2013)
17. P.M. Ashton, C.S. Saunders, G.A. Taylor, A.M. Carter, M.E. Bradley, Inertia estimation of the GB power system using synchrophasor measurements. IEEE Trans. Power Syst. **30**(2), 701–709 (2015)
18. Australian Energy Market Operator (AEMO), International Review of Frequency Control Adaptation, Final Report, NSW, Austrilia (2017)
19. ERCOT, Future Ancillary Services in ERCOT, Technical Report, Texas, USA (2013).
20. EirGrid, RoCoF Alternative & Complementary Solutions Project: Phase 2, Study Report, EirGrid, Dublin, Ireland (2016)
21. A. Kurita et al., Multiple timescale power system dynamic simulation. IEEE Trans. Power Syst. **8**(1), 216–223 (1993)
22. E. Rakhshani, D. Remon, A.M. Cantarellas, J.M. Garcia, P. Rodriguez, Virtual synchronous power strategy for multiple HVDC interconnections of multi-area AGC power systems. IEEE Trans. Power Syst. **32**(3), 1665–1677 (2017)
23. P. Rodriguez, E. Rakhshani, A. Mir Cantarellas, D. Remon, Analysis of derivative control based virtual inertia in multi-area high-voltage direct current interconnected power systems, IET Gener. Transm. Distrib. **10**(6), 1458–1469 (2016)
24. E. Rakhshani, P. Rodriguez, Inertia emulation in AC/DC interconnected power systems using derivative technique considering frequency measurement effects. IEEE Trans. Power Syst. **32**(5), 3338–3351 (2017)
25. T. Kerdphol, F.S. Rahman, M. Watanabe, Y. Mitani, Robust virtual inertia control of a low inertia microgrid considering frequency measurement effects. IEEE Access **7**(1), 57550–57560 (2019)
26. H. Bevrani, M.R. Feizi, S. Ataee, Robust frequency control in an islanded microgrid: H∞ and μ-synthesis approaches. IEEE Trans. Smart Grid **7**(2), 706–717 (2016)
27. F. Daneshfar, H. Bevrani, Load–frequency control: a GA-based multi-agent reinforcement learning. IET Gener. Transm. Distrib. **4**(1), 13 (2010)
28. Z. Li, W. Wu, M. Shahidehpour, B. Zhang, Adaptive robust tie-line scheduling considering wind power uncertainty for interconnected power systems. IEEE Trans. Power Syst. **31**(4), 2701–2713 (2016)
29. C. Luo, H.G. Far, H. Banakar, P.K. Keung, B.T. Ooi, Estimation of wind penetration as limited by frequency deviation. IEEE Trans. Energy Convers. **22**(3), 783–791 (2007)
30. N. Jaleeli, L.S. Vanslyck, Nerc's new control performance standards. IEEE Trans. Power Syst. **14**(3), 1092–1099 (1999)
31. M. Yao, R.R. Shoults, R. Keim, AGC logic based on NERC's new control performance standard and disturbance control standard. IEEE Trans. Power Syst. **15**(2), 852–857 (2000)
32. N.B. Hoonchareon, C.M. Ong, R.A. Kramer, Feasibility of decomposing $\bar{ACE}$ 1 to identify the impact of selected loads on CPS1 and CPS2. IEEE Trans. Power Syst. **17**(3), 752–756 (2002)
33. H. Bevrani, M. Watanabe, Y. Mitani, *Power System Monitoring and Control* (Wiley, New Jersey, USA, 2014)
34. NERC, *Operating Manual* (Princeton, NJ, USA, 2002)
35. P. Horacek, Securing electrical power system operation, in *Handbook of Automation* (Springer, 2009), pp. 1139–1163
36. N. Jalili, L.S. Vanslyck, Control Performance Standards and Procedures for Interconnected operation, Final Report, EPRI, USA (1997)
37. A. Ibraheem, P. Kumar, D.P. Kothari, Recent philosophies of automatic generation control strategies in power systems. IEEE Trans. Power Syst. **20**(1), 346–357 (2005)

38. C. Concordia, L.K. Kirchmayer, Tie-line power and frequency control of electric power systems. Trans. Am. Inst. Electr. Eng. Part III Power Appar. Syst. **72**, 562–572 (1953).
39. L.K. Kirchmayer, *Economic Control of Interconnected Systems* (Wiley, NY, USA, 1959)
40. T. Kerdphol, M. Watanabe, Y. Mitani, V. Phunpeng, Applying virtual inertia control topology to SMES system for frequency stability improvement of low-inertia microgrids driven by high renewables. Energies **12**(3902), 1–16 (2019)
41. IEEE Committee Report, Standard definitions of terms for automatic generation control on electric power systems. IEEE Trans Power Syst. PAS **89**, 1356–1364 (1970)
42. IEEE PES Committee Report, Dynamic models for steam and hydro-turbines in power system studies. IEEE Trans Power Syst. PAS **92**, 455–463 (1973)
43. O.I. Elgerd, C.E. Fosha, Optimum megawatt-frequency control of multiarea electric energy systems. IEEE Trans. Power Appar. Syst. PAS **89**, 556–563 (1970)
44. C.E. Fosha, O.I. Elgerd, The megawatt-frequency control problem: a new approach via optimal control theory. IEEE Trans. Power Appar. Syst. PAS **89**, 563–577 (1970)
45. IEEE PES Working Group, Hydraulic turbine and turbine control models for system dynamic studies. IEEE Trans. Power Syst. PERS **7**(1), 167–174 (1992)
46. IEEE PES Committee Report, Current operating problems associated with automatic generation control. IEEE Trans. Power Appar. Syst. **7**(3), 1106–1112 (1979)
47. J. Fang, H. Li, Y. Tang, F. Blaabjerg, On the inertia of future more-electronics power systems. IEEE J. Emerg. Sel. Top. Power Electron. **7**(4), 2130–2146 (2019)

Chapter 3
Virtual Inertia Synthesis for a Single-Area Power System

Abstract The inertia property is one of the most critical aspects to maintain the frequency stability in a single (islanded) power system. Therefore, this chapter explains the dynamic performance and frequency characteristics of a single-area system with the deployment of virtual inertia control in addition to the primary and secondary control loops. The linearized frequency response model for virtual inertia, primary, and secondary controls is presented by using the state-space representation (i.e., mathematical model of a physical system). Dynamic and static performances of the virtual inertia response model are explained in terms of small-signal (dynamic) and state-space analysis. The effects of various parameters over inertia control-based frequency response are emphasized. A dynamic model of virtual inertia control is verified by a well-tested classical load-frequency control model in the varying operating conditions of the power system. Moreover, some experimental studies with a practical virtual inertia control in the laboratory environment are also demonstrated.

Keywords Dynamic performance · Frequency control loops · Frequency deviation · Frequency response · Linearized model · Primary frequency control · Secondary frequency control · State-space model · Time delay · Virtual inertia control · Virtual inertia power

3.1 Fundamental Virtual Inertia Synthesis and Control

Recently, the integration of distributed generators (DGs) and renewable energy sources (RESs) into the traditional power system-based synchronous machines is immediately increasing according to the energy crisis, environmental concern, and economic growth. Together with the deployment of a modern (distributed) power system concept called the smart/micro-grids, such systems are suitable for integrating DGs/RESs into the distribution system [1]. Consequently, DGs/RESs develop into the highly-shared structures in modern power systems. Favorably, the consumers do not need to rely on the faraway traditional generation during a fault and they can have better power quality. On the contrary, the high DGs/RESs integration could cause critical frequency stability problems in the system as follows. Firstly, a high

T. Kerdphol et al., *Virtual Inertia Synthesis and Control*, Power Systems,
https://doi.org/10.1007/978-3-030-57961-6_3

DGs/RESs penetration curtails the number of traditional generating units, which directly contribute the initial response and reserve power for primary and secondary frequency control, resulting in larger frequency excursions and degradation of system stability and resiliency. Secondly, the DGs/RESs-based generation naturally has non-existent or low inertia and damping properties due to the deployment of power electronics interfaces (i.e., inverters/converters). This power-electronics interface has no rotating mass, which is the prominent ability in providing inertia and damping properties for slowing down the frequency change during a disturbance. The lack of system inertia and damping will result in the rising rate of change of frequency (RoCoF), leading to abrupt frequency variations with larger amplitudes and load-shedding, even at a small disturbance (see Fig. 3.1). Accordingly, the DGs/RESs-based generation may not engage in frequency control in normal system operation [2]. Therefore, penetrating DGs/RESs into the power system will undoubtedly lead to the reduction of the whole system inertia and damping, which can cause negative impacts on power system dynamics, frequency/voltage regulation as well as other operation and control issues. In the worst case, these problems might cause system instability, cascading failures, and power blackouts.

In response to the stability challenges driven by low system inertia and damping, a solution toward stabilizing a modern power system is to synthesize additional inertia and damping, virtually, allowing a high DGs/RESs participation in system operation [3–5]. Virtual inertia synthesis and control can be constructed by the short-term energy storage, power electronics converter/inverter, and advanced inertia control mechanism in the system that is called the virtual synchronous machine (VISMA) or virtual synchronous generator (VSG) concept [6, 7]. The virtual inertia control system will operate as a real synchronous machine/generator in providing virtual inertia and damping for short-time intervals. Consequently, the idea of virtual inertia can offer a fundamental for regulating a high portion of DGs/RESs in today

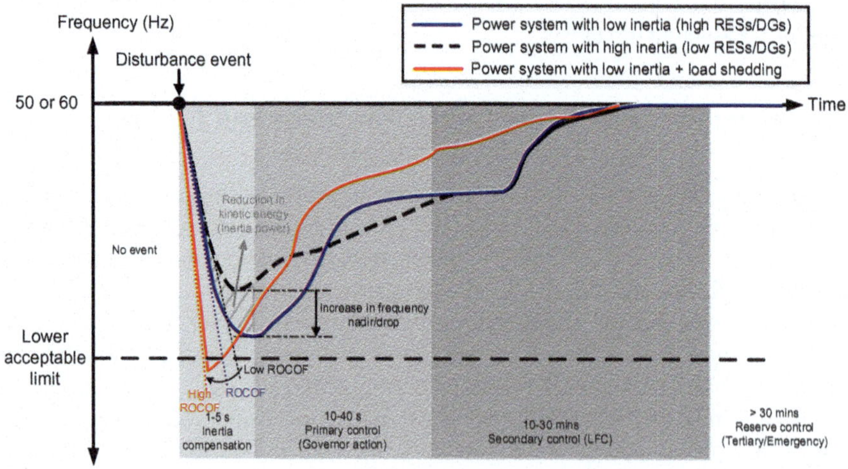

Fig. 3.1 Comparison of frequency dynamic response between in modern power systems dominated by DGs/RESs and conventional power system dominated by synchronous generators

and future power systems without compromising the system stability, reliability, and resiliency. Literature reviews, including past achievements on virtual inertia control and its applications, have been reported in [2, 8, 9]. The reports confirmed that the applications of virtual inertia control could offer uninterruptable power transfer between grid-connected and islanded operations.

The principle of virtual inertia control can be implemented either to a single RES/DG or a group of RESs/DGs. By implementing a single RES/DG, it might be suitable for individual owners of DG/RES. By implementing a group of DGs/RESs, it is easier and more economical to control in the network/grid aspect [10, 11].

Figure 3.2 displays the fundamental structure of virtual inertia control. It consists of an energy storage system (ESS), inverter, and inertia control mechanism. Then, the virtual inertia is synthesized into the system by regulating the active power via the inverter in the inverse proportion of the rotor speed. From the network point of view in regards to higher frequency noise triggered by switching of invertor's power transistors [12], it is noted that there is no difference between the electrical component of virtual inertia control and the electrical appearance of the electromechanical synchronous machine. Due to the inertia compensation principle, the virtual inertia control should absorb or inject active power; thus, the nominal state of charge (SOC) of the ESS in its system should be operated at 50% of its nominal capacity during a stationary (steady-state) circumstance. However, depending on the SOC situation, the operation of virtual inertia control can be changed due to the specified upper and lower limits (e.g., 20% for a lower limit and 80% for an upper limit). The limits can also be evaluated based on the used technology of the ESS. During such limits, the virtual inertia control is operated in the inertia control mode when the energy in the system is lacking due to the unbalance between generation and load. However, the virtual inertia control is operated in the virtual load mode when the energy in the system is excess. Finally, the emulated power from a virtual inertia control unit can be simply expressed as [7, 13]:

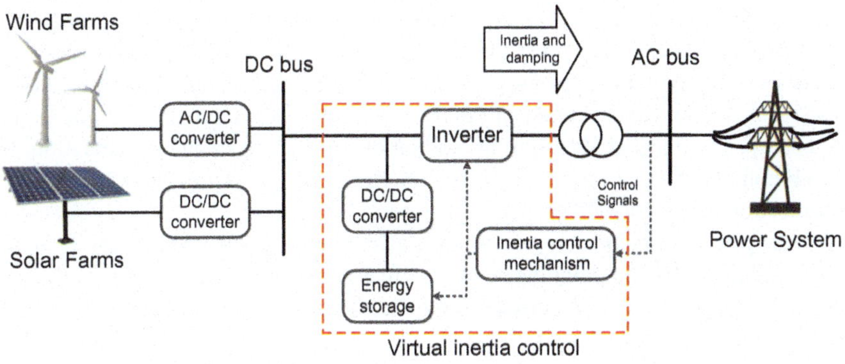

Fig. 3.2 A fundamental concept and structure of virtual inertia control

$$P_{VI} = K_{VI}\left(d\frac{\Delta f}{dt}\right) + D_{VI}(\Delta f) + P_0 \tag{3.1}$$

where,

$$K_{VI} = \frac{2HP_{Inv}}{f_0} \tag{3.2}$$

It is noted that $d\Delta f/dt$ is the rate of change of frequency (RoCoF), f_0 is the nominal frequency of the system, K_{VI} is the virtual inertia characteristic/constant, D_{VI} is the virtual damping coefficient/constant, P_{Inv} is the nominal apparent power of the inverter unit, and P_0 is the primary power that transfers to the inverter.

Generally, the initial rate of frequency variation presents the error signal with an equilibrium of zero; thus, the emulated power will be transferred only in the transient condition without restoring the system frequency to the nominal value. To deploy the frequency restoration ability for virtual inertia control, a frequency control droop must be added in terms of Eq. (3.1). Considering Eq. (3.1), the operation of virtual inertia control can be explained as three main terms as follow.

The first term, which is known as the virtual inertia, can emulate inertia behavior from a synchronous generator. This term will decrease the maximum deviation of the rotor speed and curtails system frequency nadir/overshoot after a disturbance. Regarding this term, the power will be absorbed or generated by the negative or positive initial RoCoF.

The second term, which is known as the virtual damping, can emulate the damper windings effect of a synchronous generator. The D_{VI} must be selected so that the emulated power to be equal with the nominal power of the virtual inertia control system when the system frequency oscillation occurs at the specified maximum value. This term will suppress the oscillation of system frequency after a disturbance, resulting in faster stabilizing time of the system.

The third term represents the nominal primary power that transfers to the inverter unit, generating as the constant power. In addition to K_{VI} and D_{VI}, they are negative constant and must be constant so that the virtual inertia system can exchange its maximum active power when the maximum specified frequency deviation and RoCoF occur. Increasing the K_{VI} and D_{VI} indicate that more power will be absorbed or injected for a similar amount of RoCoF and frequency deviation. By combining these three terms, the virtual inertia control system is equally effective for electromechanical synchronous generators.

Considering a practical synchronous machine, energy absorbed by the damping term is drained by the damping windings resistance. In the case of virtual inertia control, this power will be consumed by the energy storage system to balance the power of the system.

To select a suitable energy storage type for inertia control, there are some important parameters to be concerned as follows: the power of the generating unit, the maximum power of loads, averaged SOC during normal operation, operating time, and control delay [11, 14].

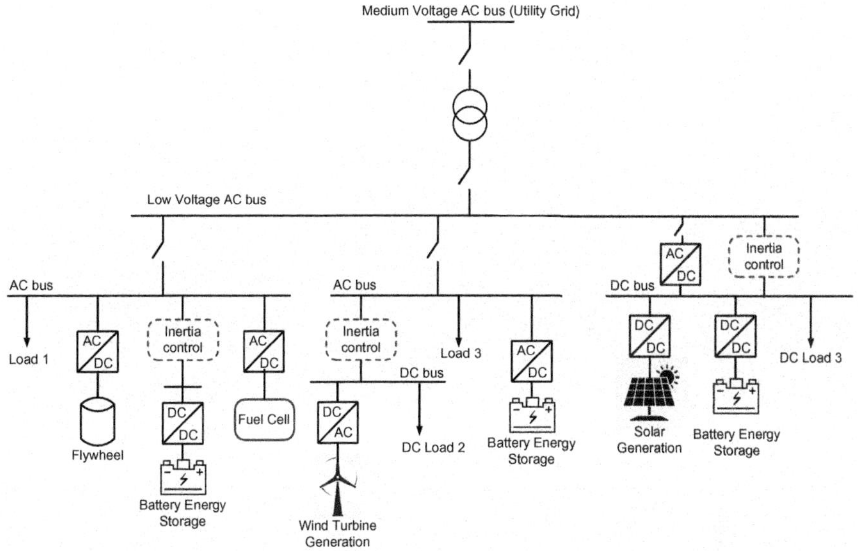

Fig. 3.3 Sufficient locations for conducting virtual inertia control

The virtual inertia control systems are usually located between the grid/AC bus and the DC bus/DC source/DG/RES, as shown in Fig. 3.3. Such systems show the grid as a synchronous machine with regards to the characteristics of inertia and damping compensation. Then, the virtual inertia control systems can effectively overcome the oscillations caused in the system due to the high share of DGs/RESs with low or no inertia and damping. Nevertheless, the demand side management may be activated in some loads via a local controller, improving effective inertia control schemes.

Figure 3.4 presents a recent overview of a practical virtual inertia control system under the VISMA project with regards to frequency and voltage stability improvement. The system was constructed at the Institute of Electrical Power Engineering and Energy Systems (IEE), TU Clausthal, Germany [15]. The system is operated as an islanded mode under the nominal frequency of 50 Hz. The size of system base is approximately 40 kVA. The system consists of two types of generation; that is 36 kW of a synchronous generator (conventional generation) and 10 kW of wind-turbine/10 kW of solar systems (RESs/DGs). The electricity is consumed by 40 kW of loads. The virtual inertia control system (VISMA) has 15 kWh installed capacity with the nominal SOC of 50%. The system has the primary and secondary control units, which is controlled by the synchronous generator. The VISMA is responsible for the inertia emulation and control. The RES units can provide a considerable amount of power to the system, but they are not participated in the frequency control. The full investigation and final report of this system can be found in [15] and [37].

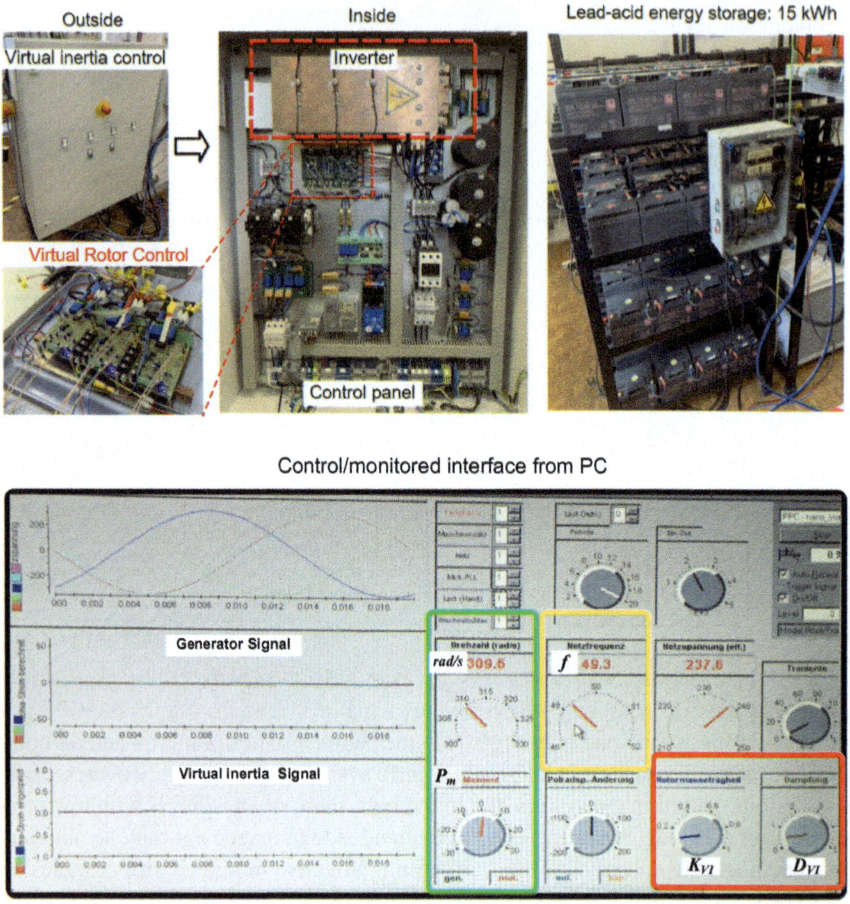

Fig. 3.4 Overview of a practical virtual inertia control system with control interface, TU Clausthal, Germany (December 3, 2019)

3.2 Droop Characteristics of Virtual Inertia Control

Traditional power systems are not always able to respond the rapid variations in voltage and frequency events, resulting in the common deployment of load-shedding with consequences on the economic aspect [11]. The virtual inertia control is provided by fast-acting storage components (e.g., ESS), which can serve the system in alleviating frequency excursions, especially from inertia reduction triggered by DGs/RESs integration, hence curtailing the load-shedding requirement.

In addition to reactive power regulation, nowadays, the modern power system is more capacitive due to the increasing reactive power generated by DGs/RESs. Subsequently, the operating voltage of the system significantly increases. To overcome this problem, the local voltage setpoint must be decreased to regulate the voltage near its nominal value.

By applying the virtual inertia control topology, active and reactive powers of the system can be properly modified, resulting in better regulation of system frequency and voltage. This concept requires the operations of virtual inertia droop characteristics, which are known as the *P-f* and *Q-V* droops. One control droop requires at least one inverter. In the case of applying both control droops, it requires parallel inverters. Then, the virtual inertia control system uses the feedback from the frequency and voltage of each RES/DG for measuring the output active and reactive powers to imitate inertia and damping properties. Thus, in the modern power system dominated by power electronics-based DGs/RESs, the frequency/voltage droop control can be created by deploying the virtual inertia control mechanism.

In a power system, when the system frequency decreases, the active power (*P*) generated by generating units should be increased to maintain the frequency, and vise versa. Similar behavior for voltage versus reactive power (*Q*) can be obtained. The feedback loop relationships for controlling active and reactive power of the system can be expressed as [13]:

$$\Delta f = f - f_0 = -R_{VI}(P - P_0) \tag{3.3}$$

$$\Delta V = V - V_0 = -R_{VIQ}(Q - Q_0) \tag{3.4}$$

where f_0 is the nominal system frequency, V_0 is the nominal system voltage, P_0 is the nominal active power, R_{VI} is the virtual inertia droop constant for controlling active power in respect to frequency regulation, R_{VIQ} is the virtual inertia droop constant for controlling reactive power in respect to voltage regulation, and Q_0 is the nominal reactive power.

It is noted that the explained droop characteristics have been determined under the situation of inductive impedance ($X \gg R$) and high inertia, representing a power system with high-voltage lines. For medium-voltage and low-voltage lines, the impedance is not dominantly inductive ($X \approx R$). For resistive lines, reactive power will depend on the rotor angle (δ), and active power will depend on voltage. Such a system recommends different droop control, known as opposite droops [16].

Over the last five years, several topologies are introduced for virtual inertia control for power system regulation objectives (e.g., frequency, voltage, and current mode control). Interested readers can study the relevant references to learn the topologies details. The virtual inertia-based frequency control has been proposed by [6, 7, 17–22]. The virtual inertia-based voltage control has been proposed by [23–31]. The virtual inertia-based current mode control has been proposed by [4, 14, 32–36]. This book is mainly focused on more flexible virtual inertia control topologies for frequency regulation issues.

3.3 Frequency Regulation for Virtual Inertia Synthesis

As discussed in Chap. 2 (a traditional power system dominated by synchronous machines), when a sudden power unbalance caused by a disturbance occurs, the power is stabilized by the natural response of the rotor or rotating mass (inertia compensation) and governors (primary control). On the contrary, in the modern power system dominated by DGs/RESs, the natural response of rotating mass (inertia compensation) is replaced by the response of converter control, which is slower than the natural response of rotating mass as shown in Fig. 3.5. Thus, immediately following a disturbance, there is no significant inertia power compensation from 1 to 5 s, resulting in rapid frequency deviations with larger amplitudes. Excessive RoCoF could occur, leading to the cascaded tripping if generation units are not designed to ride through. If system inertia falls below a certain limit, the significant loss of generation will lead to the rapid frequency drop that under frequency load-shedding would occur before, resulting in instability, cascading outages, and power blackouts. Nevertheless, if a disturbance appears between the absorbed power and the generated power, the power source voltages significantly change. Subsequently, the voltage is triggered by changes in active power. To overcome this issue, the initial frequency response should be compensated by the proposed virtual inertia control scheme.

In a virtual inertia control unit, the main control algorithm is the calculation of active/reactive power that exchanges between the inverter and the grid. Moreover, there are two other control loops for operation at a lower control level [37] as follows. Firstly, the control loop for evaluating the inverter output voltage is calculated by the main control algorithm. This control loop also implemented to obtain the active/reactive power transfer between the energy storage systems, inverter, and grid. Secondly, the control loop for controlling the DC/DC converters interfacing energy storage systems to the DC bus of the virtual inertia unit. This book directly addresses the primary control algorithm of the virtual inertia system.

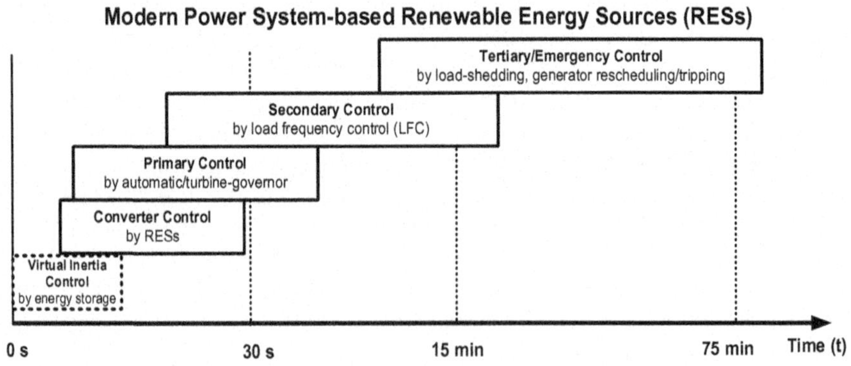

Fig. 3.5 Timescale of frequency dynamic control for modern power systems dominated by power electronics-based DGs/RESs

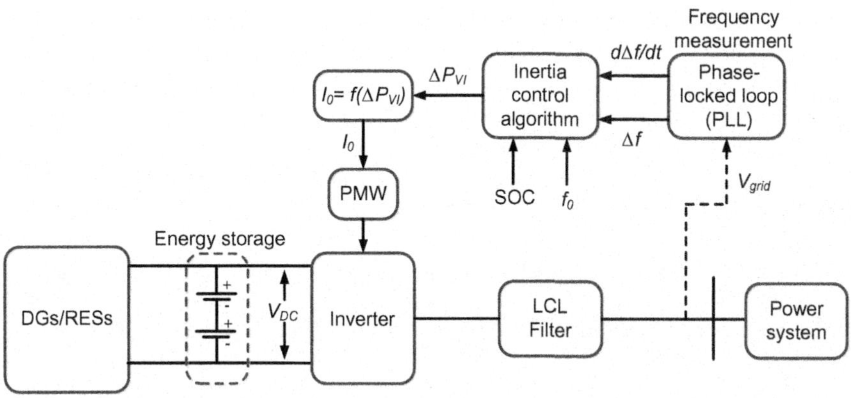

Fig. 3.6 A general frequency regulation scheme for virtual inertia control [11]

A general frequency regulation scheme for virtual inertia control is depicted in Fig. 3.6. When the system frequency and RoCoF occur larger than a set limit, the main algorithm of virtual inertia control calculates the required K_{VI} to be applied to the grid and delivers the additional power transfer (ΔP_{VI}). Considering Eq. (3.1), it can estimate the amount of power for the required inertia as [13]:

$$\Delta P_{VI} = K_{VI}\left(d\frac{\Delta f}{dt}\right) + D_{VI}(\Delta f) \tag{3.5}$$

Based on the relationship function between the transferred power from the command current signal and inverter, that is $I_0 = f(\Delta P_{VI})$, the reference current for initiating the pulse width modulation (PMW) units can be determined. This function is considering the relationship between the transferred power from the DC bus and the reference current signal. The current of DC bus at the virtual inertia unit is regulated by obtaining information data (e.g., system frequency, SOC). The frequency is calculated based on the zero-crossing algorithm. The I_0 function can be computed as [11, 13, 33]:

$$I_0 = \frac{K_{VI}\frac{d\Delta f}{dt} + D_{VI}(\Delta f)}{V_{DC}} \tag{3.6}$$

where V_{DC} is the terminal voltage between the energy storage unit and inverter.

Suddenly, after a disturbance, the required virtual inertia is promptly evaluated by the main virtual inertia algorithm. To obtain effective control, the virtual inertia unit must determine a trade between inertia and damping requirements. Also, the control performance can be enhanced by adjusting the effect of PLL in regards to frequency calculation [18]. The starting time of power delivery-based virtual inertia control is very significant to obtain a desired dynamic performance. When the system frequency and RoCoF exceed the acceptable limits (e.g., ± 0.5 Hz and ± 1 Hz/s)

[11], the operation of virtual inertia control must be immediately enabled within 1 s before the activation of load-shedding relays. To effectively reduce a frequency nadir, the amount of dynamic reserve must be determined in advance. The amplitude of frequency nadir relies on various effects, such as the amplitude of power unbalance, the available kinetic energy (inertia power) stored in the rotating mass, the sensitivity of demand to the variation in system frequency, and the activation time of primary control.

3.4 Frequency Response Model for Virtual Inertia Control

The small-signal/dynamic model is performed to study frequency regulation in the presence of transient response and active power change [37]. From the inertia control point of view, its dynamic model is mainly focused on the dynamic effects of active power change in regards to frequency regulation, which can reduce the complexity of the schematic block diagram of a closed-loop virtual inertia control in Fig. 3.6.

To design the dynamic model of virtual inertia control, dynamic characteristics from Eq. (3.5) for emulating virtual inertia and damping is combined as: (1) the first-order transfer function of inverter-based energy storage, and (2) the virtual inertia droop characteristic as shown in Fig. 3.7.

In addition to the dynamic modeling of inverters, typically, inverters have two different operating modules, acting as a voltage source or current source. By obtaining information on filter/line parameters, the output frequency and voltage, including active and reactive powers of the inverter can be regulated by local feedback to the inverter. Thus, inverters can track their reference power with a short time constant. Subsequently, the dynamic model of inverters can be modeled as the first-order transfer function [11, 38].

The dynamic equation of the virtual (emulated) inertia power (P_{VI}) can be expressed as [11, 13, 37]:

$$\Delta P_{VI}(s) = \frac{sK_{VI} + D_{VI}}{1 + sT_{INV}} \left(\frac{\Delta f(s)}{R_{VI}} \right) \tag{3.7}$$

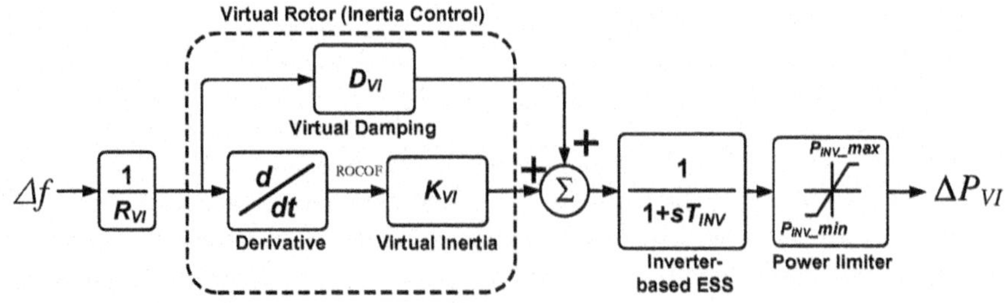

Fig. 3.7 Dynamic frequency response structure for virtual inertia control

where K_{VI} is the virtual inertia constant, D_{VI} is the virtual damping constant, T_{INV} is the time constant of inverter-based energy storage, and R_{VI} is the virtual inertia droop constant for controlling active power in respect to frequency regulation.

According to Fig. 3.7 and Eq. (3.7), the virtual inertia unit is built based on the inertia imitation technology using the derivative technique introduced in [6, 7, 17] to calculate the RoCoF (*df/dt*) for adjusting the additional power with required inertia to a setpoint of the system during the disturbance. This part will decrease the maximum deviation of the rotor speed and curtails system frequency nadir/overshoot after a disturbance.

The virtual damping unit is established for rapid stabilizing/settling time based on the system frequency deviation. This part can emulate the damper windings effect of a synchronous generator. The oscillation of system frequency after the disturbance is effectively suppressed by this damping part.

As a result, the active power from the inverter-based ESS is proportionally governed by the RoCoF. Thus, the additional power with required inertia and damping could be properly imitated into the system, improving both transient and steady-state frequency dynamics. The low-pass filter is employed to wipe out the noise problem and to obtain precise dynamics of inverter-based ESS (i.e., fast response behavior). The power limiter unit is employed to restrict the maximum power (P_{INV_max}) and minimum power (P_{INV_min}) of the inverter-based ESS output, performing the actual power response behavior of the ESS. Subsequently, the virtual inertia control unit can contribute to the system as if the DGs/RESs have inertia and damping properties similar to the traditional generation.

3.5 Frequency Analysis for Virtual Inertia Control

To perform frequency stability analysis with regards to inertia control, the small-signal/dynamic models of generation and load units should be constructed. With the approximation, an equivalent Eq. (3.7) can be used for the dynamic study of virtual inertia control in an area (isolated) power system. Subsequently, the dynamic model in Fig. 3.7 can be used as an equivalent frequency analysis model for the whole multi-inverter-based inertia control of an area power system.

Figure 3.8 shows a schematic diagram of the studied (isolated) power system dominated by DGs/RESs. Figure 3.9 shows the combined frequency response model of an area system (from Chap. 2, Fig. 2.15) with the virtual inertia control unit (from Fig. 3.7) and the additional DGs/RESs units (i.e., wind and solar power systems) [37, 39-41]. Three main frequency control schemes (i.e., virtual inertia control, primary control, and secondary control) are employed to preserve system frequency stability during a disturbance. The virtual inertia control-based ESS is in charge for providing the additional power with the required inertia and damping after a disturbance (mismatch power) at 1–5 s. The primary control unit (i.e., the governor unit in a thermal power plant) is in charge for balancing the system frequency to a new steady-state value within 10–40 s. The secondary control (also known as load

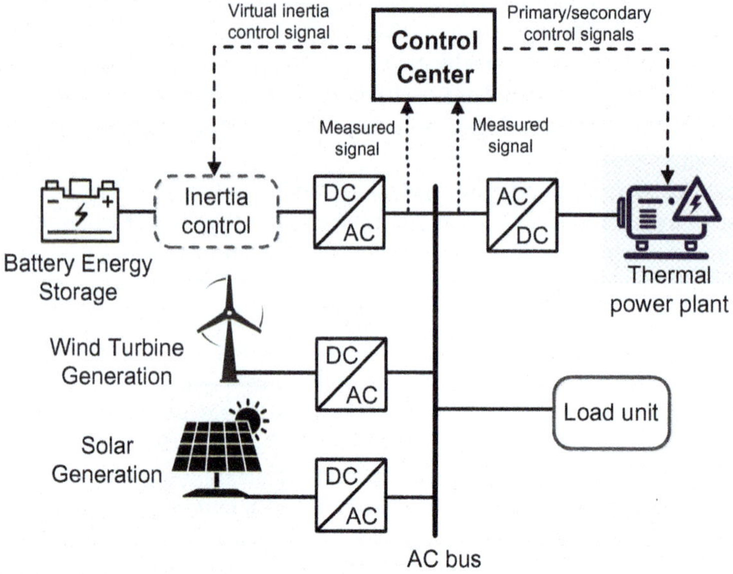

Fig. 3.8 A Schematic diagram of the modern power system-based DGs/RESs

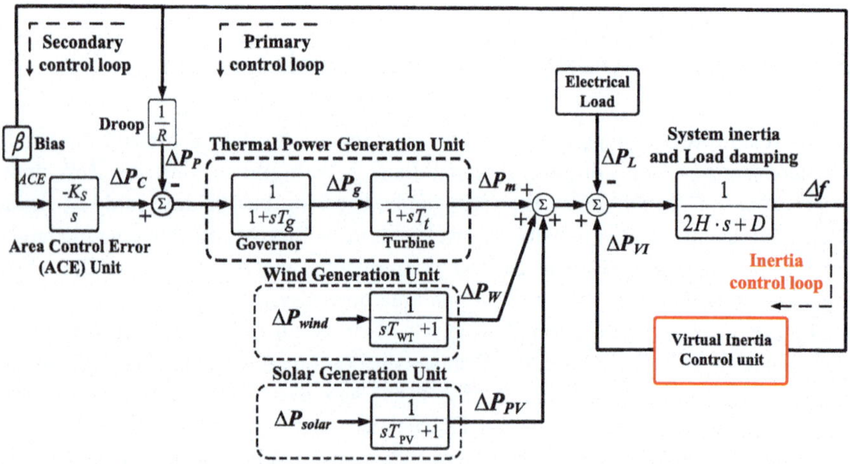

Fig. 3.9 Dynamic frequency response model of the modern power system with primary, secondary and virtual inertia control schemes

frequency control: LFC) based on area control error (ACE) is in charge for restoring the system frequency to its nominal value within 10–30 min. To make the system more practical, the DGs/RESs (wind and solar power systems) and domestic loads (industrial load and commercial-residential load) do not engage in the frequency control. Thus, they are considered as the uncertainties/disturbances to the system. Based on [6, 7, 11, 17, 38–42], the simplified model of the wind and solar systems presented in Fig. 3.9 is sufficient for frequency stability study and analysis. Hence, the simplified model used in this work is accurate enough for frequency stability study. It is noted that the simplified modeling of the system considers the influences of external inputs, control or manipulated signals, and uncontrollable disturbances in the actual mathematical models of RESs/DGs and loads. For more details about the mathematical models of RESs/DGs and loads can be found in [43].

By linearizing the dynamic effects of the load-generation blocks, including virtual inertia control, primary control, and secondary control units from Fig. 3.9, the frequency deviation of the modern power system can be expressed as [37]:

$$\Delta f(s) = \frac{1}{2Hs + D}(\Delta P_m(s) + \Delta P_W(s) + \Delta P_{PV}(s) + \Delta P_{VI}(s) - \Delta P_L(s)) \tag{3.8}$$

where

$$\Delta P_m(s) = \frac{1}{1 + sT_t}\Delta P_g(s) \tag{3.9}$$

$$\Delta P_g(s) = \frac{1}{1 + sT_g}\left(\Delta P_C(s) - \frac{1}{R}\Delta f(s)\right) \tag{3.10}$$

$$\Delta P_C(s) = \frac{K_S}{s}(\beta \cdot \Delta f(s)) \tag{3.11}$$

$$\Delta P_W(s) = \frac{1}{1 + sT_{WT}}\Delta P_{\text{wind}}(s) \tag{3.12}$$

$$\Delta P_{PV}(s) = \frac{1}{1 + sT_{PV}}\Delta P_{\text{solar}}(s) \tag{3.13}$$

where ΔP_C is the ACE action change (signal) from secondary control. ΔP_P is the control action change (signal) from primary control. ΔP_W is the generated power change from the wind system. ΔP_{wind} is the initial wind power change. ΔP_g is the generated power from the turbine system. ΔP_{solar} is the initial solar power change. ΔP_{PV} is the generated power change from the solar system. ΔP_{Ind} is the industrial/commercial load power change. ΔP_{Resi} is the residential load power. ΔP_L is the total load change of the system. ΔP_{VI} is the virtual inertia power change from Eq. (3.7).

3.6 State-Space Modeling of a Single Area Power System

In control engineering, modern control theory solves various limitations using a much deeper explination of the system dynamics called state-space representation. The state-space representation introduces the mathematical mode of a physical system, which can provide the dynamics as a set of coupled first-order differential equations in a set of internal variables (i.e., state variables), together with a set of algebraic equations that combine the state variables into physical output variables. When a suitable state-space model is evaluated, it becomes easier for designers/operators to understand or manipulate the system property. Thus, the state-space model is useful for analyzing the detail of power system stability and dynamics. Using suitable definitions and state variables as shown in Eqs. (3.15)–(3.17), the complete state-space model representation of the modern power system controlled by the virtual inertia unit can be easily calculated as [11]:

$$\dot{x} = Ax + B_1 w + B_2 u \tag{3.14}$$

Considering the dynamic block of Fig. 3.9 and Eqs. (3.7)–(3.13), the state variables and coefficients of the state-space model is evaluated as follows:

$$x^T = \left[\Delta f \ \Delta P_m \ \Delta P_g \ \Delta P_C \ \Delta P_{VI} \ \Delta P_W \ \Delta P_{PV}\right] \tag{3.15}$$

$$w^T = \left[\Delta P_{\text{wind}} \ \Delta P_{\text{solar}} \ \Delta P_L\right] \tag{3.16}$$

$$u^T = [d\Delta f/dt] \tag{3.17}$$

where x is the control output signal, w is the disturbance signal, u is the control input signal.

Equations (3.7)–(3.13) can be rewritten in the form of the first-order differential equation regarding Eq. (3.14) as:

$$s \cdot \Delta f(s) = \frac{\Delta P_m(s)}{2H} + \frac{\Delta P_W(s)}{2H} + \frac{\Delta P_{PV}(s)}{2H} + \frac{\Delta P_{VI}(s)}{2H} - \frac{\Delta P_L(s)}{2H} - \frac{D\Delta f(s)}{2H} \tag{3.18}$$

$$s \cdot \Delta P_m(s) = \frac{\Delta P_g(s)}{T_g} - \frac{\Delta P_m(s)}{T_g} \tag{3.19}$$

$$s \cdot \Delta P_g(s) = -\frac{\Delta f(s)}{RT_g} - \frac{\Delta P_g(s)}{T_g} + \frac{\Delta P_C(s)}{T_g} \tag{3.20}$$

$$s \cdot \Delta P_C(s) = K_S \cdot \beta \cdot \Delta f(s) \tag{3.21}$$

$$s \cdot \Delta P_{VI}(s) = \Delta f(s)\left(\frac{D_{VI}}{R_{VI}T_{INV}}\right) + s \cdot \Delta f(s)\left(\frac{K_{VI}}{R_{VI}T_{INV}}\right) - \frac{\Delta P_{VI}}{T_{INV}} \tag{3.22}$$

$$s\Delta P_W(s) = \frac{\Delta P_{\text{wind}}(s)}{T_{WT}} - \frac{\Delta P_W(s)}{T_{WT}} \quad (3.23)$$

$$s\Delta P_{PV}(s) = \frac{\Delta P_{\text{solar}}(s)}{T_{PV}} - \frac{\Delta P_{PV}(s)}{T_{PV}} \quad (3.24)$$

Finally, by arranging Eqs. (3.18)–(3.24) into the form of Eq. (3.14), the complete state-space equation of the system can be presented as:

$$\dot{x} = \begin{bmatrix} \frac{-D}{2H} & \frac{1}{2H} & 0 & 0 & \frac{1}{2H} & \frac{1}{2H} & \frac{1}{2H} \\ 0 & -\frac{1}{T_t} & \frac{1}{T_t} & 0 & 0 & 0 & 0 \\ \frac{-1}{RT_g} & 0 & -\frac{1}{T_g} & \frac{1}{T_g} & 0 & 0 & 0 \\ \beta \cdot K_S & 0 & 0 & 0 & 0 & 0 & 0 \\ \frac{D_{VI}}{R_{VI}T_{INV}} & 0 & 0 & 0 & \frac{-1}{T_{INV}} & 0 & 0 \\ 0 & 0 & 0 & 0 & 0 & \frac{1}{T_{WT}} & 0 \\ 0 & 0 & 0 & 0 & 0 & 0 & \frac{-1}{T_{PV}} \end{bmatrix} x + \begin{bmatrix} 0 & 0 & \frac{-1}{2H} \\ 0 & 0 & 0 \\ 0 & 0 & 0 \\ 0 & 0 & 0 \\ 0 & 0 & 0 \\ \frac{1}{T_{WT}} & 0 & 0 \\ 0 & \frac{1}{T_{PV}} & 0 \end{bmatrix} w + \begin{bmatrix} 0 \\ 0 \\ 0 \\ 0 \\ \frac{K_{VI}}{R_{VI}T_{INV}} \\ 0 \\ 0 \end{bmatrix} u \quad (3.25)$$

where,

$$A = \begin{bmatrix} \frac{-D}{2H} & \frac{1}{2H} & 0 & 0 & \frac{1}{2H} & \frac{1}{2H} & \frac{1}{2H} \\ 0 & -\frac{1}{T_t} & \frac{1}{T_t} & 0 & 0 & 0 & 0 \\ \frac{-1}{RT_g} & 0 & -\frac{1}{T_g} & \frac{1}{T_g} & 0 & 0 & 0 \\ \beta \cdot K_S & 0 & 0 & 0 & 0 & 0 & 0 \\ \frac{D_{VI}}{R_{VI}T_{INV}} & 0 & 0 & 0 & \frac{-1}{T_{INV}} & 0 & 0 \\ 0 & 0 & 0 & 0 & 0 & \frac{1}{T_{WT}} & 0 \\ 0 & 0 & 0 & 0 & 0 & 0 & \frac{-1}{T_{PV}} \end{bmatrix} \quad (3.26)$$

$$B_1 = \begin{bmatrix} 0 & 0 & \frac{-1}{2H} \\ 0 & 0 & 0 \\ 0 & 0 & 0 \\ 0 & 0 & 0 \\ 0 & 0 & 0 \\ \frac{1}{T_{WT}} & 0 & 0 \\ 0 & \frac{1}{T_{PV}} & 0 \end{bmatrix} \quad (3.27)$$

$$B_2 = \begin{bmatrix} 0 \\ 0 \\ 0 \\ 0 \\ \frac{K_{VI}}{R_{VI}T_{INV}} \\ 0 \\ 0 \end{bmatrix} \quad (3.28)$$

3.7 Simulation Results

The virtual inertia control system is implemented to a single area power system described in Fig. 3.9. The small-signal/dynamic response model is constructed using the MATLAB/Simulink software. To validate the efficiency of the proposed inertia control method, multiple disturbances are added to the studied system under various degraded situations of system inertia and load damping reductions as follows. The sudden load increase of 0.02 p.u. is applied at 2 s. Then, at 15 s, the sudden increase in solar power (0.01 p.u.) and wind power (0.01 p.u.) are simultaneously applied. The control parameters for the studied system are given in Tables 3.1 and 3.2.

Figure 3.10 investigates the system frequency response without virtual inertia operation against the degraded circumstances of system inertia and damping (reduction from 100% to 40%). With the reduction in system inertia and damping properties due to RESs/DGs penetration, it is obvious that the frequency nadir/overshoot of the system significantly increases with longer stabilizing time. The RoCoF of the system also increases, which means less time for a system operator to respond to the disturbances. These problems would be exacerbated in the system with a high share of RESs/DGs. To regulate such low inertia situations without using virtual inertia control, the under-frequency load-shedding (UFLS) may be required. Without the proper regulation, system instability, generation tripping, cascading outages, and power blackout could occur.

Table 3.1 Simulated parameters for the isolated system

Parameter	Value
Gain of Integral controller, K_s (s)	0.1
Governor time constant, T_g (s)	0.07
Turbine time constant, T_t (s)	0.37
Governor droop constant, R (Hz/p.u.)	2.6
Bias factor, β (p.u./Hz)	0.98
Virtual inertia constant, K_{VI} (p.u. s)	0.6
Virtual damping constant, D_{VI} (p.u./Hz)	0.3
Virtual inertia control droop, R_{VI} (Hz/p.u.)	2.7
Time constant of inverter-based ESS, T_{INV} (s)	1.0
Time constant of wind turbine, T_{WT} (s)	1.4
Time constant of solar system, T_{PV} (s)	1.9

Table 3.2 Inertia and load damping parameters

Percentage	System inertia, H (p.u. s)	System load damping, D (p.u./Hz)
100	0.0830	0.0160
70	0.0581	0.0112
40	0.0332	0.0064

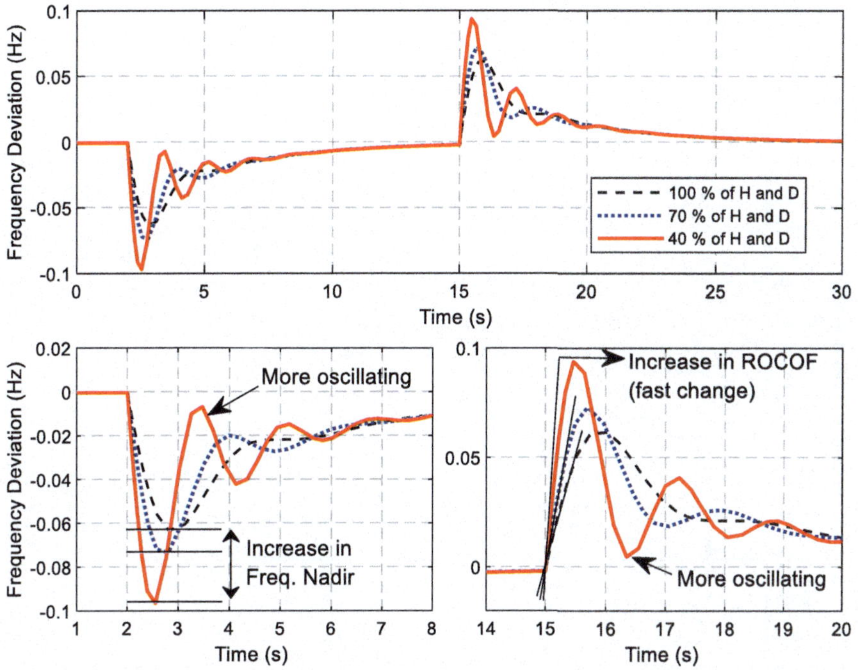

Fig. 3.10 System frequency degradation due to the inertia and load damping reduction caused by DGs/RESs penetration (without applying virtual inertia control)

By applying the virtual inertia control unit (see Fig. 3.11), the system frequency nadir/overshoot under the critical situation of low H and D (40%) could be effectively arrested after the disturbances. Figures 3.12 and 3.13 show the improvement of system frequency nadir/overshoot and RoCoF according to the deployment of virtual inertia control. It is clear that both RoCoF and stabilizing time of the system significantly improve against the reduction of H and D from 100% to 40%.

Figure 3.14 demonstrates the active power response of the conventional generating unit and inverter-based ESS unit equipped with virtual inertia control. It can be observed that the virtual inertia control unit provides a faster power response with the required inertia (see a solid line) compared with the conventional generating unit. Especially under the critical circumstance of low system inertia (40%), the virtual inertia unit could generate more additional power during the transient state, resulting in a lower frequency nadir and RoCoF.

Considering the capability of the virtual inertia control technique, it is worth to point that the system with the virtual inertia control can obtain a better performance than the system without virtual inertia control. The system can reach an optimal response in cases of numerous disturbances, especially under the critical situations of low system inertia and damping triggered by a high share of DGs/RESs. Therefore,

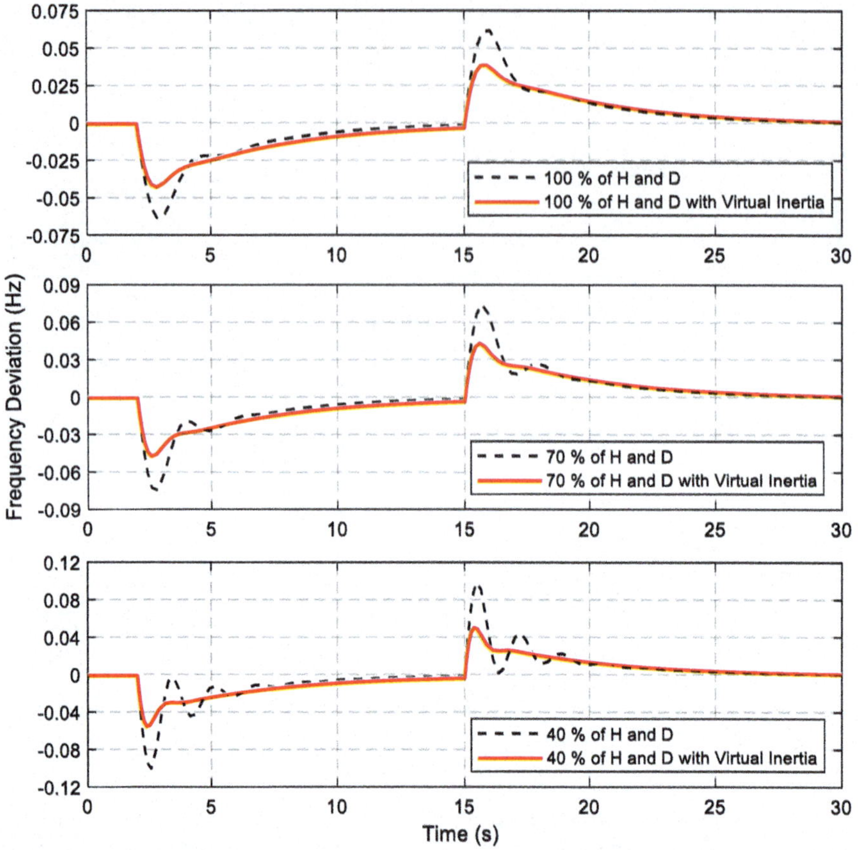

Fig. 3.11 System frequency improvement due to the deployment of virtual inertia control under the degraded inertia and load damping caused by DGs/RESs penetration

it could be confirmed that the virtual inertia control unit provides significant properties for handeling a high share of RESs/DGs penetration without compromising the system stability and resiliency.

Unlike a real synchronous machine, the parameters of virtual inertia control can be adjusted to enhance the dynamic frequency response of the system [11, 13]. In the following section, the dynamic effects of virtual inertia control parameters are fully investigated to achieve stable and robustness performances of system stability and resiliency.

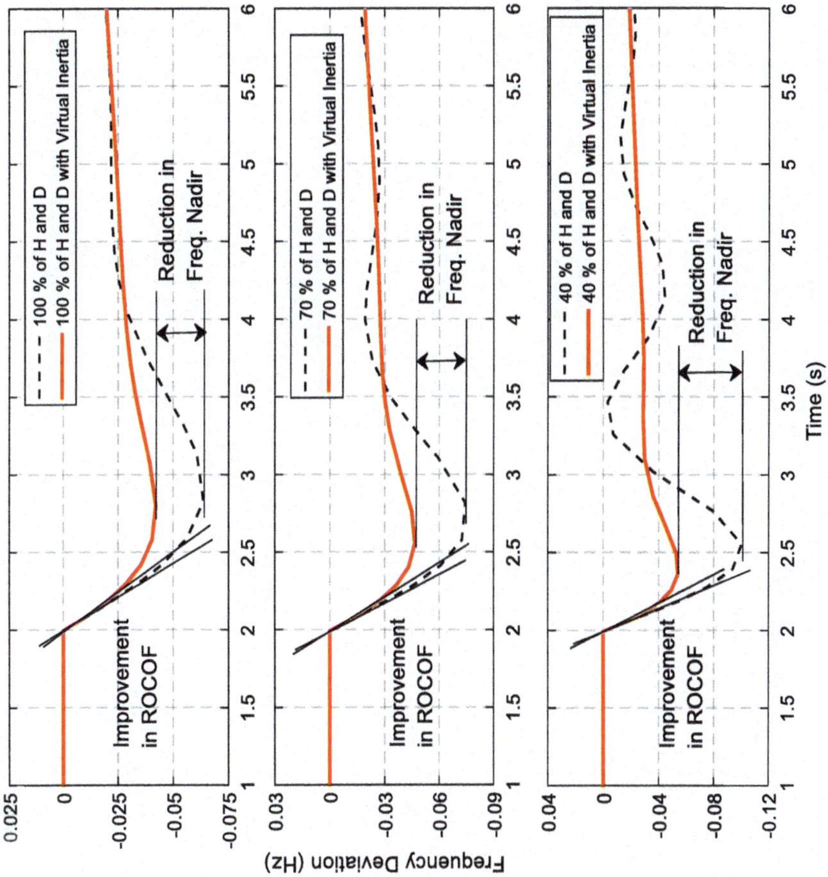

Fig. 3.12 Influence of virtual inertia control in reducing system frequency nadir and enhancing RoCoF under the degraded system inertia and load damping

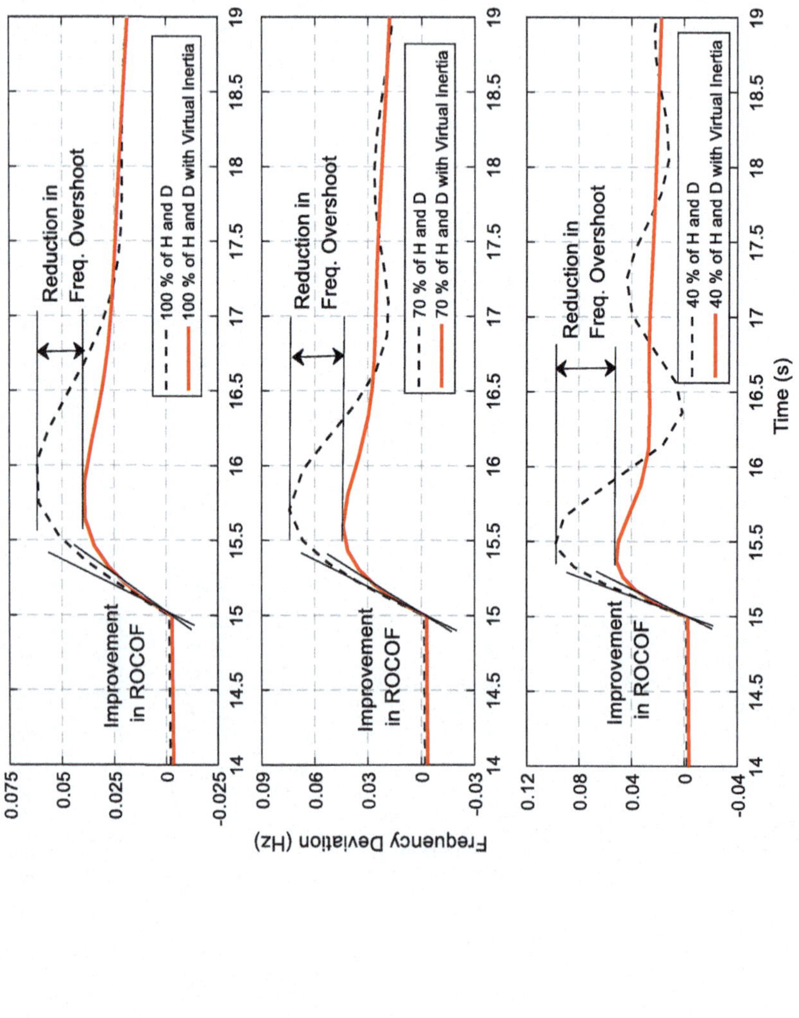

Fig. 3.13 Influence of virtual inertia control in reducing system frequency overshoot and enhancing RoCoF under the degraded system inertia and load damping

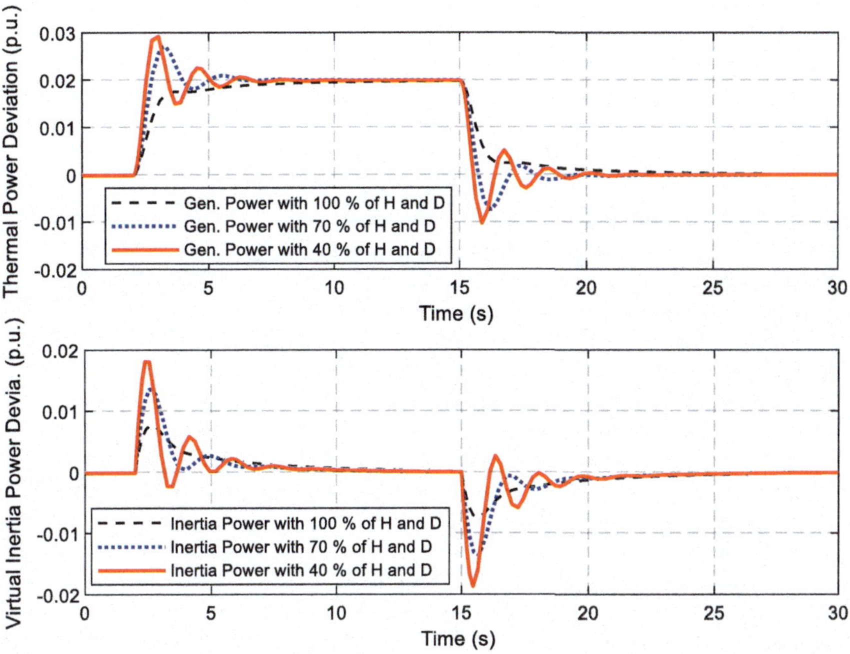

Fig. 3.14 Active power responses of the thermal generation unit and virtual inertia unit against the reduction of system inertia and load damping

3.7.1 *Effect of Virtual Inertia Control Droop*

As discussed in Sect. 3.2, the generated power with the required inertia and damping from the inverter-based ESS unit is controlled by the virtual inertia control droop, which calculates its power based on Eq. (3.3). Thus, the virtual inertia control droop characteristic could also help to improve system robustness by decreasing frequency excursions after the disturbance [11]. The control parameters of the system are given in Table 3.1.

Figure 3.15 investigates the system frequency response against the changes in the virtual inertia droop parameter (R_{VI}). The system is operated under the degraded situation of 70% of H and D. By decreasing the R_{VI}, the system can achieve better performance with lower frequency nadir/overshoot, while the stabilizing time increases.

Figure 3.16 shows active power responses of the thermal generation unit and virtual inertia unit against the variations of R_{VI}. By reducing R_{VI}, the conventional generating unit is less stressed and generates less power. On the contrary, the inverter-based ESS unit produces more inertia power with faster response time. However, as more emulated inertia power is generated with respect to the reduction of R_{VI}, the system response leads to a longer stabilizing time following the disturbances.

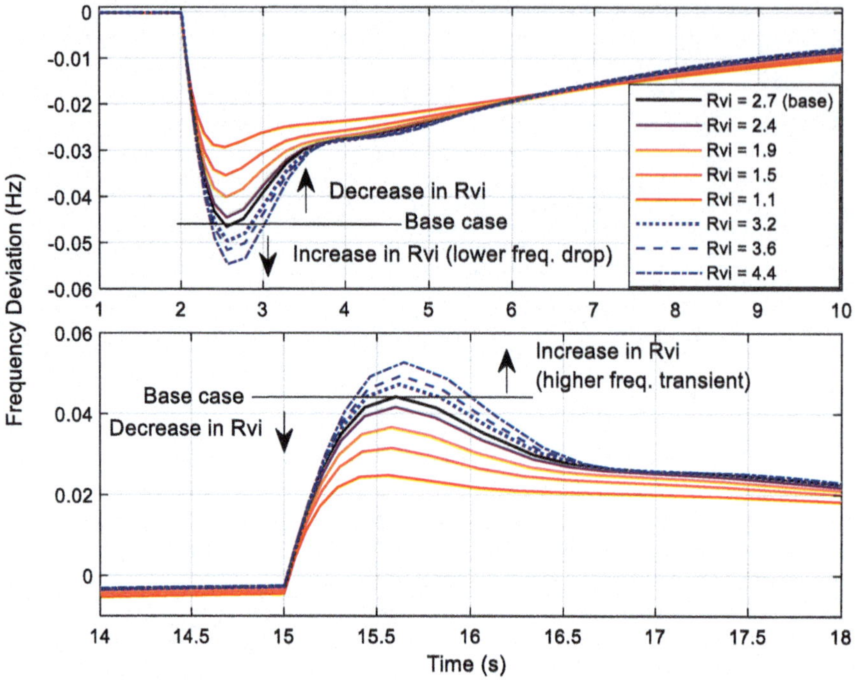

Fig. 3.15 System frequency responses under the variations of virtual inertia droop characteristic

3.7.2 Effect of Virtual Inertia Constant

In virtual inertia control, the emulated inertia power can be mainly adjusted by the virtual inertia constant (K_{VI}) to achieve better system stability and performance [22, 37].

Figure 3.17 investigates the system frequency response against the changes in virtual inertia constant. The system is operated under the degraded situation of 70% of H and D. The D_{VI} is fixed at a small value of 0.3. The other control parameters of the system are given in Table 3.1. It is found that the key factor of increasing K_{VI} is that the system frequency nadir/overshoot significantly decreases, leading to better performance and stability of the system. However, if the K_{VI} is increased too high (over 2), the system frequency would require longer time to settle, resulting in the lower damping performance. To solve this problem, the virtual damping constant may be increased for modifying suitable damping property and achieving faster stabilizing time. Figure 3.18 shows active power responses of the thermal generation unit and virtual inertia unit against the variations of K_{VI}. By increasing the K_{VI}, the conventional generating unit is less stressed and generates less power. Obviously, the inverter-based ESS unit produces more inertia power with faster response time after the disturbances.

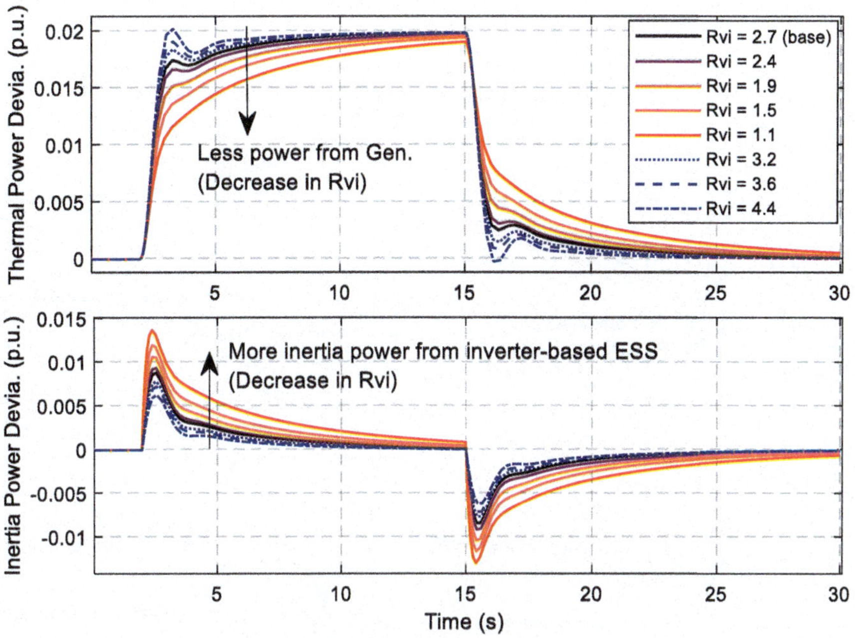

Fig. 3.16 Active power responses of the thermal generation unit and virtual inertia unit against the variations of virtual inertia droop characteristic

3.7.3 Effect of Virtual Damping

To suitably compensate the damping property and reduce the effect of long stabilizing time, the virtual damping constant can be adjusted. The virtual damping can significanlty suppress the system oscillations during multiple disturbances [15, 37].

Figure 3.19 investigates the system frequency response against the changes in virtual damping constant. The system is operated under the degraded situation of 70% of H and D. The K_{VI} is fixed at a small value of 0.6. The other control parameters of the system are given in Table 3.1. Evidently, the benefit of applying virtual damping to the system is the remarkable enhancement of stabilizing/settling performance and time. However, the side effect of increasing D_{VI} to a high value is the large overshoot effect. If the D_{VI} is increased too high (over 1.5), the large second overshoot could be observed. Figure 3.20 shows active power responses of the thermal generation unit and virtual inertia unit against the variations of D_{VI}. By increasing the D_{VI}, it is obvious that the virtual inertia unit requires more constant power after the disturbance to suppress the effect of long stabilizing time.

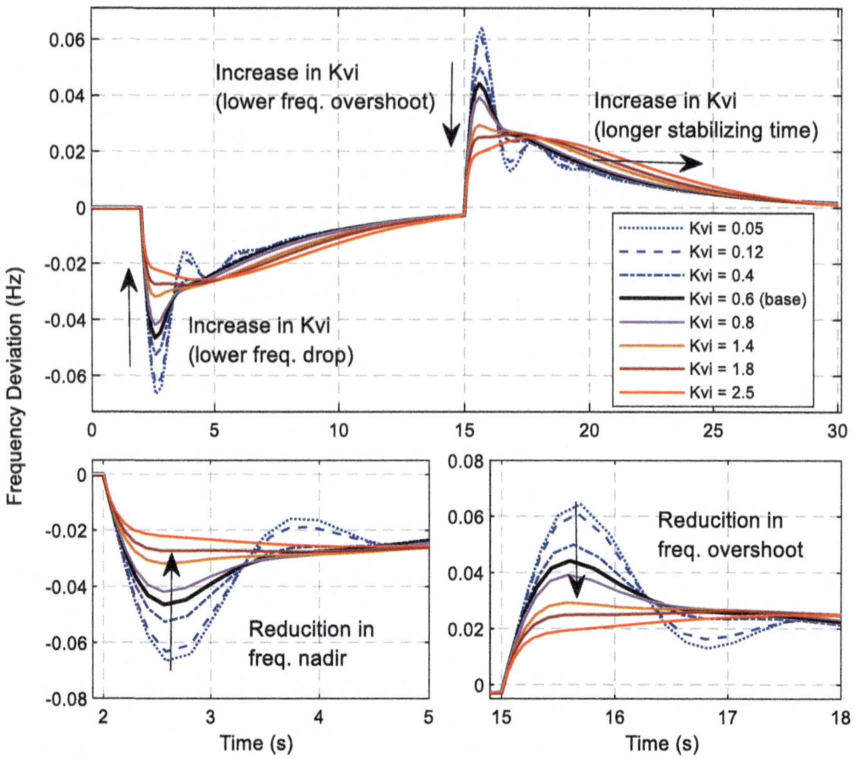

Fig. 3.17 System frequency response under the changes of virtual inertia constant

3.7.4 Effect of Time Delay

In real practice, fast response and changing devices of frequency are nearly unobservable due to delays and filters equipped in virtual inertia control process [11]. Thus, any signal measurements and filtering introduce delays, which should be analyzed. The time delay also depends on the communication channels, physical response time, physical setup, and functionality of the inverter-based ESS unit. Such time delay could cause the degradation of power system stability and performance.

Figure 3.21 analyzes the system frequency response against the variations of virtual inertia control time delay dominated by the inverter-based ESS (T_{INV}). The system is operated under the degraded situation of 70% of H and D. The other control parameters of the system are given in Table 3.2. Obviously, by increasing the T_{INV}, it yields larger frequency nadir/overshoot following the disturbances. The RoCoF of the system also increases due to the increase in T_{INV}.

Figure 3.22 shows the active power responses of the thermal generation unit and virtual inertia unit against the variations of time delay. By increasing the T_{INV}, the conventional generating unit is required to produce more power, while the virtual

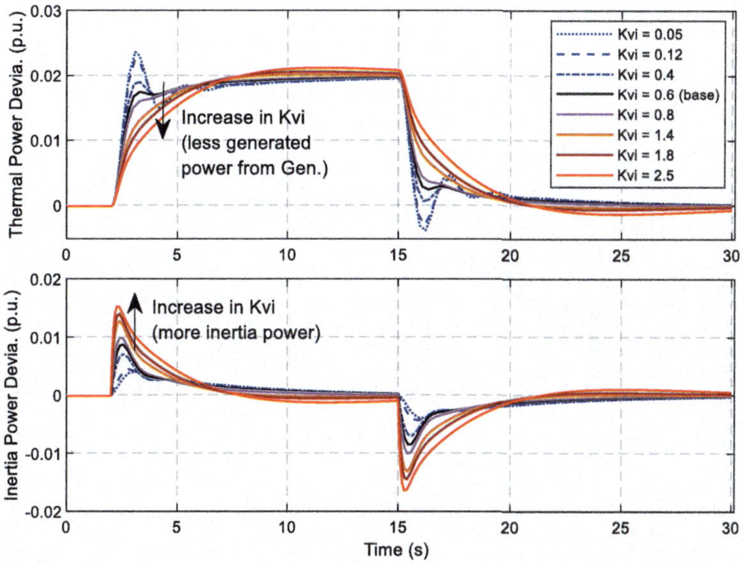

Fig. 3.18 Active power responses of the thermal generation unit and virtual inertia unit against the variations of virtual inertia constant

inertia unit generates less inertia power. By applying a high value of T_{INV}, the response time of virtual inertia control is slower than the conventional generating unit. Therefore, to get an accurate perception of virtual inertia control regarding frequency subject, it is neccessary to consider the important inherent setting and basic limitations imposted by the intertia control dynamics and design them for the sake of performance evaluation.

3.8 Summary

In this chapter, the most important concept, design framework, and topologies of virtual inertia control are fully explained. Some experimental studies with a practical virtual inertia control in the laboratory environment are also demonstrated. Then, this chapter is focused on the potential role of virtual inertia control in a single area frequency control task. The linearized frequency response model for virtual inertia, primary, and secondary controls is presented by the state-space representation. Dynamic and static performances of the virtual inertia response model are

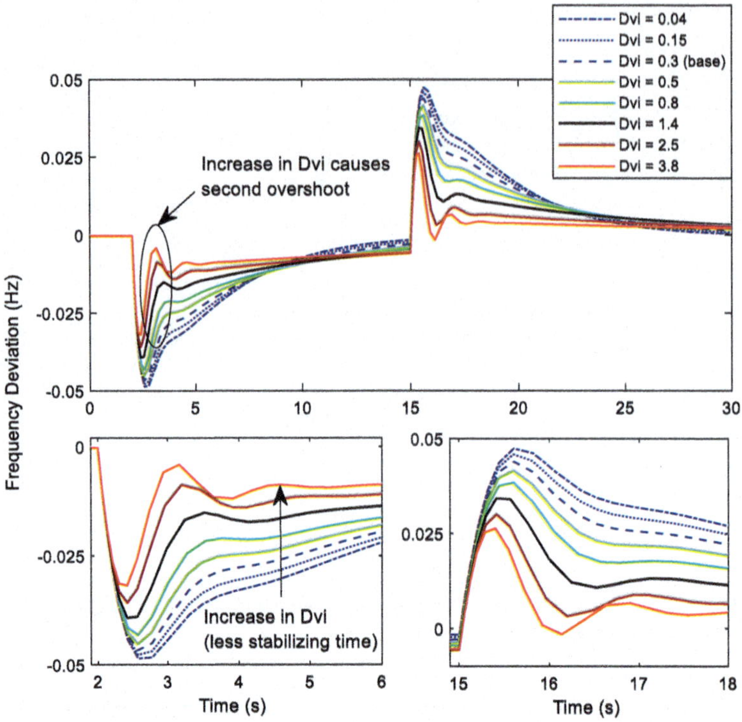

Fig. 3.19 System frequency response under the changes of virtual damping constant

explained in terms of small-signal and state-space representations. This task is validated by performing simulation results. Finally, the effects of various inherent parameters affecting the system frequency response with respect to the implementation of virtual inertia are emphasized.

Problems for Chapter 3

3.1 The inverter unit has a rated capacity of 25 MVA, 480 V, which is used for virtual inertia control. The grid frequency is 60 Hz. The system inertia constant is 2.5 MWs. Please determine the suitable virtual inertia constant for the inertia control unit.

3.2 Please construct the single-area power system based on Fig. 3.9 and Table 3.1 using MATLAB/Simulink software. The system base is 100 MVA. The system nominal frequency is 60 Hz. The gain of an integral controller is set at −0.2. The system inertia and damping are reduced to 40% of its nominal values from Table 3.2. The 5 MW of wind system and 5 MW of solar system are connected to the system at 5 s. Then, if the step load (ΔP_L) of 20 MW is added at the time $t = 10$ s, please determine the following.

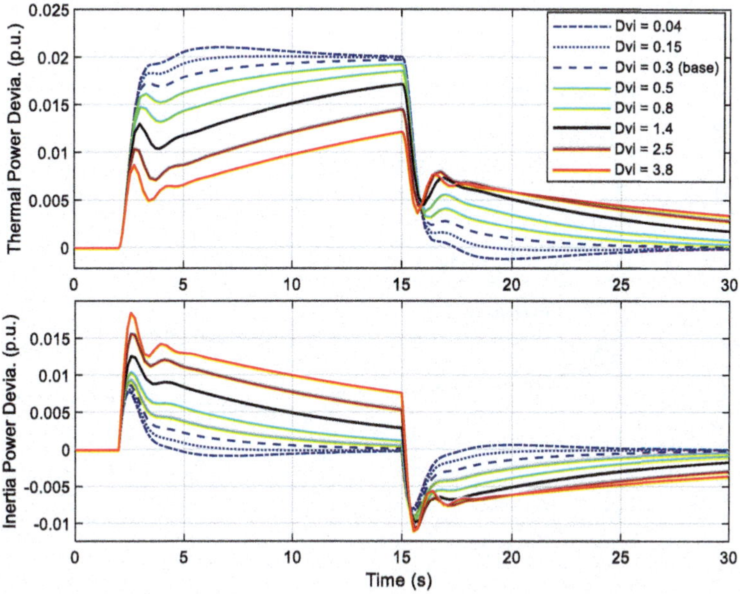

Fig. 3.20 Active power responses of the thermal generation unit and virtual inertia unit against the variations of virtual damping constant

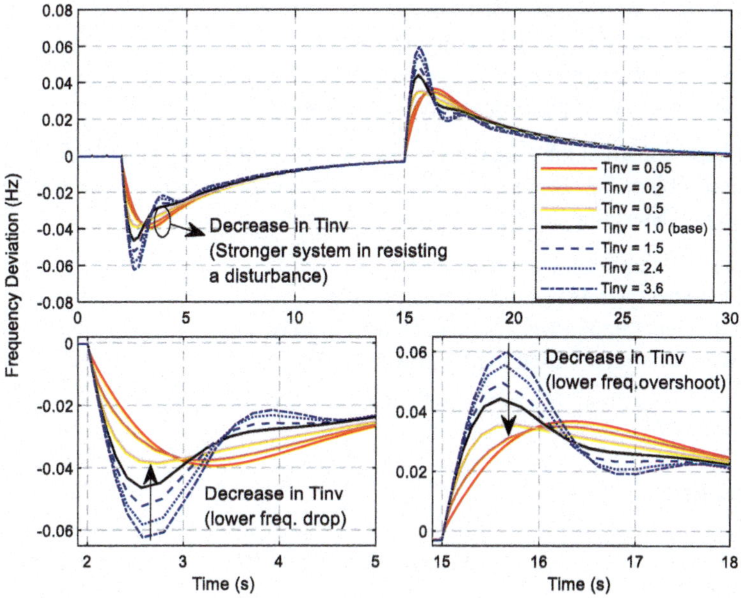

Fig. 3.21 System frequency response under the changes of time delay dominated by the inverter-based ESS

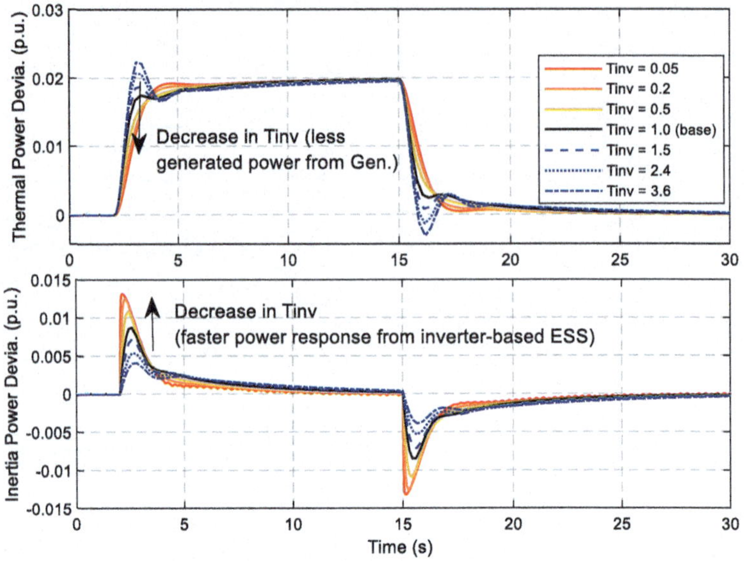

Fig. 3.22 Active power responses of the thermal generation unit and virtual inertia unit against the variations of time delay dominated by the inverter-based ESS

(1) System frequency nadir in Hz
(2) Steady-state system frequency in Hz
(3) Generated power of the generator in MW
(4) Generated virtual inertia power in MW

3.3 Please construct the single-area power system with virtual inertia control based on Problem 3.2. If the system is operated without the secondary control unit, please determine the following.

(1) System frequency nadir in Hz
(2) Steady-state system frequency in Hz
(3) Generated power of the generator in MW
(4) Generated virtual inertia power in MW

References

1. D.E. Olivares et al., Trends in microgrid control. IEEE Trans. Smart Grid **5**(4), 1905–1919 (2014)
2. M. Dreidy, H. Mokhlis, S. Mekhilef, Inertia response and frequency control techniques for renewable energy sources: a review. Renew. Sustain. Energy Rev. **69**(1), 144–155 (2017)
3. H.P. Beck, R. Hesse, Virtual synchronous machine, in Proc. *International Conference on Electrical Power Quality and Utilisation*, 1-6 (2007)

4. J. Driesen, K. Visscher, Virtual synchronous generators, in Proc. *IEEE Power and Energy Society General Meeting (IEEE PES GM)*, 1-6 (2008)
5. Q.C. Zhong, G. Weiss, Synchronverters: inverters that mimic synchronous generators. IEEE Trans. Ind. Electron. **58**(4), 1259–1267 (2011)
6. P. Rodriguez, E. Rakhshani, A. Mir Cantarellas, D. Remon, Analysis of derivative control based virtual inertia in multi-area high-voltage direct current interconnected power systems. IET Gener. Transm. Distrib. **10**(6), 1458–1469 (2016)
7. E. Rakhshani, D. Remon, A.M. Cantarellas, J.M. Garcia, P. Rodriguez, Virtual synchronous power strategy for multiple HVDC interconnections of multi-area AGC power systems. IEEE Trans. Power Syst. **32**(3), 1665–1677 (2017)
8. H. Bevrani, T. Ise, Y. Miura, Virtual synchronous generators: a survey and new perspectives. Int. J. Electr. Power Energy Syst. **54**, 244–254 (2014)
9. S. D'Arco, J.A. Suul, O.B. Fosso, A virtual synchronous machine implementation for distributed control of power converters in SmartGrids. Electr. Power Syst. Res. **122**(1), 180–197 (2015)
10. U. Tamrakar, D. Shrestha, M. Maharjan, B. Bhattarai, T. Hansen, R. Tonkoski, Virtual inertia: current trends and future directions. Appl. Sci. **7**(654), 1–29 (2017)
11. H. Bevrani, *Robust Power System Frequency Control*, 2nd edn. (Springer, New York, USA, 2014)
12. Y. Chen, R. Hesse, D. Turschner, H.P. Beck, Improving the grid power quality using virtual synchronous machines, in Proc. *International Conference on Power Engineering, Energy and Electrical Drives,* 1-6 (2011)
13. H. Bevrani, I. Francois, T. Ise, *Microgrid Dynamics and Control* (IEEE-Wiley, Hoboken, NJ, 2017)
14. M. Albu, A. Nechifor, D. Creanga, Smart storage for active distribution networks. Estimation and measurement solutions, in Proc. *IEEE International Instrumentation and Measurement Technology Conference,* 1-6 (2010)
15. T. Kerdphol, F.S. Rahman, M. Watanabe, Y. Mitani, D. Turschner, H.P. Beck, Extended virtual inertia control design for power system frequency regulation, in Proc. *IEEE PES GTD Grand International Conference and Exposition Asia*, 97-101 (2019)
16. K. De Brabandere, B. Bolsens, J. Van den Keybus, A. Woyte, J. Driesen, R. Belmans, A voltage and frequency droop control method for parallel inverters. IEEE Trans. Power Electron. **22**(4), 1107–1115 (2007)
17. E. Rakhshani, P. Rodriguez, Inertia emulation in AC/DC interconnected power systems using derivative technique considering frequency measurement effects. IEEE Trans. Power Syst. **32**(5), 3338–3351 (2017)
18. T. Kerdphol, F.S. Rahman, M. Watanabe, Y. Mitani, Robust virtual inertia control of a low inertia microgrid considering frequency measurement effects. IEEE Access **7**(1), 57550–57560 (2019)
19. T. Kerdphol, F.S. Rahman, Y. Mitani, M. Watanabe, S. Kufeoglu, Robust virtual inertia control of an islanded microgrid considering high penetration of renewable energy. IEEE Access **6**(1), 625–636 (2017)
20. T. Kerdphol, M. Watanabe, K. Hongesombut, Y. Mitani, Self-adaptive virtual inertia control-based fuzzy logic to improve frequency stability of microgrid with high renewable penetration. IEEE Access **7**(1), 76071–76083 (2019)
21. T. Kerdphol, F.S. Rahman, Y. Mitani, Virtual inertia control application to enhance frequency stability of interconnected power systems with high renewable energy penetration. Energies **11**(4), 981–997 (2018)
22. T. Kerdphol, M. Watanabe, Y. Mitani, V. Phunpeng, Applying virtual inertia control topology to SMES system for frequency stability improvement of low-inertia microgrids driven by high renewables. Energies **12**(3902), 1–16 (2019)
23. J. Alipoor, Y. Miura, T. Ise, Stability assessment and optimization methods for microgrid with multiple VSG units. IEEE Trans. Smart Grid **9**(2), 1462–1471 (2018)
24. T. Shintai, Y. Miura, T. Ise, Oscillation damping of a distributed generator using a virtual synchronous generator. IEEE Trans. Power Deliv. **29**(2), 668–676 (2014)

25. J. Alipoor, Y. Miura, T. Ise, Power system stabilization using virtual synchronous generator with alternating moment of inertia. IEEE J. Emerg. Sel. Top. Power Electron. **3**(2), 451–458 (2015)
26. J. Alipoor, Y. Miura, T. Ise, Distributed generation grid integration using virtual synchronous generator with adoptive virtual inertia, in Proc. *IEEE Energy Conversion Congress Exposition*, 4546–4552 (2013)
27. H. Hlaing, J. Liu, Y. Miura, H. Bevrani, T. Ise, Enhanced Performance of a Stand-Alone Gas-Engine Generator Using Virtual Synchronous Generator and Energy Storage System. IEEE Access. **7**(1), 2169–3536 (2019)
28. N.B. Sharifmuddin, T. Ise, T. Suwa, Comparative study of series and parallel schemes for stabilization of a microgrid integrated with a solar thermal cogeneration system using virtual synchronous generator control, in Proc. *International Energy Sustainable Conference*, 1–6 (2016)
29. J. Liu, Y. Miura, T. Ise, Comparison of dynamic characteristics between virtual synchronous generator and droop control in inverter-based distributed generators. IEEE Trans. Power Electron. **31**(5), 3600–3611 (2016)
30. J. Alipoor, Y. Miura, T. Ise, Voltage sag ride-through performance of virtual synchronous generator, in Proc. *International Power Electronics Conference*, 3298–3305 (2014)
31. V. Karapanos, S. de Haan, K. Zwetsloot, Real time simulation of a power system with VSG hardware in the loop, in Proc. *Annual Conference of IEEE Industrial Electronics Society*, 3748–3754 (2011)
32. V. Karapanos, S. de Haan, K. Zwetsloot, Real time simulation of a power system with VSG hardware in the loop, in Proc. *Annual Conference of IEEE Industrial Electronics Society*, 3748–3754 (2011)
33. M. Albu, M. Calin, D. Federenciuc, J. Diaz, The measurement layer of the Virtual Synchronous Generator operation in the field test, in Proc. *IEEE International Workshop on Applied Measurements for Power Systems*, 1-5 (2011)
34. T. Vu Van et al., Virtual synchronous generator: An element of future grids, in Proc. *IEEE PES Innovative Smart Grid Technologies Conference Europe*, 1–7 (2010)
35. M. Albu et al., Measurement and remote monitoring for virtual synchronous generator design, in Proc. *IEEE International Conference on Applied Measurements for Power Systems*, 1-6 (2010)
36. V. Van Thong et al., Virtual synchronous generator: laboratory scale results and field demonstration, in Proc. *IEEE PowerTech: Innovative Ideas Toward the Electrical Grid of the Future*, 1-6 (2009)
37. T. Kerdphol, F.S. Rahman, M. Watanabe, Y. Mitani, D. Turschner, H.P. Beck, Enhanced virtual inertia control based on derivative technique to emulate simultaneous inertia and damping properties for microgrid frequency regulation. IEEE Access **7**(1), 14422–14433 (2019)
38. P. Kundur, *Power System Stability and Control* (McGraw-Hill, New York, 1994)
39. H. Bevrani, M.R. Feizi, S. Ataee, Robust frequency control in an Islanded microgrid: $H\infty$ and μ-synthesis approaches. IEEE Trans. Smart Grid **7**(2), 706–717 (2016)
40. J. Pahasa, I. Ngamroo, PHEVs bidirectional charging/discharging and SoC control for microgrid frequency stabilization using multiple MPC. IEEE Trans. Smart Grid **6**(2), 526–533 (2015)
41. J. Pahasa, I. Ngamroo, Coordinated control of wind turbine blade pitch angle and PHEVs using MPCs for load frequency control of microgrid. IEEE Syst. J. **10**(1), 97–105 (2016)
42. M. Hajiakbari Fini, M. E. Hamedani Golshan, Determining optimal virtual inertia and frequency control parameters to preserve the frequency stability in islanded microgrids with high penetration of renewables. Electr. Power Syst. Res. **154**(1), 13–22 (2018)
43. C. Bordons, F. Garcis-Torres, M.A. Ridao, *Model Predictive Control of Microgrids*, (Springer, Switzerland, 2020)

Chapter 4
Multiple-Virtual Inertia Synthesis for Interconnected Systems

Abstract In the previous chapter, the regulation of the interchange power and the implementation of multiple virtual inertia controls are not considered. In real practice, most of the power systems are interconnected to enable the power transfer between each area in the system for more efficient use of the available resources. With the increasing penetration of renewable energy sources (RESs)/distributed generators (DGs), the inertia of some areas will decrease and could lead to power system oscillations. In this regard, the regional inertia property is critical to managing the oscillations and avoiding system instability. To overcome this problem, this chapter investigates the coordination of multiple virtual inertia control systems to improve the frequency stability and responses of the inertia power and tie-line power in the interconnected power systems with RESs/DGs. The dynamic performance and frequency characteristics of an interconnected system with multiple virtual inertia, primary, and secondary control loops are explained. A dynamic response model of the interconnected system is linearized and investigated in detail. The role of interchange power between control areas is briefly described. The effects of control parameters affecting the system frequency response with respect to the implementation of multiple virtual inertia units in the interconnected system are discussed via the state-space representation.

Keywords Dynamic performance · Frequency control loops · Frequency deviation · Interconnected power systems · Linearized model · Primary control · Secondary control · State-space model · Tie-line power control · Uncertainty

4.1 Introduction to Interconnected Systems

In real practice, most power systems are interconnected via a tie-line or transmission line to help each other in regulating and exchanging interchange power, improving system stability, reliability, and resiliency [1]. An interconnected power system consists of control areas that are connected by tie-lines or transmission lines with high voltage [2]. The trend of measured frequency in each region is an important index of the power unbalance in the interconnected area (not in the single area). Nowadays, frequency control in such a system is gaining more concerns according

T. Kerdphol et al., *Virtual Inertia Synthesis and Control*, Power Systems,
https://doi.org/10.1007/978-3-030-57961-6_4

to the changing structure, increasing size, and complexity with respect to DGs/RESs penetration. Especially, the integration of numerous DGs/RESs in interconnected systems could provoke the massive lack of system inertia and damping. Since the DGs/RESs have exchanged power to multi-area systems using power electronic interfaces (e.g., inverters/converters), the interfaces of power electronics naturally diminish the whole system inertia and damping property. Compared with traditional generation (based on synchronous machines) [3, 4], the power electronic interfaces directly result in the lower the voltage and frequency stabilization. Subsequently, the decline of sufficient inertia will be one of the primary restrictions of the interconnected systems with DGs/RESs worldwide. By rising DGs/RESs levels, the sufficient inertia and damping of the interconnected system may be inadequate, and this creates the dynamic problems to regulate system frequency and voltage, leading to undesirable effects on system stability and resiliency [5, 6]. Therefore, the virtual inertia control can play a crucial function as effective ancillary service in reinforcing power exchanges with the desirable inertia and damping property for interconnected power systems dominated by DGs/RESs [7–12]. Recently, a few research works have applied to the virtual inertia control designs into the interconnected systems [13–16].

In this chapter, the power system is concerned as a group of control areas connected via transmission/tie lines (high voltage). Each area would control its own inertia, frequency, and loads as well as deliver tie-line power exchange with the desirable inertia and damping to its neighbors.

This section demonstrates an extended operation of virtual inertia control to enhance the dynamic and stability of interconnected systems. The virtual inertia control unit is implemented to the studied interconnected system, augmenting suitable system inertia with damping, frequency performance, and preventing instability and failures during DGs/RESs integration. Thus, the basic structure of an interconnected system is re-established by integrating the control loop of virtual inertia for advanced-level frequency control analysis. The primary goal of this chapter is to introduce a new control method of the inertia control-based frequency analysis for multi-area power systems. Accordingly, the proposed inertia control system can efficiently provide to superior exploitation of DGs/RESs in the interconnected systems, while preserving the system robustness of operation. The contribution of this chapter is to design a suitable structure for the pre-evaluation of dynamic effects of an interconnected system, including inertia synthesis based on high-level control for power system applications. Eventually, the proposed inertia control system can augment the frequency stability of the interconnected system when the DGs/RESs is greatly integrated at partial or low load consumption. Compared to traditional interconnected systems [3, 4, 17, 18], the interconnected system with inertia control provides superior dynamic performance, and stability to today's and future's power system, which is intended to penetrate higher renewable energy towards 100% without compromising system stability, and thus, avoiding cascading failures and power blackouts.

4.2 Modeling of Multiple-Virtual Inertia Control

In this section, the study of an interconnected power system with DGs/RESs integration is constructed in Fig. 4.1. The system contains two control areas, which are connected by a tie line (high-voltage AC line). By integrating the DGs/RESs (e.g., solar and wind systems) into the interconnected system, the power electronic interfaces naturally create a low moment of inertia and absence of frequency stabilization to the system. The virtual inertia control units in both regions are greatly expected to help in compensating power unbalance with the required inertia as well as incorporating the secondary control or when the secondary control is not available.

To perform frequency study and analysis, a low-order structure can be exploited to create the generation-load dynamic behavior of the interconnected system. The small-signal or dynamic structure of the system is depicted in Fig. 4.2. The control parameters of the system are given in Table 4.1. To receive a precise behavior of the actual interconnected system, this chapter analyzes the essential restrictions from the physical dynamics of the generation-load systems, including inherent aspects. The crucial physical restriction of the thermal generation is the rate of change of power generation due to the mechanical movement limit. The physical dynamics of the thermal generation can be employed using the maximum and minimum valve gate opening/closing for the turbine system and generation rate constraint (GRC). The V_U and V_L are the maximum and minimum constraints, which regulate the rate of the valve-gate closing/opening speed of the turbine unit [2]. The governor dead band of the non-reheat thermal generating unit is used as 20% p.u.MW/min for Area 1, and 25% p.u.MW/min for Area 2. This chapter assumed that the synchronous generating units, which have the similar component or behavior, are demonstrated through the aggregated structure. The solar and wind power sources as well as loads

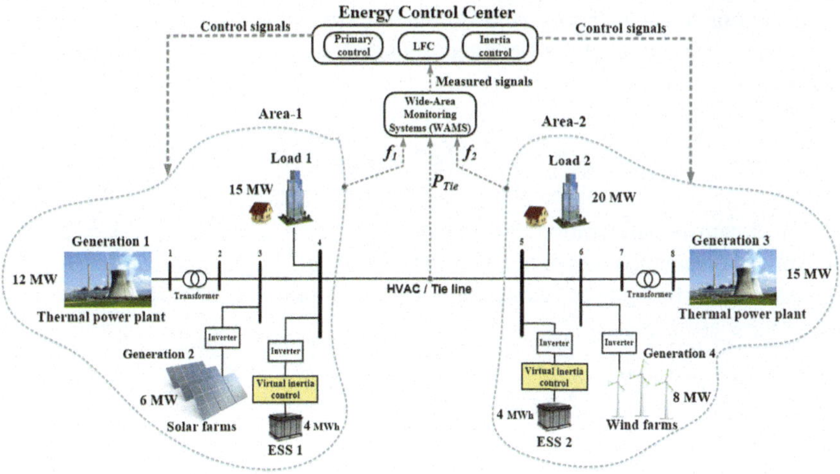

Fig. 4.1 Schematic structure of the interconnected system with virtual inertia control units

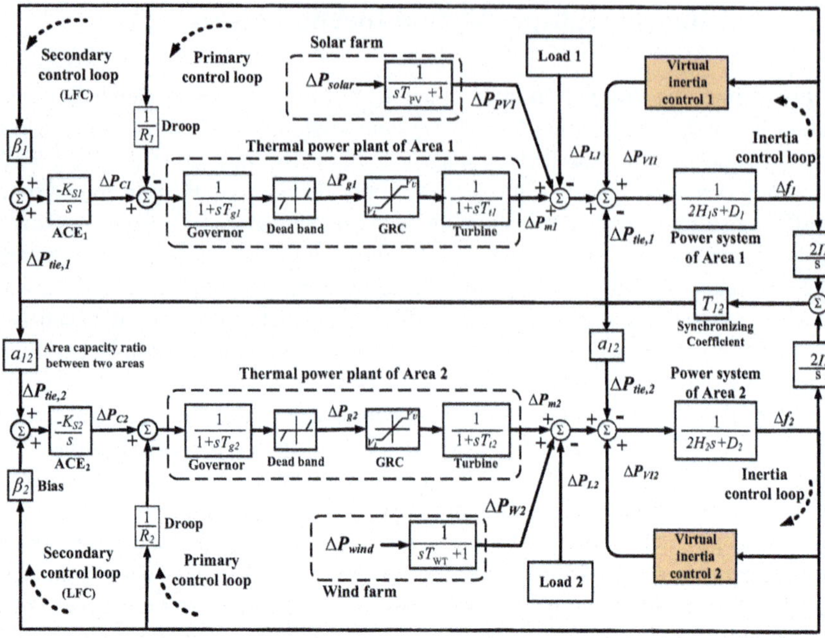

Fig. 4.2 Small-signal/dynamic structure of the interconnected system with different inertia control droops

are analyzed as the external disturbance to the interconnected system. The DGs/RESs in other systems may have a high-order dynamic response. However, the low-order dynamic structures presented in this chapter are suitable enough to investigate the frequency control issue [2, 19, 20].

In addition to the interconnected system, the area control error (ACE) is modified as a linear combination of the weighted frequency deviation and tie-line power flow, as depicted below [2].

$$\Delta P_{Cj} = \frac{K_{Sj}}{s}\left[\beta_j \cdot \Delta f_j\right] + \Delta P_{Tie_j} \tag{4.1}$$

The dynamic configuration of tie-line power is fully described in Chap. 2. Alternatively, to create the interconnections between N areas in an interconnected power system, the tie-line power change between the area-j and the rest of the areas is represented as [2, 21]:

$$\Delta P_{Tie,j}(s) = \sum_{\substack{j=1\\ j\neq k}}^{y} \Delta P_{Tie,jk}(s)$$

Table 4.1 Control parameters for the interconnected system

Parameters	Area 1	Area 2
Frequency bias factor, B (p.u.MW/Hz)	0.3483	0.3827
Area control error gain, K_S (s)	0.3	0.2
Governor time constant, T_g (s)	0.08	0.06
Turbine time constant, T_t (s)	0.4	0.44
Droop constant, R (Hz/p.u.MW)	3	2.73
System inertia constant, H (p.u.MW s)	0.083	0.1010
Damping coefficient, D (p.u.MW/Hz)	0.015	0.016
Virtual inertia constant, K_{VI} (p.u. s)	1.3	1.5
Virtual damping constant, D_{VI} (p.u.MW/Hz)	0.2	0.2
Virtual inertia time constant, T_{INV} (s)	1	1.2
Inverter-based ESS power capacity, T_{INV_max}, T_{INV_min} (p.u.MW)	±0.2	±0.3
Virtual inertia control droop, R_{VI} (Hz/p.u.MW)	2.5	2.7
Solar system time constant, T_{PV} (s)	1.3	–
Wind turbine time constant, T_{WT} (s)	–	1.5
Maximum limit of valve gate, V_U (p.u.MW)	0.4	0.5
Minimum limit of valve gate, V_L (p.u.MW)	−0.4	−0.5
System base (MW)	1500	
Synchronizing coefficient, T_{12} (p.u.MW/Hz)	0.08	
Area capacity ratio between two areas, α_{12}	−0.6	

$$= \frac{2\pi}{s}\left[\sum_{\substack{j=1\\ j\neq k}}^{y} T_{jk}\Delta f_j - \sum_{\substack{j=1\\ j\neq k}}^{y} T_{jk}\Delta f_k\right] \tag{4.2}$$

$$\Delta P_{Tie,jk} = \alpha_{kj}\Delta P_{Tie,kj} \tag{4.3}$$

$$\Delta\alpha_{kj} = -\frac{P_{rk}}{P_{rj}} \tag{4.4}$$

where T_{jk} is the synchronizing coefficient between areas, $\Delta P_{Tie,jk}$ is the tie-line power transfer between the j area and the k area. P_{rk} and P_{rj} are the rated power in the k area and the j area, respectively.

By linearizing the dynamic effects of the load-generation blocks, including virtual inertia control, primary control, and secondary control units from Fig. 4.2, the frequency deviation of the control area 1 can be expressed as [16]:

$$\Delta f_1(s) = \frac{1}{2H_1 s + D_1}\left(\Delta P_{m1}(s) + \Delta P_{PV1}(s) + \Delta P_{VI1}(s) - \Delta P_{L1}(s) - \Delta P_{Tie,1}(s)\right) \tag{4.5}$$

where,

$$\Delta P_{m1}(s) = \frac{1}{1 + sT_{t1}}\Delta P_{g1}(s) \tag{4.6}$$

$$\Delta P_{g1}(s) = \frac{1}{1 + sT_{g1}}\left(\Delta P_{C1}(s) - \frac{1}{R_1}\Delta f_1(s)\right) \tag{4.7}$$

$$\Delta P_{C1}(s) = \frac{K_{S1}}{s}(\beta_1 \cdot \Delta f_1(s)) + \Delta P_{Tie,1} \tag{4.8}$$

$$\Delta P_{PV1}(s) = \frac{1}{1 + sT_{PV1}}\Delta P_{solar}(s) \tag{4.9}$$

$$\Delta P_{VI1}(s) = \frac{sK_{VI1} + D_{VI1}}{1 + sT_{INV1}}\left(\frac{\Delta f_1(s)}{R_{VI1}}\right) \tag{4.10}$$

$$\Delta P_{Tie,1} = \Delta P_{Tie,12} = T_{12} \cdot \frac{2\pi}{s}(\Delta f_1 - \Delta f_2) \tag{4.11}$$

Similarly, by linearizing the dynamic effects of the load-generation blocks, including virtual inertia control, primary control, and secondary control units from Fig. 4.2, the frequency deviation of the control area 2 can be expressed as:

$$\Delta f_2(s) = \frac{1}{2H_2 s + D_2}\left(\Delta P_{m2}(s) + \Delta P_{W2}(s) + \Delta P_{VI2}(s) - \Delta P_{L2}(s) - \Delta P_{Tie,2}(s)\right) \tag{4.12}$$

where,

$$\Delta P_{m2}(s) = \frac{1}{1 + sT_{t2}}\Delta P_{g2}(s) \tag{4.13}$$

$$\Delta P_{g2}(s) = \frac{1}{1 + sT_{g2}}\left(\Delta P_{C2}(s) - \frac{1}{R_2}\Delta f_2(s)\right) \tag{4.14}$$

$$\Delta P_{C2}(s) = \frac{K_{S2}}{s}(\beta_2 \cdot \Delta f_2(s)) + \Delta P_{Tie,2} \tag{4.15}$$

$$\Delta P_{W2}(s) = \frac{1}{1 + sT_{W2}}\Delta P_{wind}(s) \tag{4.16}$$

$$\Delta P_{VI2}(s) = \frac{sK_{VI2} + D_{VI2}}{1 + sT_{INV2}} \left(\frac{\Delta f_2(s)}{R_{VI2}} \right) \tag{4.17}$$

$$\Delta P_{Tie,2} = \Delta P_{Tie,21}(s) = \alpha_{12} \Delta P_{Tie,12} \tag{4.18}$$

where ΔP_m is the generated power change from the thermal generation, ΔP_g is the signal change from the turbine unit, ΔP_{VI} is the emulated power from the virtual inertia control unit, ΔP_W is the generated power change from the wind farm unit, ΔP_{PV} is the generated power change from the solar farm unit, ΔP_C is the secondary control change (signal).

4.3 State-Space Modeling of Interconnected Systems

As discussed in Chap. 3, a state-space model, which represents a mathematical structure of a physical system can be useful for analyzing the detail of interconnected system stability and dynamics. When a suitable state-space model is determined, it is easier for designers/engineers to understand or manipulate the property of the interconnected system without examining its specific physical structure. Using suitable state variables and definitions, the complete state-space model representation of the interconnected system controlled by multiple virtual inertia units can be calculated as [2, 21]:

$$\dot{x} = Ax + B_1 w + B_2 u \tag{4.19}$$

Considering the dynamic block diagram of Fig. 4.2 and Eqs. (4.5)–(4.18), the state variables and coefficients of the state-space representation in Eq. (4.19) is evaluated as follows:

$$x^T = \begin{bmatrix} x_1 \; x_2 \end{bmatrix} \tag{4.20}$$

$$x_1 = \begin{bmatrix} \Delta f_1 \; \Delta P_{m1} \; \Delta P_{g1} \; \Delta P_{C1} \; \Delta P_{VI1} \; \Delta P_{PV1} \; \Delta P_{Tie,1} \end{bmatrix} \tag{4.21}$$

$$x_2 = \begin{bmatrix} \Delta f_2 \; \Delta P_{m2} \; \Delta P_{g2} \; \Delta P_{C2} \; \Delta P_{VI2} \; \Delta P_{W2} \; \Delta P_{Tie,2} \end{bmatrix} \tag{4.22}$$

$$w^T = \begin{bmatrix} \Delta P_{wind} \; \Delta P_{solar} \; \Delta P_{L1} \; \Delta P_{L2} \end{bmatrix} \tag{4.23}$$

$$u^T = \begin{bmatrix} d\Delta f_1/dt \; d\Delta f_2/dt \end{bmatrix} \tag{4.24}$$

where x is the control output signal, w is the disturbance signal, u is the control input signal.

The state matrix A is partitioned as follows:

$$A = \begin{bmatrix} A_{11} & A_{12} \\ A_{21} & A_{22} \end{bmatrix}_{(14 \times 14)} \tag{4.25}$$

Based on Eqs. (4.5)–(4.18), the A_{11}, A_{12}, A_{21}, and A_{22} can be determined as:

$$A_{11} = \begin{bmatrix} -D_1/2H_1 & 1/2H_1 & 0 & 0 & 1/2H_1 & 1/2H_1 & -1/2H_1 \\ 0 & -1/T_{t1} & 1/T_{t1} & 0 & 0 & 0 & 0 \\ -1/T_{g1}R_1 & 0 & -1/T_{g1} & 1/T_{g1} & 0 & 0 & 0 \\ K_{S1}\beta_1 + 2\pi T_{12} & 0 & 0 & 0 & 0 & 0 & 0 \\ D_{VI1}/R_{VI1}T_{INV1} & 0 & 0 & 0 & -1/T_{INV1} & 0 & 0 \\ 0 & 0 & 0 & 0 & 0 & 1/T_{PV} & 0 \\ 2\pi T_{12} & 0 & 0 & 0 & 0 & 0 & 0 \end{bmatrix}_{(7 \times 7)} \tag{4.26}$$

$$A_{12} = \begin{bmatrix} 0 & 0\,0\,0\,0\,0\,0 \\ 0 & 0\,0\,0\,0\,0\,0 \\ 0 & 0\,0\,0\,0\,0\,0 \\ -2\pi T_{12} & 0\,0\,0\,0\,0\,0 \\ 0 & 0\,0\,0\,0\,0\,0 \\ 0 & 0\,0\,0\,0\,0\,0 \\ -2\pi T_{12} & 0\,0\,0\,0\,0\,0 \end{bmatrix}_{(7 \times 7)} \tag{4.27}$$

$$A_{21} = \begin{bmatrix} 0 & 0\,0\,0\,0\,0\,0 \\ 0 & 0\,0\,0\,0\,0\,0 \\ 0 & 0\,0\,0\,0\,0\,0 \\ 2\pi T_{12} & 0\,0\,0\,0\,0\,0 \\ 0 & 0\,0\,0\,0\,0\,0 \\ 0 & 0\,0\,0\,0\,0\,0 \\ \alpha_{12}2\pi T_{12} & 0\,0\,0\,0\,0\,0 \end{bmatrix}_{(7 \times 7)} \tag{4.28}$$

$$A_{22} = \begin{bmatrix} -D_2/2H_2 & 1/2H_2 & 0 & 0 & 1/2H_2 & 1/2H_2 & -1/2H_2 \\ 0 & -1/T_{t2} & 1/T_{t2} & 0 & 0 & 0 & 0 \\ -1/T_{g2}R_2 & 0 & -1/T_{g2} & 1/T_{g2} & 0 & 0 & 0 \\ K_{S2}\beta_2 - 2\pi T_{12} & 0 & 0 & 0 & 0 & 0 & 0 \\ D_{VI2}/R_{VI2}T_{INV2} & 0 & 0 & 0 & -1/T_{INV2} & 0 & 0 \\ 0 & 0 & 0 & 0 & 0 & 1/T_W & 0 \\ -\alpha_{12}2\pi T_{12} & 0 & 0 & 0 & 0 & 0 & 0 \end{bmatrix}_{(7 \times 7)} \tag{4.29}$$

The state matrix B_1 is partitioned as follows:

$$B_1 = \begin{bmatrix} B_{11} & B_{12} \end{bmatrix}_{(14 \times 4)} \tag{4.30}$$

Based on Eqs. (4.5)–(4.18), the B_{11} and B_{12} can be determined as:

$$B_{11}^T = \begin{bmatrix} 0\,0\,0\,0\,0 & 0 & 0\,0\,0\,0\,0\,0\,1/T_{W2} & 0 \\ 0\,0\,0\,0\,0\,1/T_{PV1} & 0\,0\,0\,0\,0\,0 & 0 & 0 \end{bmatrix}_{(2\times 14)} \tag{4.31}$$

$$B_{12}^T = \begin{bmatrix} -1/2H_1 & 0\,0\,0\,0\,0\,0 & 0 & 0\,0\,0\,0\,0\,0 \\ 0 & 0\,0\,0\,0\,0\,0 & -1/2H_2 & 0\,0\,0\,0\,0\,0 \end{bmatrix}_{(2\times 14)} \tag{4.32}$$

The state matrix B_2 is partitioned as follows:

$$B_2 = \begin{bmatrix} C_{11} & C_{12} \end{bmatrix}_{(14\times 2)} \tag{4.33}$$

Based on Eqs. (4.5)–(4.18), the C_{11} and C_{12} can be determined as:

$$C_{11}^T = \begin{bmatrix} 0\,0\,0\,0\,K_{VI1}/R_{VI1}T_{VI1}\,0\,0\,0\,0\,0\,0\,0\,0\,0 \end{bmatrix}_{(1\times 14)} \tag{4.34}$$

$$C_{12}^T = \begin{bmatrix} 0\,0\,0\,0\,0\,0\,0\,0\,0\,0\,0\,K_{VI2}/R_{VI2}T_{VI2}\,0\,0 \end{bmatrix}_{(1\times 14)} \tag{4.35}$$

4.4 Multiple Virtual Inertia Control Droops

As mentioned in Chap. 3, by applying the virtual inertia control topology, active and reactive power of the power system can be properly modified, resulting in regulation of system frequency and voltage. This concept requires the operations of virtual inertia droop characteristics, which can be built in the forms of the *P-f* and *Q-V* droops. One control droop requires at least one inverter. In the case of applying both control droops, it requires parallel inverters. Then, the virtual inertia control system uses the feedback from the frequency and voltage of each RESs/DGs for measuring the output active and reactive powers to imitate inertia and damping properties [2]. Thus, in the modern power system dominated by power electronics-based DGs/RESs, the frequency/voltage droop control can be enabled by integrating the virtual inertia control system.

In this section, we are interested in the virtual inertia control with respect to frequency regulation. Thus, each area of the interconnected system in Fig. 4.2 will be equipped with one *P-f* droop for inertia control. This means each area has a different inertia control droop. The main control parameter of R_{VI} in this virtual droop can be adjusted to emulate more or less inertia power to the system regarding the setting of K_{VI}, D_{VI}, and T_{INV} (as discussed in Chap. 3). Figure 4.3 shows the concept of multiple virtual inertia control units with different *P-f* droop characteristics. Both inertia control units are operating at a unique nominal frequency of the interconnected system with the different inertia powers. With the changes in DGs/RESs or loads in each area, the virtual inertia control units would help each other by increasing or decreasing their power until reaching a new common operating frequency of the interconnected system. As expressed in Fig. 4.3, the amount of emulated power

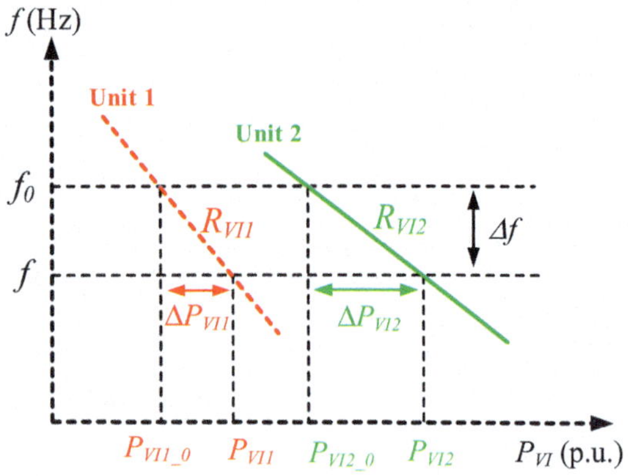

Fig. 4.3 DGs/RESs or load tracking by multiple virtual inertia control units with different *P*-*f* droops

from each inertia control unit to compensate the disturbance depends on the virtual inertia droop characteristic.

4.4.1 Sensitivity Analysis for Multiple Inertia Control Units

In this section, the dynamic influences of multiple-virtual inertia control are investigated for the interconnected system. In this section, the primary goal is to analyze the dominant effect of multiple virtual inertia control gains (i.e., K_{VI1}, K_{VI2}) in providing the desirable inertia service over the system stability and performance. It is focused on the main function of virtual inertia control; that is, the imitation of inertia power, and the virtual damping in both areas is set as the constant of 0.1 p.u.MW/Hz. Thus, the eigenvalue trajectory of the interconnected system is evaluated for a wide range of parameter variations of K_{VI1} and K_{VI2}. In this section, we investigated only the effect of multiple-virtual inertia constant, which is the main structure in producing the emulated inertia power into the system. It is noted that other control parameters (i.e., virtual droop and virtual damping) could also affect the emulation of virtual inertia control in the interconnected system. In a future work, the investigation of those control parameters can be evaluated for advanced analysis of interconnected systems.

The impact of numerous variations on the virtual inertia constant in the area 1 (K_{VI1}) is investigated, as shown in Fig. 4.4. By rising the K_{VI1}, the system reaches better stability as the negative real part of most eigenvalues situated far away in the left side of the s-plane. Clearly, the interconnected system with virtual inertia control can significantly enhance its dynamic stability and performance by approaching the eigenvalues to a greater area in the left-half side of the s-plane. On the contrary, when the K_{VI1} is rised too high, the 6th mode (λ_6) and 7th mode (λ_7) will have adverse effects on the system performance by approaching back to the right side of the s-plane, leading to the unstable and more oscillatory area. Hence, based on

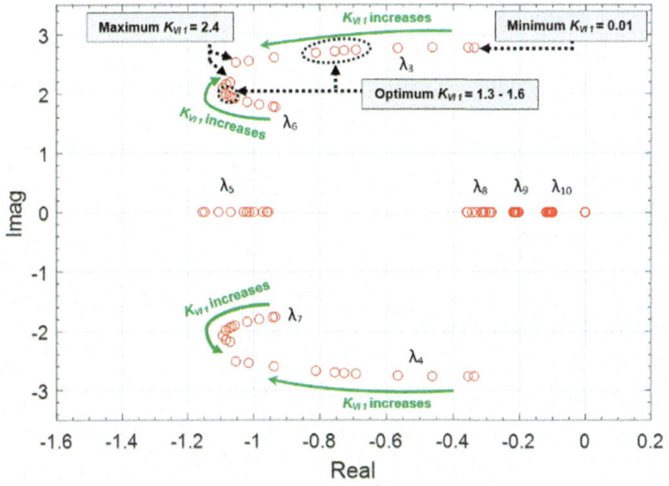

Fig. 4.4 Eigenvalue trajectory of dominant poles over changes of virtual inertia constant in Area 1 (with $K_{VI2} = 0.03$)

the trajectory in Fig. 4.4, the optimum values of K_{VI1}, which could not degrade the system dynamic and stability, are evaluated in a range of 1.3–1.6.

Similarly, the impact of various variations on the virtual inertia constant in the area 2 (K_{VI2}) is analyzed with observation in Fig. 4.5. Simulation results reveal that by rising the K_{VI2}, the system could achieve better stability with less deviation. This is because most of the eigenvalues approach far away to the left side of the s-plane.

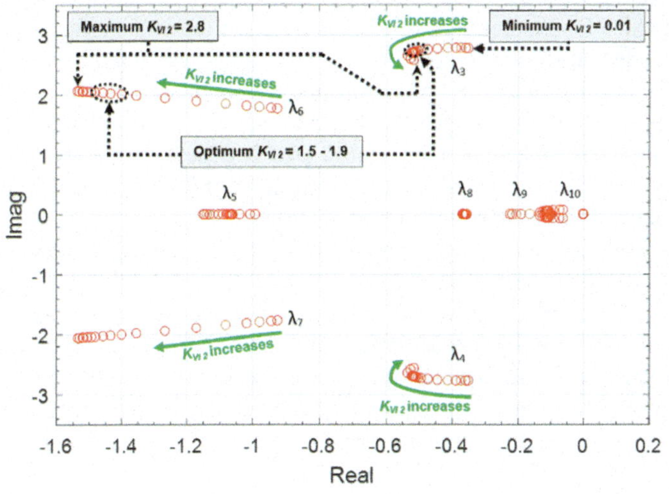

Fig. 4.5 Eigenvalue trajectory of dominant poles over changes of virtual inertia control gain in Area-2 (With $K_{VI1} = 1.54$)

However, by integrating a higher K_{VI2}, it can be seen that several modes (i.e., λ_3, λ_4) reach to the right of the s-plane, which can degrade the dynamic performance of the system. By investigating the trajectory in Fig. 4.5, the optimal values of K_{VI2}, which could not decay the system stability, are determined in a range of 1.5–1.9.

Hence, before applying the virtual inertia constant, the eigenvalue analysis should be performed to evaluate a suitable trade-off between inertia control response and dynamic stability of the system.

4.5 Simulation Results

This section describes the simulation results of the multiple virtual inertia control units in the interconnected system. The study is performed in the MATLAB/Simulink environment. The results are divided into two subsections: the first section shows the main effectiveness of multiple virtual inertia units for the interconnected system; the second section validates the robustness of multiple virtual inertia units under a high share of DGs/RESs and load disturbances.

4.5.1 Efficacy of Multiple-Virtual Inertia Control

The efficiency of the multiple virtual inertia units over the dynamic and performance of the interconnected system is verified in this section. The studied system is constructed based on Fig. 4.2 and Table 4.1. It is noted that the conventional generating unit and ESS-based inertia capacities in the area 1 are smaller than the area 2. Then, a sudden load increase of 0.1 p.u. is added to the system via the area 2. The study is divided into four technical cases. Case 1 represents the interconnected system without the installation of virtual inertia control in both areas. In Case 2, the virtual inertia control unit is only installed in the control area 1. In Case 3, the virtual inertia control unit is only installed in the control area 2. Case 4 represents the concept of multiple-virtual inertia control, which has installed the virtual inertia control units in both control areas of the interconnected system.

Figure 4.6 shows the frequency response of the interconnected system after the disturbance. Without virtual inertia control units (i.e., Case 1), it can be seen that the system frequency response has the largest frequency nadir of −0.25 Hz in both areas. Thermal generation-based synchronous units in both areas corporately respond the disturbance, while the synchronous unit in the area 1 generates more power and delivers through its tie-line to help the area 2 in regulating its disturbance, see Fig. 4.7. Consequently, the tie-line power flow of 0.07 p.u. can be observed in this case, see Fig. 4.8.

By applying virtual inertia control in the area 1 (i.e., Case 2), the frequency nadir of the area 1 reduces to −0.17 Hz, while the frequency nadir of the area 2 remains the same as the case 1 due to no virtual inertia support in its area.

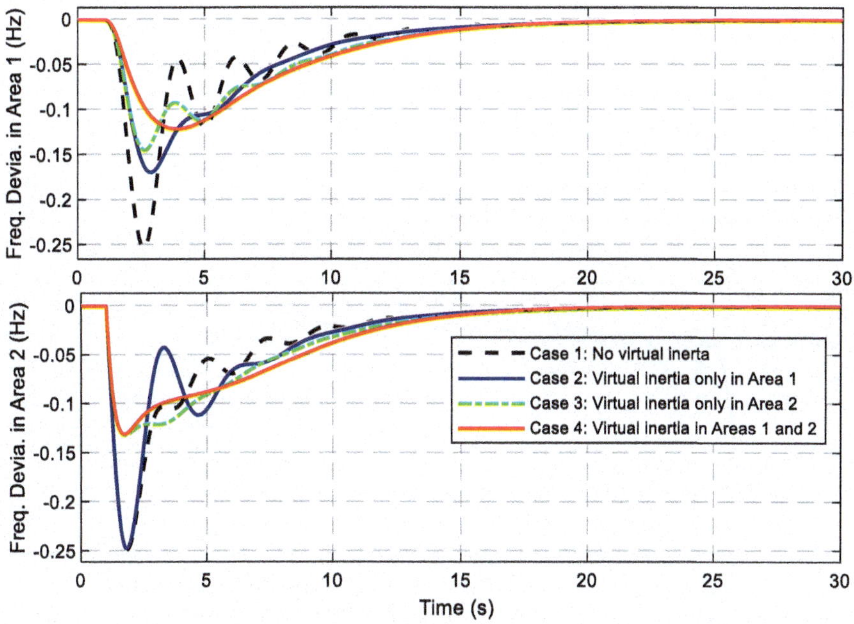

Fig. 4.6 Frequency response of the interconnected system

By applying virtual inertia control in the area 2 (i.e., Case 3, the area of the disturbance), it could immediately respond the disturbance, and thus, the frequency nadir of the area 2 significantly reduces. Obviously, if the disturbance is immediately arrested, it could prevent the degraded effect in another area, leading to a lower frequency nadir in the area 1. Nevertheless, the tie-line power flow is reduced. Thus, it is recommended that the virtual inertia control unit should be placed on the area, where a disturbance frequently occurs.

By applying multiple virtual inertia control units in both areas (i.e., Case 4), it is obvious that the lowest frequency nadir with less oscillation can be observed in both areas. Clearly, the tie-line power is significantly reduced with lower oscillation. This means that the requirement of power transfer among the areas is less. Moreover, the thermal generating units in both areas are less stressed due to less power generation.

4.5.2 *Stability Analysis Under Continuous Disturbances*

In this part, the robustness of the multiple-virtual inertia control strategy is tested by various nature of DGs/RESs and load changes with continuous changes. To perform a more drastic situation, the system inertia (H) and load damping (D) property in both areas are decreased to 50% of its nominal values. The disturbance power output from DGs/RESs and loads in both areas is depicted in Fig. 4.9.

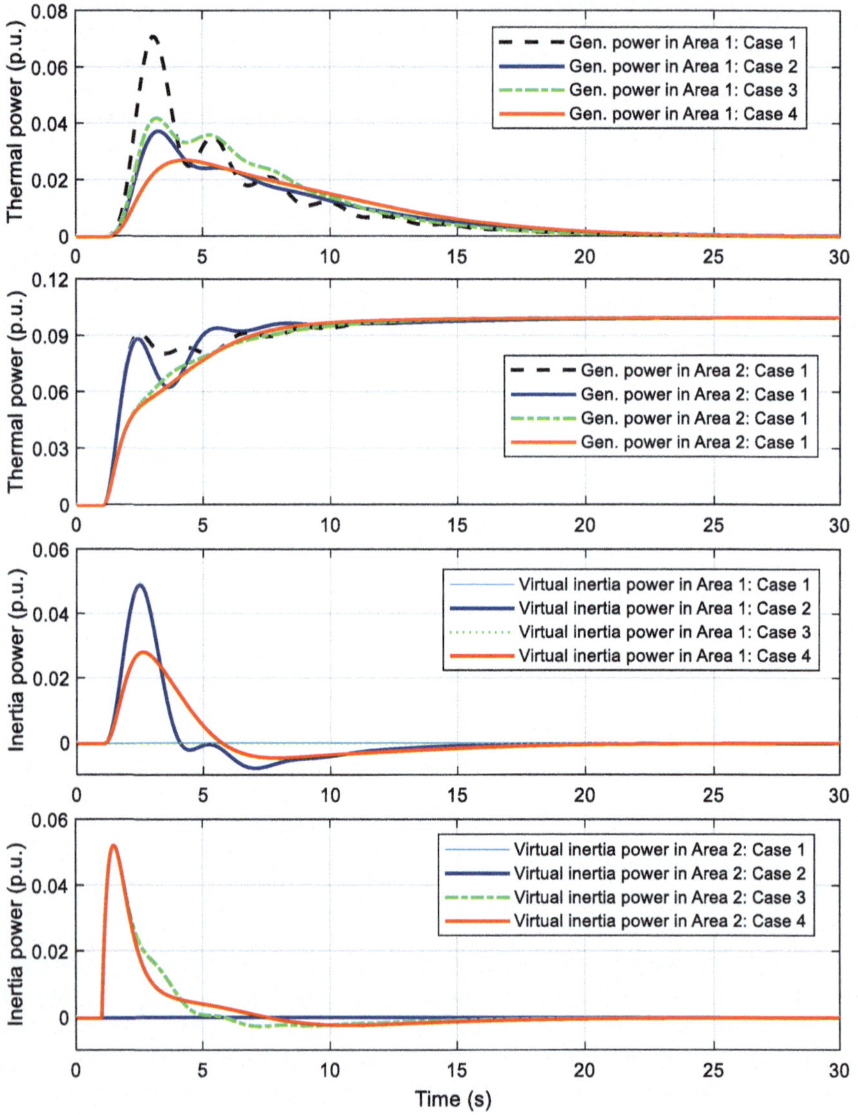

Fig. 4.7 Active power response of the interconnected system

Figure 4.10 shows the system frequency response without virtual inertia control. It can be seen that the system fluctuates severely with larger oscillations triggered by the severe situation of low system inertia and damping. Following the significant disturbance at 35 s caused by the sudden drop of solar irradiation, it is obvious that the system without virtual inertia control could not maintain its stability and turns into unstable. If the interconnected system did not perform any protection scheme

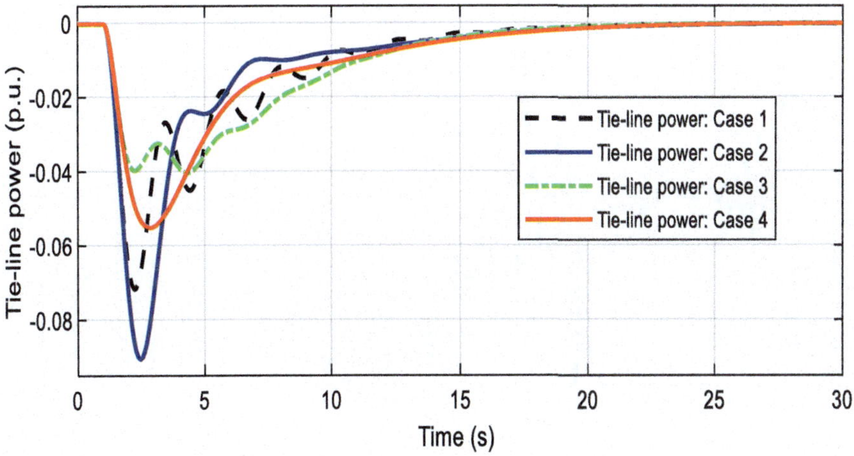

Fig. 4.8 Tie-line power response of the interconnected system

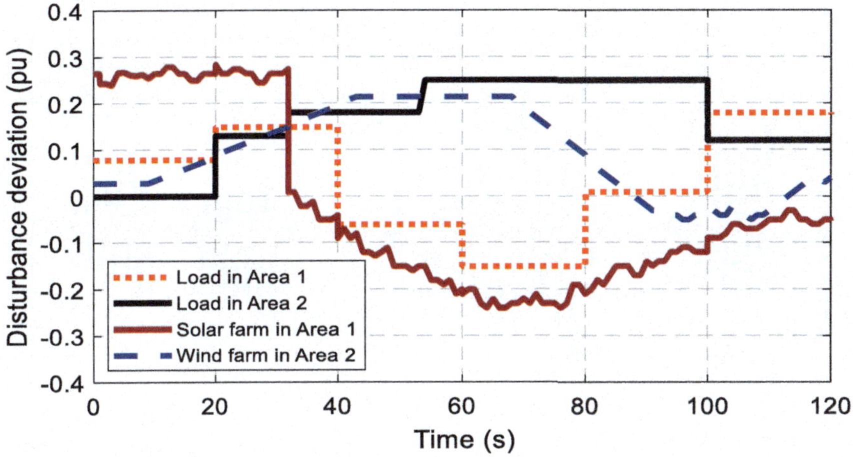

Fig. 4.9 Multiple disturbance power outputs

(e.g., load shedding) within the specific timescale, it would lead to the cascading outages and system collapse and wide-area power blackout.

By integrating virtual inertia control in the area 2, the interconnected system could maintain better stability with lower frequency peak/drop and oscillations (see Fig. 4.11) compared with the system without virtual inertia control. In addition, the virtual inertia control unit could help the conventional generating units of both areas in stabilizing the disturbances under the presence of system inertia and damping reduction. From Fig. 4.12, it is obvious that the tie-line power flow is still oscillating due to the lack of virtual inertia support in the area 1.

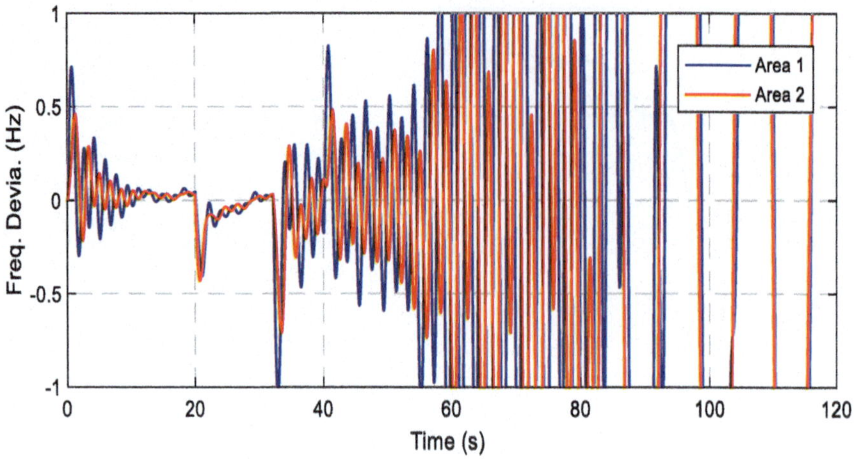

Fig. 4.10 Frequency response of the interconnected system without virtual inertia control against the disturbances and inertia/damping reductions

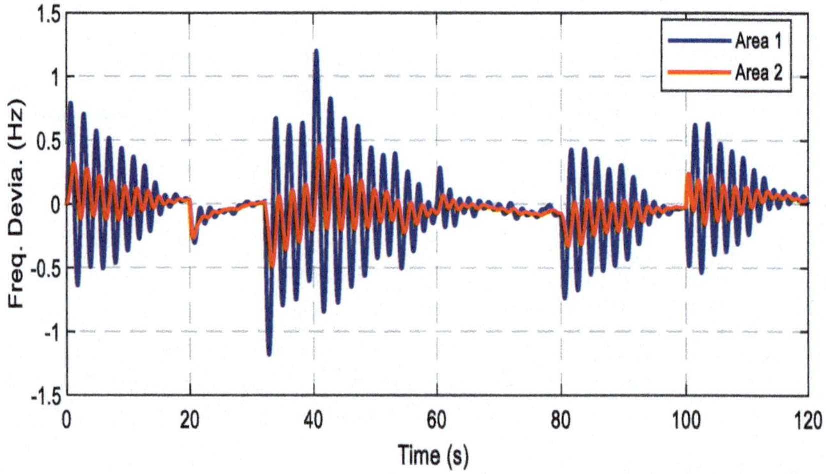

Fig. 4.11 Frequency response of the interconnected system with virtual inertia control in one area against the disturbances and inertia/damping reductions

By applying multiple virtual inertia control units in the system, the frequency stability and robustness of the interconnected system significantly improve, as shown in Fig. 4.13. The system could properly maintain its stability within the acceptable frequency operating standard of ±0.1 Hz suggested by the national electric market (NEM) [22], which is in line with the mainland frequency operating standards for interconnected systems. Also, the tie-line power flow is significantly reduced with less oscillation, as depicted in Fig. 4.14. Therefore, it is ensured that the multiple-

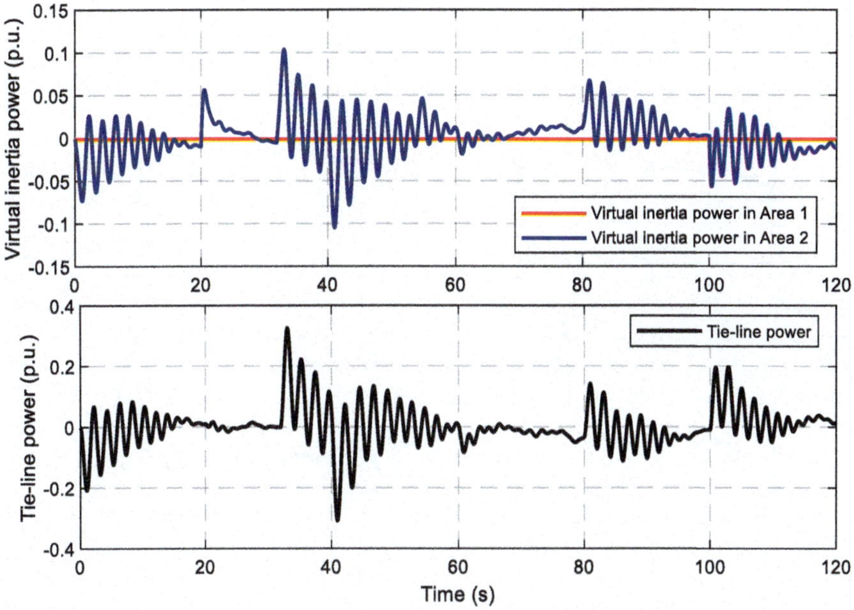

Fig. 4.12 Virtual inertia and tie-line power responses of Fig. 4.11 against the disturbances and inertia/damping reductions

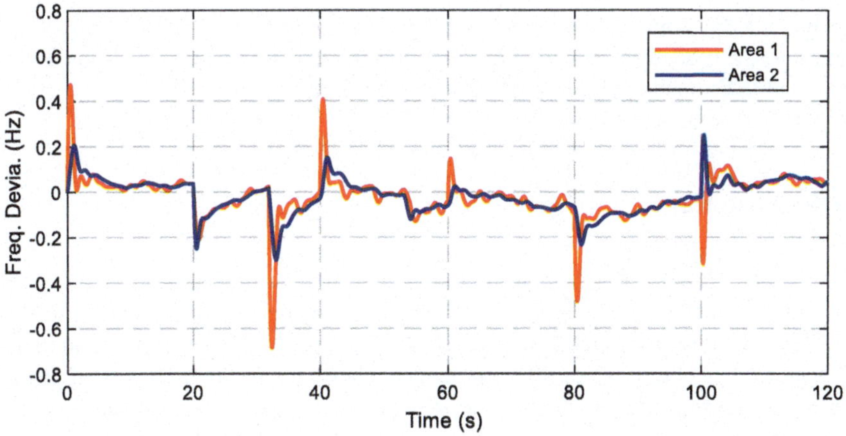

Fig. 4.13 Frequency response of the interconnected system with virtual inertia control in both areas against the disturbances and inertia/damping reductions

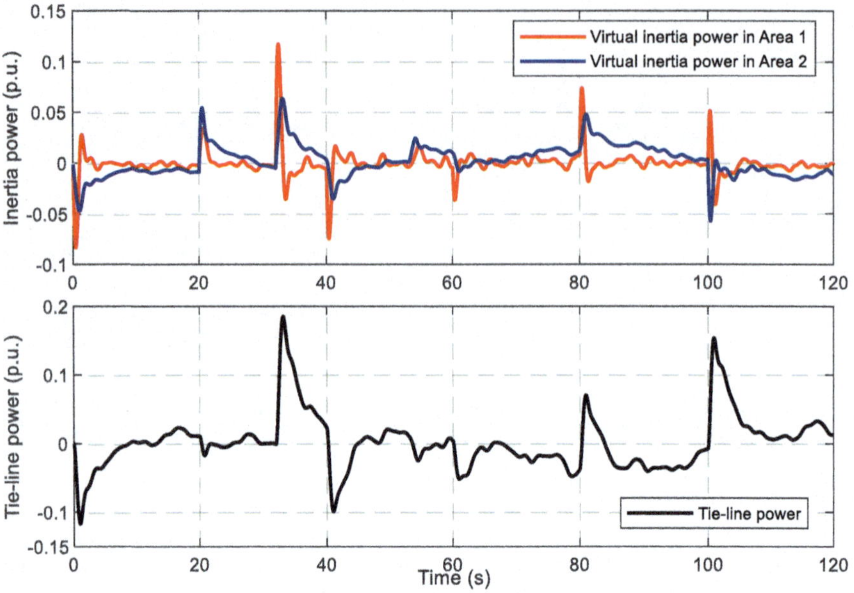

Fig. 4.14 Virtual inertia and tie-line power responses of Fig. 4.13 against the disturbances and inertia/damping reductions

virtual inertia control strategy is able to reduce frequency nadir/overshoot and improve the robustness of interconnected system response under numerous changes in DGs/RESs, loads, system inertia and system damping.

4.6 Summary

This chapter investigates the coordination of multiple virtual inertia control systems to improve the frequency stability and response of the inertia power and tie-line power in the interconnected systems with DGs/RESs. The dynamic performance and frequency characteristics of an interconnected system with multiple virtual inertia, primary, and secondary control loops are explained. A dynamic response model of the interconnected system is linearized and analyzed in detail. The effects of control parameters affecting the system frequency response with respect to the implementation of multiple virtual inertia systems in the interconnected system are investigated and discussed via the state-space model representation. A comprehensive sensitivity analysis of main inertia control parameters is conducted to show the actual impacts of inertia emulation over stability of the interconnected system. Finally, the efficiency and robustness of the multiple-virtual inertia control strategy are tested and verified through the interconnected system model against various changes in DG/RESs, loads, system inertia and damping.

Problems for Chapter 4

4.1 Please construct the interconnected power system based on Fig. 4.2 and Table 4.1 using MATLAB/Simulink software. The system inertia and damping are reduced to 60% of its nominal values in both areas. The nominal system frequency is 50 Hz. The system base is 1000 MVA. At 5 s, the 30 MW of wind farm is connected to Area 1 and 20 MW of solar system is connected to Area 2. Then, if the step load (ΔP_L) increment of 100 MW occurs in Area 2 at the time $t = 25$ s, please determine the following values in both areas.

(1) System frequency nadir in Hz
(2) Steady-state system frequency in Hz
(3) Generated power of the generator in MW
(4) Generated virtual inertia power in MW
(5) Maximum tie-line power flow in MW

4.2 Based on Problem 4.1. If the system is operated without the secondary control unit in both areas, please determine the following values in both areas.

(1) System frequency nadir in Hz
(2) Steady-state system frequency in Hz
(3) Generated power of the generator in MW
(4) Generated virtual inertia power in MW
(5) Maximum tie-line power flow in MW

References

1. Z. Li, W. Wu, M. Shahidehpour, B. Zhang, Adaptive robust tie-line scheduling considering wind power uncertainty for interconnected power systems. IEEE Trans. Power Syst. **31**(4), 2701–2713 (2016)
2. H. Bevrani, *Robust Power System Frequency Control*, 2nd Edi. (Springer, New York, USA, 2014)
3. T.H. Mohamed, H. Bevrani, A.A. Hassan, T. Hiyama, Decentralized model predictive based load frequency control in an interconnected power system. Energy Convers. Manag. **52**(2), 1208–1214 (2011)
4. M.Z. Bernard, T.H. Mohamed, Y.S. Qudaih, Y. Mitani, Decentralized load frequency control in an interconnected power system using Coefficient Diagram Method. Int. J. Electr. Power Energy Syst. **63**(1), 165–172 (2014)
5. H. Bevrani, M. Watanabe, Y. Mitani, *Power System Monitoring and Control* (IEEE-Wiley, New Jersey, USA, 2014)
6. S. Küfeoğlu, M. Lehtonen, Macroeconomic assessment of voltage sags. Sustainability **8**(12), 1–12 (2016)
7. Y. Chen, R. Hesse, D. Turschner, H.P. Beck, Investigation of the virtual synchronous machine in the Island mode, in Proc. *IEEE PES Innovative Smart Grid Technologies Conference Europe*, 1–6 (2012)
8. T. Kerdphol, F.S. Rahman, M. Watanabe, Y. Mitani, D. Turschner, H.P. Beck, Extended virtual inertia control design for power system frequency regulation, in Proc. *2019 IEEE PES Grand International Conference and Exposition Asia*, 97–101 (2019)

9. H.P. Beck, R. Hesse, Virtual synchronous machine, in Proc. *International Conference on Electrical Power Quality and Utilisation*, 1–5 (2007)
10. Y. Hirase, K. Sugimoto, K. Sakimoto, T. Ise, Analysis of resonance in microgrids and effects of system frequency stabilization using a virtual synchronous generator. IEEE J. Emerg. Sel. Top. Power Electron. **4**(4), 1287–1298 (2016)
11. J. Alipoor, Y. Miura, T. Ise, Power system stabilization using virtual synchronous generator with alternating moment of inertia. IEEE J. Emerg. Sel. Top. Power Electron. **3**(2), 451–458 (2015)
12. J. Alipoor, Y. Miura, T. Ise, Stability assessment and optimization methods for microgrid with multiple VSG units. IEEE Trans. Smart Grid **9**(2), 1462–1471 (2018)
13. E. Rakhshani, P. Rodriguez, Inertia emulation in AC/DC interconnected power systems using derivative technique considering frequency measurement effects. IEEE Trans. Power Syst. **32**(5), 3338–3351 (2017)
14. P. Rodriguez, E. Rakhshani, A. Mir Cantarellas, D. Remon, Analysis of derivative control based virtual inertia in multi-area high-voltage direct current interconnected power systems. IET Gener. Transm. Distrib. **10**(6), 1458–1469 (2016)
15. E. Rakhshani, D. Remon, A.M. Cantarellas, J.M. Garcia, P. Rodriguez, Virtual synchronous power strategy for multiple HVDC interconnections of multi-area AGC power systems. IEEE Trans. Power Syst. **32**(3), 1665–1677 (2017)
16. T. Kerdphol, F.S. Rahman, Y. Mitani, Virtual inertia control application to enhance frequency stability of interconnected power systems with high renewable energy penetration. Energies **11**(4), 981–997 (2018)
17. H. Bevrani, P.R. Daneshmand, Fuzzy logic-based load-frequency control concerning high penetration of wind turbines. IEEE Syst. J. **6**(1), 173–180 (2012)
18. M. Saejia, I. Ngamroo, Alleviation of power fluctuation in interconnected power systems with wind farm by SMES with optimal coil size. IEEE Trans. Appl. Supercond. **22**(3), 5701504 (2012)
19. D.J. Lee, L. Wang, Small-signal stability analysis of an autonomous hybrid renewable energy power generation/energy storage system part I: time-domain simulations. IEEE Trans. Energy Convers. **23**(1), 311–320 (2008)
20. H. Bevrani, F. Habibi, P. Babahajyani, M. Watanabe, Y. Mitani, Intelligent frequency control in an AC microgrid: online PSO-based fuzzy tuning approach. IEEE Trans. Smart Grid **3**(4), 1935–1944 (2012)
21. P. Kundur, *Power System Stability and Control* (McGraw-Hill, New York, USA, 1994)
22. Australian Energy Market Operator (AEMO), International review of frequency control adaptation (Sydney, Australia, 2017)

Chapter 5
Application of PI/PID Control for Virtual Inertia Synthesis

Abstract In the previous chapters, to deal with the disturbances including high integration of distributed generators (DGs)/renewable energy sources (RESs), the virtual inertia constant, which is the crucial factor in emulating additional inertia power into the system, is fixed at one value. In the application of virtual inertia control, improper selection of its control value may result in a higher frequency deviation, slower recovery time, and instability. To overcome this problem, in this chapter, the basic proportional-integral (PI) or proportional-integral-derivative (PID) controllers, which are widely used in the real-practice in the industrial systems, are applied to the virtual inertia control to generate proper virtual inertia constant for imitating the effective inertia power and improving system frequency stability. This chapter provides the synthesis of a new decentralized PI/PID-based virtual inertia control to evaluate the virtual inertia power under different levels of RESs/DGs penetration and load disturbances. The uses of the PI/PID controllers for frequency stability enhancement are briefly discussed. Then, the optimal setting of PI/PID parameters using the classical and modern tuning techniques are described in detail to obtain the sufficient virtual inertia constant with respect to the additional power, assuring stable grid operation. Finally, the proposed method is tested in a control area power system with different levels of RESs/DGs, loads, and system inertia and damping reduction.

Keywords PI/PID control · Dynamic control · Frequency control · Inertia control · Internal-Model-Control (IMC) · Parameter tuning · Virtual inertia constant · Virtual inertia synthesis

5.1 Introduction to PI/PID Control

Over a century, the proportional-integral (PI) and proportional-integral-derivative controls (PID) are extensively integrated into industrial automatic control applications due to their functional simplicity. Additionally, more than 90% of the control loops in process control are the PID type with the most loop using PI control. Due to its simplicity and intuitiveness regarding satisfactory performance, the controllers can offer a wide range of processes [1]. Subsequently, they become the standard

T. Kerdphol et al., *Virtual Inertia Synthesis and Control*, Power Systems,
https://doi.org/10.1007/978-3-030-57961-6_5

controllers in industrial settings. The controllers are also involved with the technology progress. Up to now, they are often applied in digital platforms rather than with electrical or pneumatic components. The controllers are usually equipped on all kinds of control devices as a functional system in programmable logic controllers, or distributed control systems, or stand-alone controllers [2]. The main control actions of the controllers can be simply divided as follows: the action of proportional to the error (based on its current value), the action of proportional to the integral of the error (based on its past value), and the action of proportional to the derivative of the error (based on its future value).

The new potentialities of such controllers have been supported by the advancement of the software packages, including digital technology, leading to the rapid growth of the research related to PI/PID controllers. The wide utilization of controllers is also ensured by the fact that they provide the essential components for more complicated control schemes [3]. This advantage is useful when the fundamental control law is not adequate to procure the desired performance, or the sophisticated control issue is of concern [4]. The PI/PID controllers serve in numerous changes in technology. They started with pneumatic control via direct digital control. In the present, logic, selector, block function, and sequence are combined with the PI/PID controllers. Several complicated regulations and control strategies, including start-up and shutdown strategies, could be designed via the controllers. This fundamental serves as the essential solutions for smooth transient, appropriate regulation, stable and secure operation. In advanced control techniques, such as model predictive control (MPC), the PI/PID controller is included as a basic block at the regulatory level. Moreover, the computing power of microprocessors offers extra features (e.g., model switching and automatic tuning gain scheduling) for the PI/PID controllers [5, 6]. Nevertheless, the PI/PID controllers are also gaining attention in modern applications, such as autonomous robots, unmanned aerial vehicles, and driverless cars [7].

In the industry practice regarding frequency control, the utilization of PI controllers is more popular than those of PID controllers due to less complexity. Currently, some control techniques have been implemented to the design of PI/PID controllers for solving the frequency control problems as follows. In [8–12], the authors proposed the PI control design approach with a combination of robust and frequency control theorems. In [13], the authors presented the sequential decentralized PI approach with a combination of a robust theorem. In [14–19], the decentralized PID methods have been applied to enrich frequency stability of power systems. In [20, 21], the authors proposed the fractional-order PID concept for load frequency control. In [22], the PI/PID controllers have been utilized for digital decentralized frequency problems.

In the virtual inertia synthesis with respect to frequency control, the application of PI/PID controllers may be extended to the main function of the system, that is, the emulation of inertia power [23]. Thus, the PI/PID controller designs can be implemented in the tuning process of virtual inertia control. In the previous chapters, to deal with the disturbances and high integration of RESs/DGs, the virtual inertia constant, which is the key factor in emulating additional power with the required inertia and damping into the system, is fixed at one value. The improper selection of virtual inertia control parameter may result in a higher frequency deviation, slower recovery time, and instability. To solve this issue, this chapter proposes the design of PI/PID controllers to help the virtual inertia unit in selecting the suitable virtual inertia constant for emulating the sufficient inertia power against various changes in loads, generations, system inertia, and damping.

5.2 Fundamental Feedback Control

A feedback control concept is a powerful and common tool used in the design of a control system. In this section, the control system objective is to acquire a desirable response and performance for a studied/controlled system. This idea can be completed with an open-loop system, where the controller defines the input signal to the process on the fundamental of the reference signal. For a closed-loop system, the controller defines the input signal to the process using the measurement of the output, known as the feedback signal [1, 7].

Thus, the feedback control is important to maintain the process variable close to the desired value under the presence of changes (disturbances) of process dynamics [24]. The development of feedback control techniques has a significant effect on several engineering fields. Figure 5.1 depicts the concept of the feedback control system. The overall performance of the control system relies on the suitable selection of each component [25]. Focusing on the objectives of the controller design, the sensor, and actuator dynamic in Fig. 5.1 are ignored. Thus, the dynamic structure of the control system is represented as Fig. 5.2. $P(s)$ is the process, $C(s)$ is the controller, y_s is the reference signal (desired process output), $F(s)$ is the feedforward filter, y is the process output, u is the controlled output, n is the measurement noise signal, and d is the disturbance signal [1].

Ignoring the measurement noise signal, the control error (e) can be easily computed in the time domain as [1, 7]:

$$e(t) = y_s(t) - y(t) \tag{5.1}$$

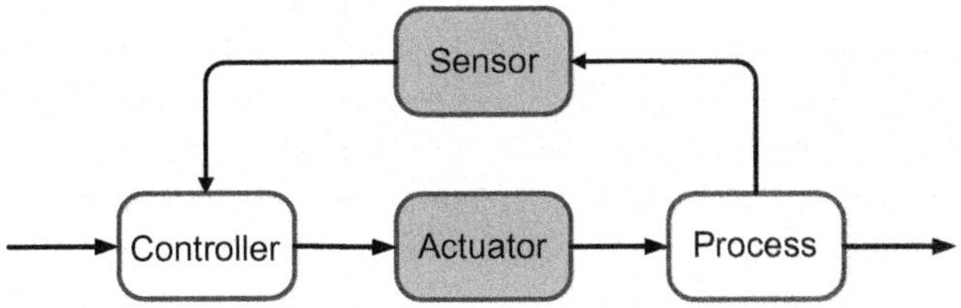

Fig. 5.1 Schematic diagram of a feedback control loop

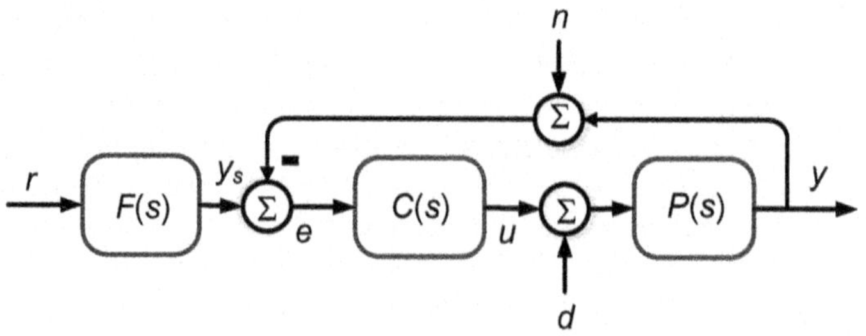

Fig. 5.2 Dynamic structure of a feedback control loop

5.3 Actions of PI/PID Control

The PI or PID contol is a form of feedback control. They employ the feedback concept in the control mechanism. In general, the PI controller consists of two control actions (i.e., proportional and integral actions), while the PID controller consists of three control actions (i.e., proportional, integral, and derivative actions). These actions are briefly explained as follows.

5.3.1 Proportional Action

The proportional control relates to the current value of its control error. The action of proportional control (u_p) is proportional to the current control error due to the expression as [1, 7]:

$$u_p(t) = K_p e(t) = K_p(y_s(t) - y(t)) \tag{5.2}$$

where K_p is the proportional control gain.

The transfer function of the proportional (P) controller can be defined in a form of Laplace domain as:

$$C(s) = K_p \tag{5.3}$$

The main advantage of the proportional controller is the ability in offering a less control variable when the control error is small, preventing excessive control attempts. The drawback of integrating a single proportional controller is that it generates a steady-state error. To solve such a problem, the addition of a bias named u_b (i.e., a reset unit) is added, which modifies the action of P control as [1, 7]:

$$u_p(t) = K_p e(t) + u_b \tag{5.4}$$

It is noted that the value of u_b can be determined as constant or adjusted manually until the steady-state error is diminished to zero. Usually, the u_b can be defined as [1]:

$$u_b = \frac{u_{b\_max} + u_{b\_min}}{2} \tag{5.5}$$

where the u_{b_max} and u_{b_min} are the maximum and minimum values of u_b, respectively.

In commercial practice, the proportional gain is likely substituted by the proportional band (PB), which represents the error range. This range results in a full range variation of the control variable. The PB equation can be defined as [1]:

$$\mathrm{PB} = 100/K_p \tag{5.6}$$

5.3.2 Integral Action

The integral control relates to the past value of its control error. The action of integral control is proportional to the integral of the control error as [1, 5, 7]:

$$u_i(t) = K_i \int_0^t e(t)\, dt = K_i \int_0^t (y_s(t) - y(t))\, dt \tag{5.7}$$

where K_i is the integral control gain.

It should be noted that the integral action deals with the past values of the control error. The transfer function of the integral controller is defined in a form of Laplace domain as:

$$C(s) = \frac{K_i}{s} \tag{5.8}$$

where

$$K_i = \frac{K_P}{T_i} \tag{5.9}$$

By applying the pole at the origin of the complex plane, it enables the drop to zero for the steady-state error when the step reference signal is added, or the step disturbance appears [5]. Additionally, the integral action automatically sets the correct value of u_b in Eq. (5.4); thus, the steady-state error is zero. The resulting transfer function is called as the PI control and expressed as [1]:

$$C(s) = K_p + \frac{K_i}{s} = K_p\left(\frac{1}{sT_i} + 1\right) \tag{5.10}$$

where T_i is the controller integral time. By adjusting the T_i, it affects both the proportional and integral control actions. By adjusting the K_p, it affects both control actions.

The utilization of a proportional action in combination with an integral action (i.e., a PI controller) resolves several issues of the oscillatory response and the steady-state error compared to a pure proportional controller.

5.3.3 Derivative Action

The derivative control deals with the predicted future value of its control error. The action of derivative control is defined as [1, 5, 7]:

$$u_d(t) = K_d\frac{de(t)}{dt} = K_d\frac{d(y_s(t) - y(t))}{dt} \tag{5.11}$$

where K_d is the derivative control gain and defined as:

$$K_d = K_pT_d \tag{5.12}$$

The transfer function of the derivative controller can be defined in a form of Laplace domain as:

$$C(s) = sK_d = sK_pT_d \tag{5.13}$$

where T_d is the controller derivative time.

To understand better detail for the derivative action, the Taylor series development of the control error at the time T_d should be studied as [1, 7]:

$$e(t + T_d) \cong e(t) + T_d\frac{de(t)}{dt} \tag{5.14}$$

By considering the control law regarding this expression, it results in the combination of a proportional-derivative (PD) control. The transfer function can be rewritten as [1]:

$$C(s) = K_p\left(T_d\frac{de(t)}{dt} + e(t)\right) \tag{5.15}$$

Considering the above equation, the control variable at time t is based on the predicted value of the control error at time $t + T_d$; thus, the derivative action is known as the anticipatory control. By adjusting the T_d, it affects both proportional and derivative control actions. Similarly, by adjusting the K_p, it affects both control actions. Consequently, the derivative action has the important capability in enhancing the control performance as it directly predicts a false trend of the control error and reacts for it. For more details about the control actions in this section, interested readers can be found in [1, 5, 7].

5.4 Structures of PI/PID Control

The fundamental structure of a PI controller contains two control actions (i.e., proportional and integral actions), while the structure of a PID controller contains three control actions (i.e., proportional, integral, and derivative actions). In this section, the modeling structures of PI/PID controllers are briefly described as follows.

5.4.1 *Modeling of PI Controller*

For the PI controller, the proportional and integral control actions are additive, and the PI structure is universally applied to all controllers. By considering Eq. (5.10), the combined structure of the proportional and integral control actions is modeled, as depicted in Fig. 5.3.

5.4.2 *Modeling of PID Controller*

Unlike the PI controller, the PID controller combines the proportional, integral, and derivative actions, which is represented in several forms. From Fig. 5.4, the ideal

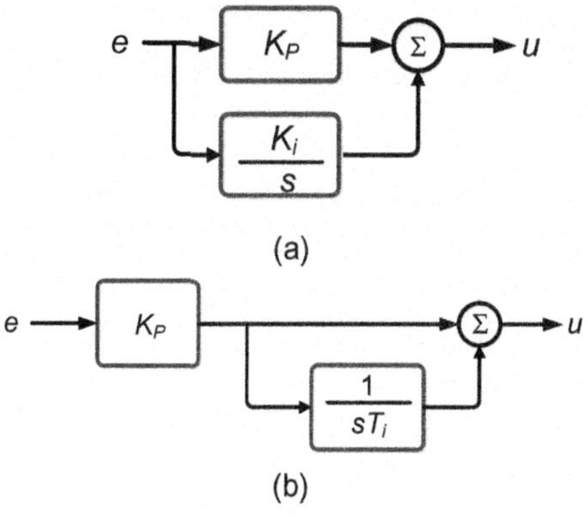

Fig. 5.3 Dynamic structures of a PI controller; **a** in the form of K_p and K_i, **b** in the form of K_p and T_i

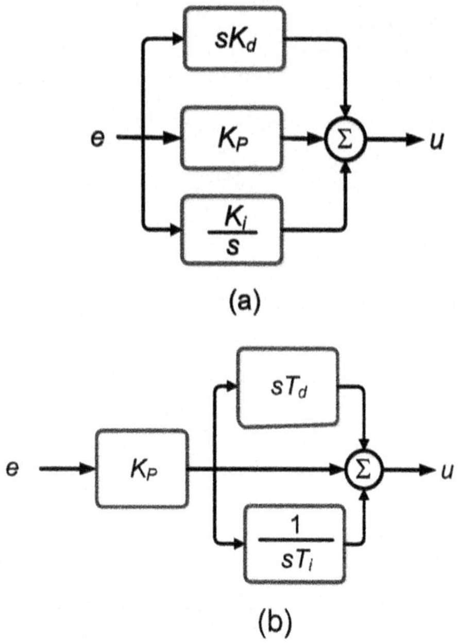

Fig. 5.4 Dynamic structures of a PID controller; **a** in the form of K_p, K_i, and K_d, **b** in the form of K_p, T_i and T_d

form of the PID control can be expressed as [1, 7]:

$$C(s) = K_p + \frac{K_i}{s} + sK_d \tag{5.16}$$

Alternatively, it can be represented in the form of K_p, T_i, and T_d as [1, 7]:

$$C(s) = K_p\left(1 + \frac{1}{sT_i} + sT_d\right) \tag{5.17}$$

Representing in the time domain (t), Eq. (5.17) can be rewritten in terms of the controlled output $u(t)$ of the PID controller as:

$$u(t) = u_p(t) + u_i(t) + u_d(t) \tag{5.18}$$

$$u(t) = \left(K_p(y_s(t) - y(t))\right) + \left(\frac{K_p}{T_i}\int_0^t (y_s(t) - y(t)dt)\right) + \left(K_pT_d\frac{d(y_s(t) - y(t))}{dt}\right) \tag{5.19}$$

Example 5.1 The controlled system has the setpoint/reference signal $y_s(t) = 1$ for $t \geq 0$ and the process output $y(t) = 0$ for $t \geq 0$. The PID control gains are given as $K_p = 1$, $T_i = 4.5$, and $T_d = 0.3$. Please determine the control output of the PID controller for $t \geq 0$.

Solution 5.1

Considering Eq. (5.2), the control output of the proportional unit is computed as:

$$u_p(t) = K_p(y_s(t) - y(t))$$
$$u(t) = 1(1 - 0) = 1 \text{ for } t \geq 0$$

Considering Eq. (5.7), the control output of the integral unit is computed as:

$$u_i(t) = \frac{1}{4.5}\int_0^t (1 - 0dt)$$
$$u_i(t) = \frac{t}{4.5} \text{ for } t \geq 0$$

Considering Eq. (5.11), the control output of the derivative unit is computed as:

$$u_d(t) = K_pT_d\frac{d(y_s(t) - y(t))}{dt}$$
$$u_d(t) = 1(0.3)\frac{d(1 - 0)}{dt} = 0 \text{ for } t \geq 0$$

By summing all control units as Eq. (5.19), the control output of the PID controller for $t \geq 0$ can be determined as:

$$u(t) = 1 + \frac{t}{4.5}$$

Example 5.2 The controlled system has the setpoint $y_s(t) = 1$ for $t \geq 0$, $y_s(t) = 0$ for $t \leq 0$ and the process output $y(t) = 0$. The PID control gains are given as $K_p = 1$, $T_i = 4.5$, and $T_d = 0.3$. Please determine the control output of the PID controller for $t \geq 0$.

Solution 5.2

Considering Eq. (5.2), the control output of the proportional unit is computed as:

$$u_p(t) = K_p(y_s(t) - y(t))$$
$$u(t) = 1(1 - 0) = 1 \text{ for } t \geq 0$$

Considering Eq. (5.7), the control output of the integral unit is computed as:

$$u_i(t) = \frac{1}{4.5}\int_0^t (1 - 0dt)$$
$$u_i(t) = \frac{t}{4.5} = 0 \text{ for } t \geq 0$$

Considering Eq. (5.11), the control output of the derivative unit is computed as:

$$u_d(t) = K_p T_d \frac{d(y_s(t) - y(t))}{dt}$$

$$u_d(t) = 1(0.3)\frac{d(1-0)}{dt} = 0 \text{ for } t \geq 0$$

$$u_d(t) = \infty = 0 \text{ for } t = 0$$

By summing all control units as Eq. (5.19), the control output of the PID controller can be determined as:

$$u(t) = 1 + \frac{t}{4.5} \text{ for } t \geq 0$$

and,

$$u(t) = \infty \text{ for } t = 0$$

5.5 Tuning Rules for PI/PID Control

According to the widespread utilization of PI/PID controllers, several approaches for tuning its performance have been presented over the years. In this section, the most popular classical and modern methods for achieving the suitable control gains of PI/PID controllers are briefly described.

5.5.1 *Classical Tuning*

- **Trial and Error Approach**
  This approach is implemented when there is no systematic method to track during the design control process. This method is based on the experience in adjusting the suitable control gains of K_p, K_i, and K_d for achieving the desired time response with respect to speed and closed-loop stability. It is noted that this technique needs some experience to adjust the control gain for achieving the desired performance [1, 7, 26]. Typically, by increasing each control gain, the system performance is affected, as depicted in Table 5.1.

Table 5.1 System performance after increasing the control gains [1]

Parameter	Stability	Steady-state error	Speed
K_p	Decreased	Decreased	Increased
K_i	Increased	Eradicated	Decreased
K_d	Increased	Increased	Increased

- **Ziegler-Nichols Step Response Approach**
  This approach has been widely used in its original form regarding the changes (disturbances). It is based on a step response of the open-loop stable system [7, 26]. The tuning process can be described as follows.

  1. Evaluate the step response of the open-loop system.
  2. Plot a tangent with the maximum slope regarding the step response, see Fig. 5.5.
  3. Determine L, which represents the distance from the intersection of the vertical axis and slope to the starting point of the step response.
  4. Determine A, which represents the distance from the intersection of the vertical axis and slope to the horizontal axis.
  5. Calculate the PI/PID control gains from the following equations [1, 7].

$$K_p = \frac{1.2}{A} \tag{5.20}$$

$$K_i = \frac{0.6}{LA} \tag{5.21}$$

$$K_d = \frac{0.6L}{A} \tag{5.22}$$

- **Ziegler-Nichols Frequency Response Approach**
  This approach evaluates a proportional controller regarding the closed-loop system. The goal is to determine the maximum frequency where the phase of

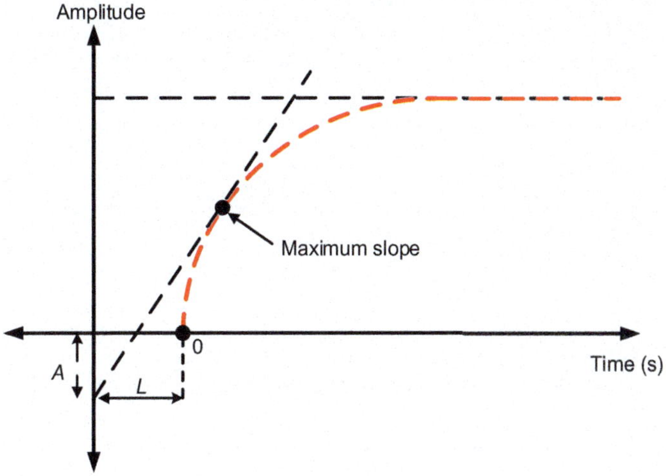

Fig. 5.5 Step response-based Ziegler-Nichols method

the process is at 180° [1, 7]. That indicates the maximum gain where the system achieves its stability border. The tuning process can be described as follows.

1. Link a proportional (P) controller to the studied system in the presence of a closed-loop configuration.
2. Gradually rise the P gain until the output starts oscillating. This gain is collected as the maximum control gain (K_m).
3. Measure the oscillating period. This period is collected as the maximum period (T_m).
4. Calculate the PI/PID control gain as follows [1, 7]:

$$K_p = 0.6K_m \tag{5.23}$$

$$K_i = \frac{1.2K_m}{T_m} \tag{5.24}$$

$$K_d = 0.075K_mT_m \tag{5.25}$$

This technique is assumed that the plant is based on a first-order transfer function with a delay.

5.5.2 *Modern Tuning*

- **Pole Placement Technique**

  This approach is a control design technique based on the fact of the transfer function of the system [27]. The goal is to evaluate the closed-loop pole positions on the complex plane by adjusting the gains of a controller. This technique can be applied to the plant that has a first or second-order transfer function [1]. For higher-order systems, the transfer function must be approximated and reduced to the first or second order.

  For the first-order plant, the system is defined as [7]:

$$P(s) = \frac{K}{sT + 1} \tag{5.26}$$

  where T is the time constant, and K is the gain of the system.

  For PI control, the equation of control actions can be represented as [1]:

$$C(s) = K_p\left(\frac{1}{sT_i} + 1\right) \tag{5.27}$$

Then, the closed-loop transfer function of the system is found as:

$$G(s) = \frac{P(s)C(s)}{1 + P(s)C(s)} \tag{5.28}$$

It can be represented in terms of the second-order as:

$$Q(s) = s^2 + s\left(\frac{1 + K \cdot K_p}{T}\right) + \frac{K \cdot K_p}{TT_i} \tag{5.29}$$

The second-order equation is presented in terms of the natural frequency (ω_n) and relative damping (ζ) as:

$$Q(s) = s^2 + 2\omega_n \zeta s + \omega_n^2 \tag{5.30}$$

Thus, the suitable K_p and T_i for a PI controller can be computed as:

$$K_p = \frac{2\omega_n \zeta T - 1}{K} \tag{5.31}$$

$$T_i = \frac{2\omega_n \zeta T - 1}{\omega_n^2 T} \tag{5.32}$$

For the second-order plant without zeros, the system is defined as:

$$P(s) = \frac{K}{(sT_1 + 1)(sT_2 + 1)} \tag{5.33}$$

For PID control, the equation of control actions can be represented as:

$$C(s) = \frac{K_p}{sT_i}\left(T_i T_d s^2 + T_i s + 1\right) \tag{5.34}$$

It can be represented in terms of the third-order as:

$$Q(s) = s^3 + \left(\frac{KK_pT_d}{T_1T_2} + \frac{1}{T_2} + \frac{1}{T_i}\right)s^2 + \left(\frac{KK_p}{T_1T_2} + \frac{1}{T_1T_2}\right)s + \frac{KK_d}{T_iT_1T_2} \tag{5.35}$$

The third-order equation is presented in terms of the natural frequency (ω_n), decay rate parameter (α), and relative damping (ζ) as:

$$Q(s) = \left(s^2 + 2\omega_n \zeta s + \omega_n^2\right)(s + \alpha\omega_n) \tag{5.36}$$

Thus, the suitable K_p, T_i, and T_d for a PID controller can be computed as:

$$K_p = \frac{T_1 T_2 \omega_n^2 (2\alpha\zeta + 1) - 1}{K} \tag{5.37}$$

$$T_i = \frac{T_1 T_2 \omega_n^2 (2\alpha\zeta + 1) - 1}{T_1 T_2 \alpha \omega_n^3} \tag{5.38}$$

$$T_d = \frac{T_1 T_2 \omega_n (2\zeta + \alpha) - T_1 - T_2}{T_1 T_2 \omega_n^2 (2\alpha\zeta + 1) - 1} \tag{5.39}$$

- **Time Domain Optimization Technique**
  This approach calculates the gains of the controller based on the numerical optimization technique, where a goal is set. There are three main goal functions that can be chosen based on the desired system performance as follow.
  The integral time-weighted absolute error (ITAE) is applied to integrate the absolute error multiplied by time as a weight. The ITAE function can be defined as [1, 7]:

$$J(V) = \int_0^\infty t|e(V,t)|dt \tag{5.40}$$

  where V is the vector with the controller gains, and $e(V, t)$ is the error signal of the studied system.
  The integral absolute error (IAE) is applied to integrate the absolute error without weights. This function is expressed as:

$$J(V) = \int_0^\infty |e(V,t)|dt \tag{5.41}$$

  The integral square error (ISE) is implemented to only integrate the square of the error. This function is represented as:

$$J(V) = \int_0^\infty e(V,t)^2 dt \tag{5.42}$$

  Finally, the controller gains can be found after minimizing a selected goal function to achieve a desired performance in the closed-loop system.
- **Cohen-Coon Technique**
  This implies an open-loop tuning technique, which used similar procedures as the Ziegler-Nichols step response technique. Figure 5.6 depicts the step response of the system, where the control parameters of K_{CC}, L, and T are calculated. The varibles L and T can be found in Fig. 5.6, which indicates the distances regarding the step response. The Cohen-Coon gain (K_{CC}) is obtained by computing the ratio between the increase of the control signal and the increase of amplitude output as [1, 7]:

$$K_{CC} = \frac{\Delta y}{\Delta u} \tag{5.43}$$

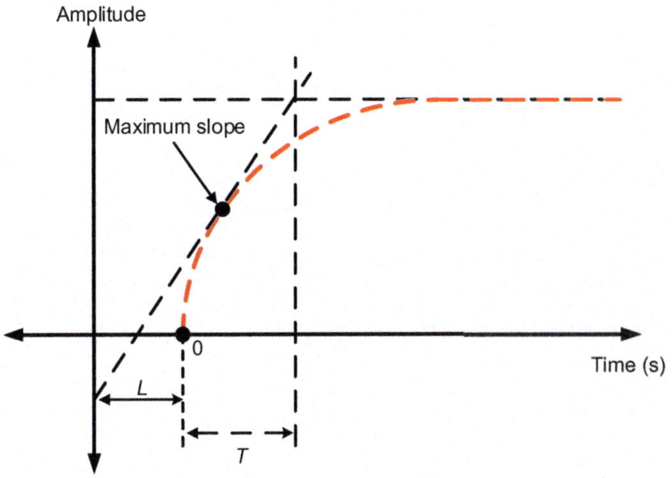

Fig. 5.6 Step response-based Cohen-Coon technique

Then, the controller gains can be computed as:

$$K_p = \frac{1}{K_{CC}}\left(\frac{1.35T}{L} + 0.25\right) \tag{5.44}$$

$$T_i = L\left(\frac{\frac{0.46L}{T} + 2.5}{\frac{0.61L}{T} + 1}\right) \tag{5.45}$$

$$T_d = L\left(\frac{0.37}{\frac{0.19L}{T} + 1}\right) \tag{5.46}$$

- **Internal Model Control (IMC) Technique**
  This method is evaluated when the system is in a stable condition. As the controller internally contains a system model, this method is called the internal model control [17, 28]. Figure 5.7 shows the structure of a closed-loop system. $G'(s)$ is the estimation of the system $G(s)$. $G_L(s)$ is a low pass filter. $G''(s)$ is the inverse of $G'(s)$. $\bar{y}$ is the estimation of the controlled variable. The controller's goal is to reject the zeros and poles of the original system $G(s)$ by merging it in parallel with $G'(s)$. The objective of $G_L(s)$ is to create the system less sensitive to the modeling error. Then, $C(s)$ can be expressed as [1, 7]:

$$C(s) = \frac{G_L(s)G''(s)}{1 - G_L(s)G''(s)G'(s)} \tag{5.47}$$

Then, this technique is integrated to the PI/PID controller. In the case of the first-order system with time delay, it can be expressed as [7]:

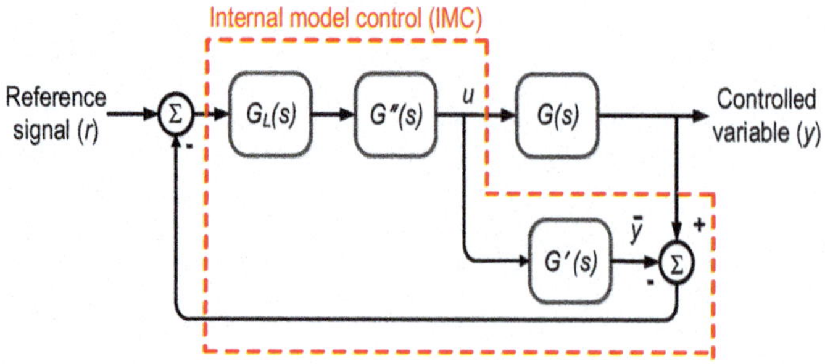

Fig. 5.7 Closed-loop system with IMC

$$P(s) = \frac{K}{sT + 1}\left(e^{-sL}\right) \tag{5.48}$$

$$G''(s) = \frac{sT + 1}{K} \tag{5.49}$$

$$G_L(s) = \frac{1}{sT_f + 1} \tag{5.50}$$

where K is the integrator gain, L is the delay time, T and T_f are the time constants of the systems.

By applying a first-order Padé approximation for the time delay, it can be expressed as [7]:

$$e^{-sL} \approx \frac{1 - sL/2}{1 + sL/2} \tag{5.51}$$

Later, the controller form is obtained as:

$$C(s) \approx \frac{\left(1 + s \cdot \frac{L}{2}\right)(s \cdot T + 1)}{s \cdot K\left(L + T_f + s \cdot T_f \cdot \frac{L}{2}\right)} \approx \frac{\left(1 + s \cdot \frac{L}{2}\right)(s \cdot T + 1)}{s \cdot K\left(L + T_f\right)} = \frac{K_i + sK_p + s^2 K_d}{s} \tag{5.52}$$

Finally, the controller gains can be computed as [7]:

$$K_i = \frac{1}{(L + T_f)K} \tag{5.53}$$

$$K_p = \frac{L + 2T}{(L + T_f)2K} \tag{5.54}$$

$$K_d = \frac{LT}{(L + T_f)2K} \tag{5.55}$$

5.6 Modeling of PI/PID-Based Virtual Inertia Control

In this section, the synthesis of a new decentralized PI/PID-based virtual inertia control is proposed. The PI/PID controller can be implemented to utilize the main function of virtual inertia control, that is, the emulation of inertia power. The emulation of inertia power is proportional to the rate of change of frequency (RoCoF) and virtual inertia constant (K_{VI}). The improper selection of virtual inertia control parameter may result in a higher frequency deviation, slower recovery time, and instability. To emulate the sufficient inertia power, the K_{VI} should effectively respond to a wide range of disturbances (e.g., changes in RESs/DGs, loads, system inertia and damping) without threatening the system stability.

By obtaining the suitable value of K_{VI} for responding to a wide range of disturbances, the PI/PID controller can participate in the parameter tuning process. Figure 5.8 shows the structure of a PI/PID controller, which is modified to combine with the dynamic structure of virtual inertia control. The input of the PI/PID controller is the RoCoF or derivative of frequency (*df/dt*), and the output is the controlled value of K_{VI}. The main objective of this controller is to use the well-known PI/PID feedback control technique to calculate the fixed (robust) value of K_{VI}. Once the optimal PI/PID control gains are obtained, the decentralized PI/PID-based virtual inertia controller is ready in hand, and no additional computational is required.

In the control area *j*, the RoCoF performs the input signal of the PI/PID controller to be used by the virtual inertia control.

The control equation for the PI controller can be expressed as:

$$u_j = \Delta P_{Cj} = K_{pj} ROCOF_j + K_{ij} \int ROCOF_j \tag{5.56}$$

The control equation for the PID controller can be expressed as:

$$u_j = \Delta P_{Cj} = K_{pj} ROCOF_j + K_{ij} \int ROCOF_j + K_{dj} \frac{d}{dt} ROCOF_j \tag{5.57}$$

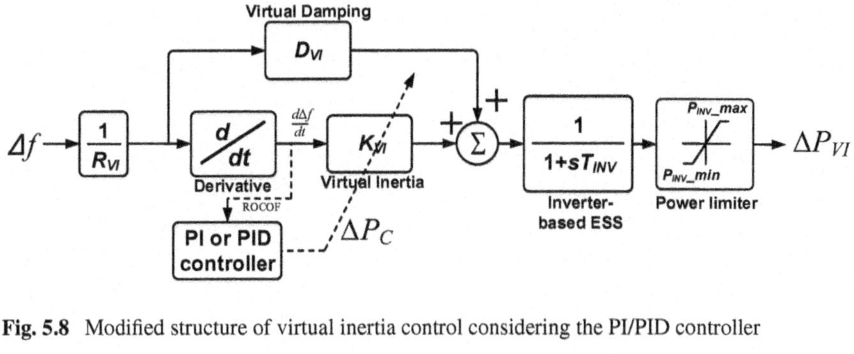

Fig. 5.8 Modified structure of virtual inertia control considering the PI/PID controller

where K_p, and K_i, and K_d are the gain constants of proportional, integral, and derivative actions, respectively.

Thus, by expanding the system description to include the RoCoF, its proportional, integral, and derivative as a measured output, the PI/PID-based virtual inertia controller can satisfy the performance requirements. Equations (5.56) and (5.57) can be changed to the feedback control form as:

$$u_j = K_j y_j \tag{5.58}$$

Then, Eq. (5.56) can be rewritten as:

$$u_j = \left[K_{pj}\ K_{ij} \right] \begin{bmatrix} ROCOF_j \\ \int ROCOF_j \end{bmatrix} \tag{5.59}$$

Also, Eq. (5.57) can be rewritten as:

$$u_j = \left[K_{pj}\ K_{ij}\ K_{dj} \right] \begin{bmatrix} ROCOF_j \\ \int ROCOF_j \\ \frac{d}{dt} ROCOF_j \end{bmatrix} \tag{5.60}$$

After that, y_j from Eq. (5.58) can be augmented in the form of the PI controller as Eq. (5.61). Figure 5.9 displays the PI structure and control variables.

$$y_j = \left[ROCOF_j\ \int ROCOF_j \right] \tag{5.61}$$

Similarly, y_j from Eq. (5.58) can be augmented in the form of the PID controller as Eq. (5.62). Figure 5.10 displays the PID structure and control variables.

$$y_j = \left[ROCOF_j\ \int ROCOF_j\ \frac{d}{dt} ROCOF_j \right] \tag{5.62}$$

Finally, considering $G_j(s)$ as the linear time-invariant model for the given control area j with the obtained state-space model from Chap. 3. The general equation can be expressed as [29]:

$$\dot{x}_j = A_j x_j + B_{1j} w_j + B_{2j} u_j \tag{5.63}$$

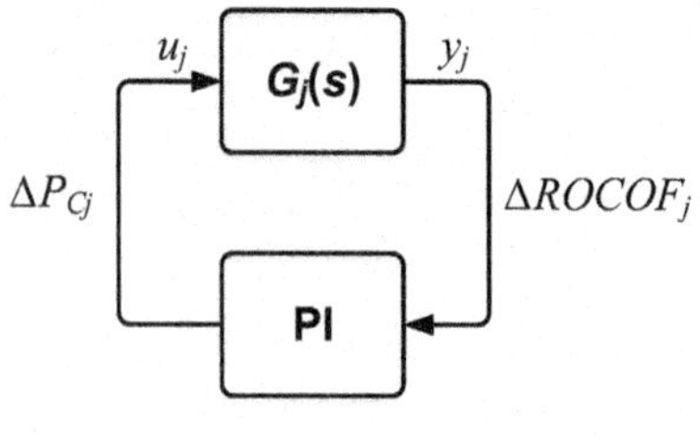

Fig. 5.9 Structure of PI-based virtual inertia control

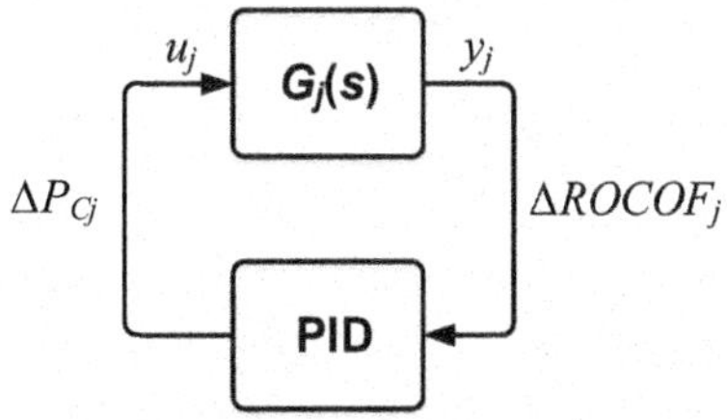

Fig. 5.10 Structure of PID-based virtual inertia control

$$y_j = K_{VIj} u_j \tag{5.64}$$

The next section will explain the process of tuning PI/PID control gains for achieving a suitable $K_{VI.}$

5.7 MATLAB-Based PI/PID Tuning Approach

The tuning of PI/PID control gains for simple applications may be an easy task. On the contrary, in complex applications with various disturbances, finding a set of control gains is a complex task for ensuring optimal system performance. Conventionally, the PI/PID controllers are tuned manually or tuned based on the classical rule-based approach. The manual/classical tuning approaches are iterative, hence, time-consuming [26]. These approaches may result in some reimbursements to the hardware of the system in real-world applications. Moreover, these classical approaches have some limitations as they typically appropriate for a specific type of dynamic systems (e.g., low-order plants or stable plants).

To deal with this problem, the MATLAB-based tuning algorithm offers a simple software interface for tuning the control gains, automatically (i.e., Simulink PID tuner). This algorithm could track an optimal solution to achieve the designed requirements [30, 31]. The MATLAB-based tuning approach offers several features, as follows:

- Tuning single-input and single-output (SISO) of the PI/PID controllers
- Tuning multiple loops of PI/PID controllers
- Tuning controller gains with interactive interfaces
- Tuning multiple controllers in a batch mode
- Identifying a plant model from test data of the input-output
- Modeling controllers in Simulink using controller blocks
- Modeling controllers in MATLAB using controller objectives.

5.7.1 Optimal PI Control Gains

The IMC tuning method is a useful feature in the MATLAB-based tuning algorithm. The method considers the model disturbances (uncertainties) and allows the designer to control a tradeoff between the system robustness and control system performance with respect to the modeling errors and perturbations. In this section, the IMC tuning method is used for determining optimal PI control gains. The main procedures are briefly described as follows:

1. Open a 'Block Parameter' of a PI controller (from 'Simulink Library Browser') and select a 'Tune' tab for initializing the tuning, see Fig. 5.11.
2. The next step is to determine the desirable performance of the PI controller, see Fig. 5.12. To reduce system frequency oscillation, drag the 'Response Time' slider (i.e., upper row in the red circle) to the right. This can increase the response speed of the controller (i.e., open loop bandwidth). To minimize transient behavior, drag the 'Transient Behavior' slider (i.e., lower row in the red circle) to the right until the system response has lower transient amplitude. It is noted that moving this transient slider toward the right results in more robust control.
3. When the PI performance is met the desired criteria, click an 'Update Block' tab to export the designed PI controller into the associated power system model or

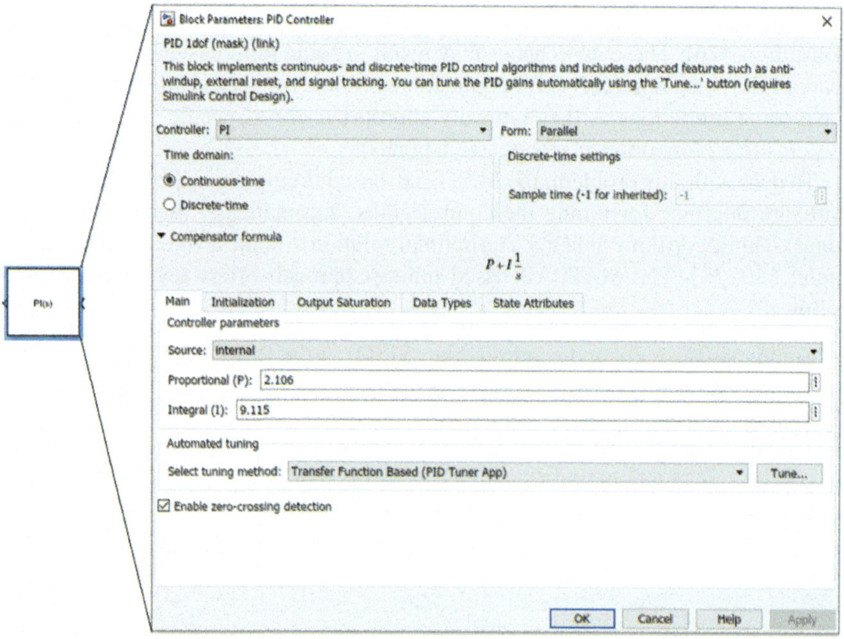

Fig. 5.11 MATLAB interface for tuning the PI controller

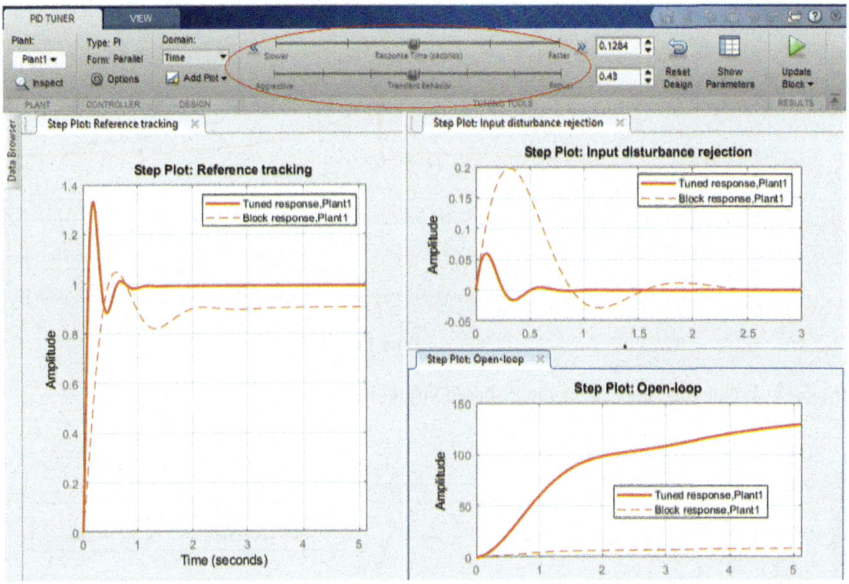

Fig. 5.12 PI performance tuning

MATLAB Workspace. In this study, the optimal control gains of K_p and K_i are obtained as 2.908 and 2.254, respectively.

The PI-based virtual inertia control system was designed and implemented in a single area system, as described in Fig. 3.9. The system model is constructed using the MATLAB/Simulink® software. To validate the efficiency of the proposed method, multiple disturbances are added to the system under various degraded situations of system inertia and load damping reductions as follows. The sudden load increase of 0.02 p.u. is applied at 2 s. Then, at 15 s, the sudden increase in solar power (0.01 p.u.) and wind power (0.01 p.u.) are simultaneously applied. The initial control parameters for the studied system are given in Table 3.1. The system inertia and damping are operated at 100% of its values (given in Table 3.2).

By applying the optimal control gains of K_p and K_i, the PI-based virtual inertia controller achieves better system performance and stability. Obviously, the system frequency nadir and overshoot are significantly reduced with less stabilizing time, see Fig. 5.13.

Figure 5.14 demonstrates the active power response of the conventional generating unit and inverter-based ESS unit equipped with PI-based virtual inertia control. It can be observed that the proposed PI-based virtual inertia control provides a faster power response with the required inertia (see the red line) compared with the traditional inertia control. Particularly, the virtual inertia unit could generate more additional power during the contingency, resulting in a lower frequency nadir and overshoot.

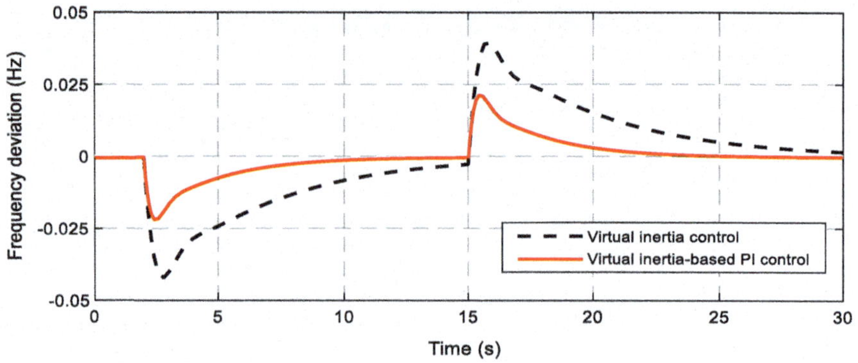

Fig. 5.13 Frequency response of the PI-based virtual inertia control

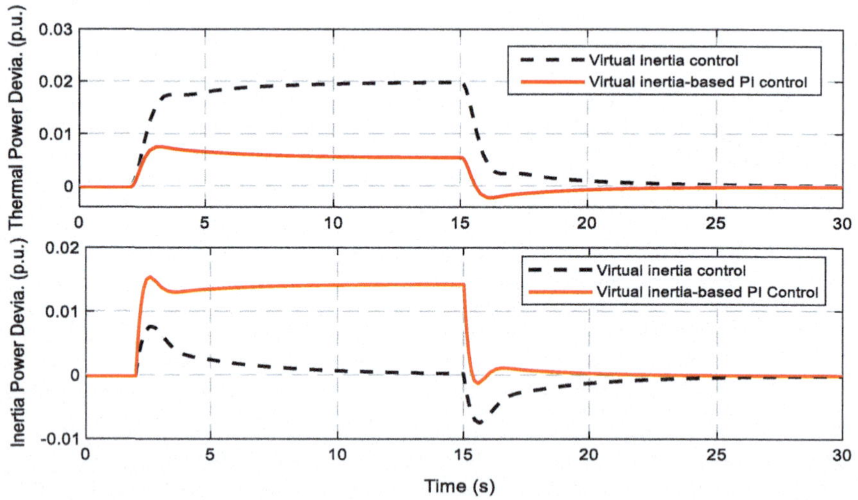

Fig. 5.14 Power responses (deviations) of the PI-based virtual inertia control

5.7.2 *Optimal PID Control Gains*

In this section, the IMC tuning method is used for determining optimal PID control gains. The main procedures are briefly described as follows:

1. Open a 'Block Parameter' of a PID controller (from 'Simulink Library Browser') and select a 'Tune' tab for initializing the tuning, see Fig. 5.15.
2. The next step is to determine the desirable performance of the PID controller, see Fig. 5.16. To reduce system frequency oscillation, drag the 'Response Time' slider (i.e., upper row in the red circle) to the right. This can improve the response speed of the controller (i.e., open loop bandwidth). To minimize transient behavior, drag the 'Transient Behavior' slider (i.e., lower row in the red

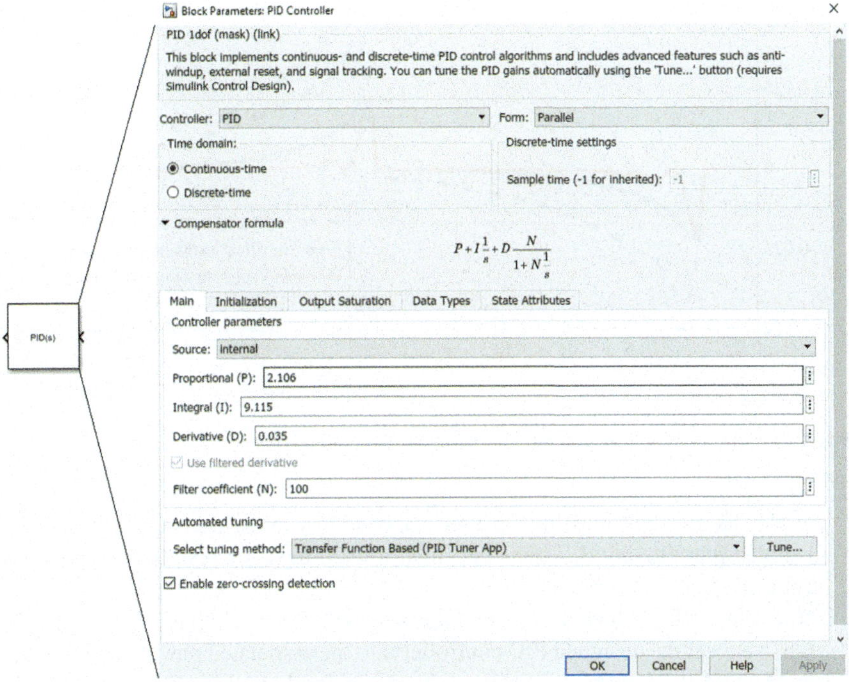

Fig. 5.15 MATLAB interface for tuning the PID controller

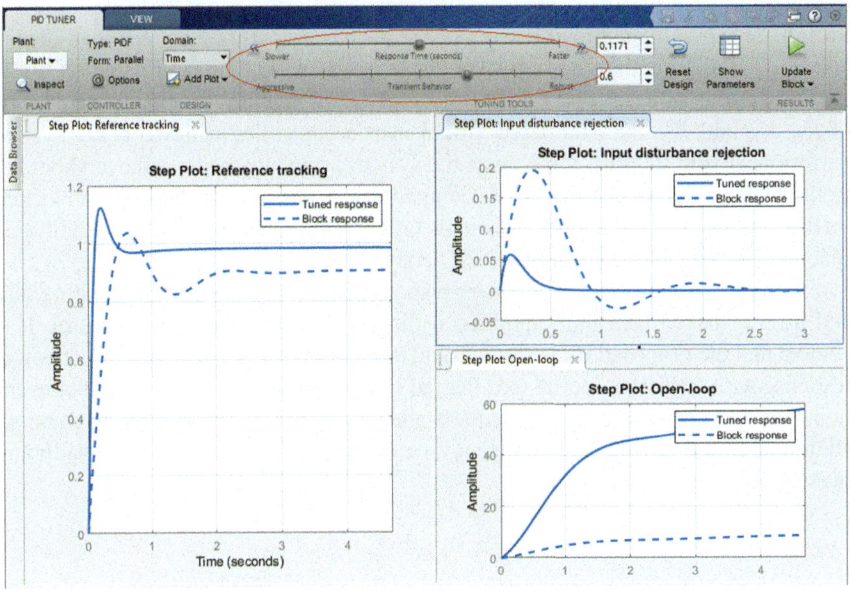

Fig. 5.16 PID performance tuning

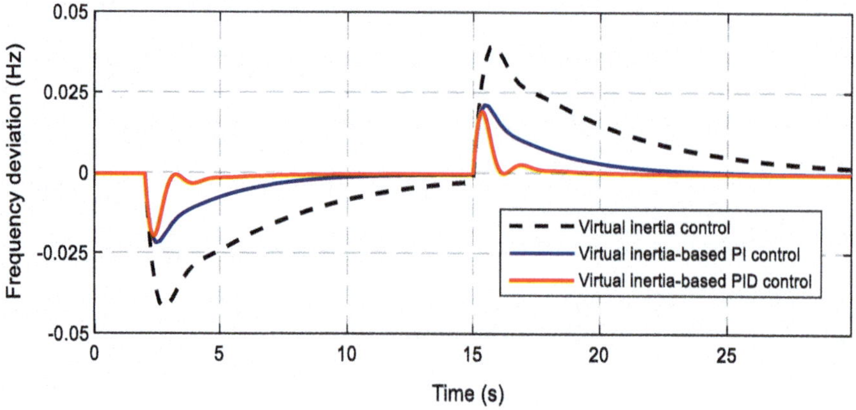

Fig. 5.17 Frequency response of the PID-based virtual inertia control

circle) to the right until the system response has lower transient amplitude. It is noted that moving this transient slider toward the right results in more robust control.

3. When the PID performance is met the desired criteria, click an 'Update Block' tab to export the designed PID controller into the associated power system model or MATLAB Workspace. In this study, the optimal control gains of K_p, $K_{i,}$ and K_d are obtained as 2.106, 9.115, and 0.035, respectively.

It is noted that this tuning process is designed and tested under the same setting of Sect. 5.7.1. To validate the efficiency of the proposed method, the step load disturbance of −0.02 p.u. and the step RES disturbance of +0.02 p.u. are applied to the studied system as described in Sect. 5.7.1. By applying the optimal control gains of K_p, K_i, and K_d, the PID-based virtual inertia controller achieves better system performance and stability. Obviously, the system frequency nadir and overshoot are significantly reduced, see Fig. 5.17. Compared with the case of the PI-based virtual inertia control, the system response yields faster stabilizing time due to the additional function of derivative control in the PID controller.

Figure 5.18 shows the active power response of the conventional generating unit and inverter-based ESS unit equipped with PID-based virtual inertia control. It is evident that the proposed PID-based virtual inertia control provides the fastest power response with sufficient inertia (see the red line) compared with the PI and conventional inertia control systems. Particularly, the virtual inertia unit could generate more additional power during the contingency, resulting in the lowest frequency nadir and overshoot.

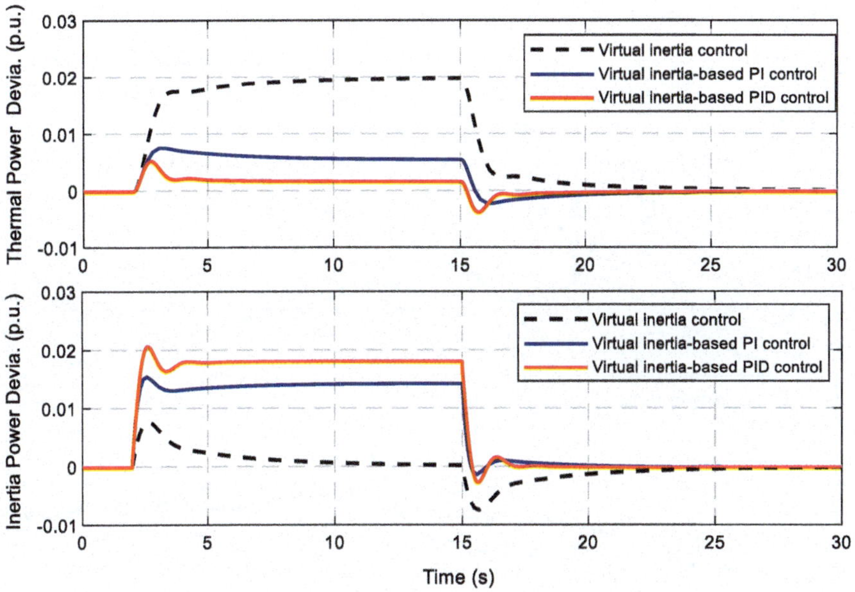

Fig. 5.18 Power responses (deviations) of the PID-based virtual inertia control

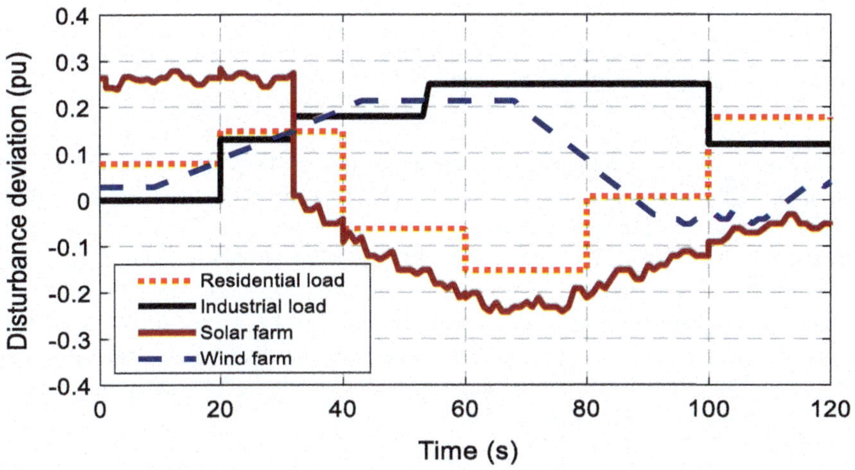

Fig. 5.19 Multiple disturbances in RESs and loads penetration

5.8 Simulation Results

In this section, the performance of the proposed PI/PID-based virtual inertia control system is tested under high power uncertainties of RESs and loads, see in Fig. 5.19. The studied system is created based on Fig. 3.9 and Table 3.1.

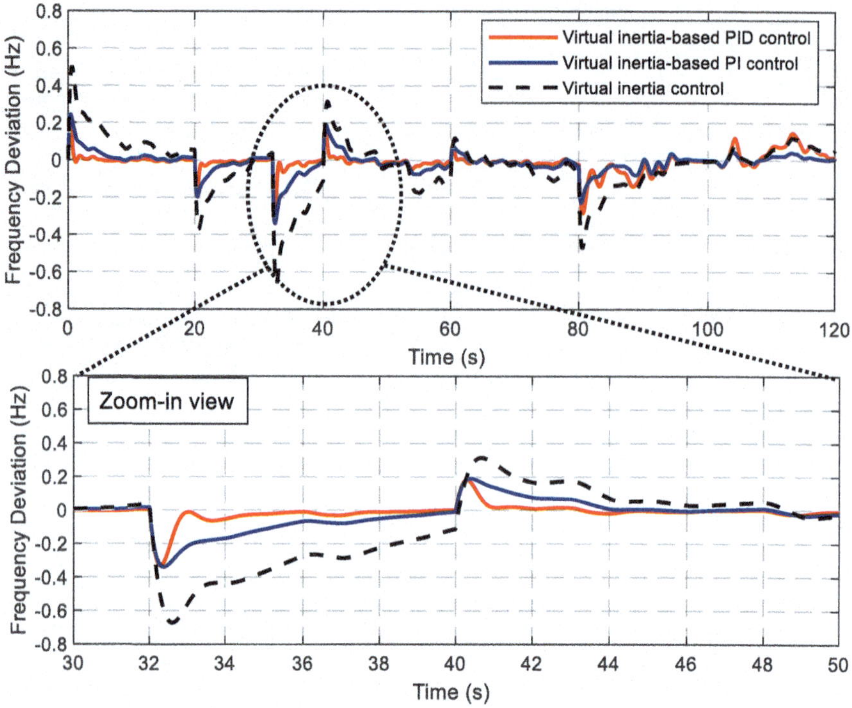

Fig. 5.20 Frequency responses of the PI and PID-based virtual inertia controllers under the condition of high system inertia and damping

Figure 5.20 shows the system performance under 100% of system inertia and damping reduction (i.e., normal operation). Without the deployment of the PI/PID controller (i.e., dash line), it yields the largest frequency nadir and overshoots with longer stabilizing time. By applying the PI-based virtual inertia control (i.e., blue line), the frequency nadir and overshoot are significantly reduced with less stabilizing time. By applying the PID-based virtual inertia control (i.e., red line), the system response rapidly returns to the nominal state after the disturbance. This indicates the ability of fast stabilizing time due to the deployment of the derivative control function in the PID controller.

Then, the system inertia and damping are lowered to 70% of its nominal values to test the robustness of the proposed PI/PID-based virtual inertia control. Figure 5.21 shows the system performance under the degraded condition of low system inertia and damping. It is clear that the proposed PI/PID-based virtual inertia control could resist multiple disturbances in RESs/DGs and loads penetration, even in the degraded condition of low system inertia and damping. Evidently, the frequency nadir and overshoot are immediately arrested after the disturbance, resulting in lower values with less stabilizing time. In the case of the conventional virtual inertia control (a

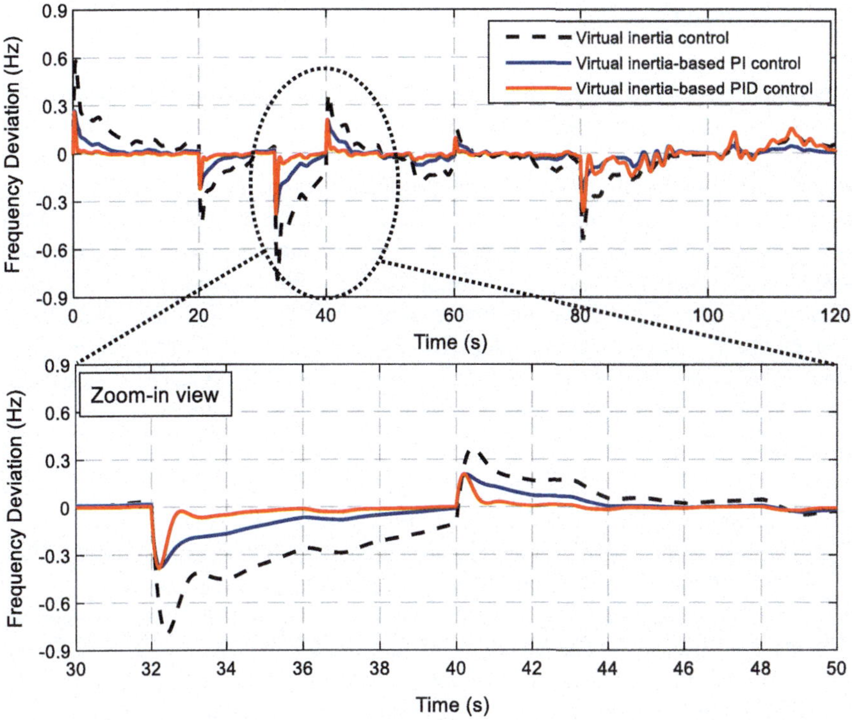

Fig. 5.21 Frequency responses of the PI and PID-based virtual inertia controllers under the critical condition of low system inertia and damping

non-optimal K_{VI}), it yields longer response time, leading to slower actions in dealing with the disturbances (see the dash line in Fig. 5.21).

In summary, the conventional virtual inertia control with non-optimal virtual inertia control constant (K_{VI}) fails to meet the desired frequency control objective in the system. It yields the larger frequency nadir and overshoot with longer stabilizing time under the wide range of operation. Obviously, the deployment of PI/PID controllers introduces more effective control approaches and more stable control tools for achieving an optimal K_{VI} parameter. It is clear that the proposed PI/PID-based virtual inertia control could provide a reasonable approximation in estimating a suitable value of K_{VI}. Thus, the PI/PID-based virtual inertia control system can achieve quite better system stability and performance with robustness regarding a wide range of uncertainties and disturbances.

5.9 Summary

The decentralized PI/PID-based virtual inertia control approaches have been developed to evaluate the powerful virtual inertia power under a wide range of RESs/DGs, and load penetration. The uses of the PI/PID controllers for frequency stability enhancement are briefly discussed. Then, the optimal setting of PI/PID parameters using the classical and modern tuning techniques was described in detail to obtain sufficient virtual inertia constant with respect to the additional power, assuring stable grid operation. Finally, the proposed approaches were implemented to a control area system using a nonlinear simulation with different levels of RESs/DGs, loads, system inertia, and system damping reductions. It was shown that the designed controllers could provide stable performance under a wide range of area load uncertainties without threatening the system stability and resiliency.

Problems

1.1 Please explain the roles of control actions of PI and PID controllers.

1.2 Please draw the dynamic structure of a PID controller for a unit step error. Then, explain the influences of the changing parameters over the response.

1.3 The controlled system has the setpoint $y_s(t) = 1$ for $t \geq 0$ and the process output $y(t) = 0$ for $t \geq 0$. The PID control gains are given as $K_p = 2.5$, $T_i = 7$, and $T_d = 0.5$. Please determine the control output of the PID controller for $t \geq 0$.

1.4 The controlled system has the setpoint $y_s(t) = 1$ for $t \geq 0$, $y_s(t) = 0$ for $t < 0$, and the process output $y(t) = 0$. The PID control gains are given as $K_p = 3$, $T_i = 6.5$, and $T_d = 0.4$. Please determine the control output of the PID controller for $t \geq 0$.

1.5 Please construct the single-area power system based on Fig. 3.9 and Table 3.1 using MATLAB/Simulink software. The system inertia and damping are set as 100% of its nominal values (see Table 3.2). The system base is 500 MW. The nominal frequency is 60 Hz. At time of 15 s, the step load increment (ΔP_L) of 50 MW occurs. During the load increment, the wind and solar systems are generated constant power to the system. Please determine the following.

(1) System frequency nadir without the virtual inertia control

(2) System frequency nadir with the virtual inertia control

(3) PI control gains (K_p and K_I) for virtual inertia control with the objective of maintaining frequency nadir within −0.1 Hz

(4) System frequency nadir after applying the designed PI-based virtual inertia control

(5) PID control gains (K_p, K_I, and K_D) for virtual inertia control with the multi-objectives of maintaining frequency nadir within −0.08 Hz and stabilizing time within 3 s after the disturbance

(6) System frequency nadir after applying the designed PID-based virtual inertia control.

1.6 From Problem 1.5, the system inertia and damping are massively lowered to 40% of its nominal values (see Table 3.2). Then, at the time of 10 s, the integrations of 50 MW wind power and 40 MW solar power are applied to the system. During the wind and solar integration, the load is consumed constant power from the system. Please determine the following.

(1) System frequency overshoot without the virtual inertia control
(2) System frequency overshoot with the virtual inertia control
(3) PI control gains (K_p and K_I) with the objective of maintaining frequency overshoot within +0.08 Hz
(4) System frequency overshoot after applying the designed PI-based virtual inertia control
(5) PID control gains (K_p, K_I, and K_D) with the multi-objectives of maintaining frequency overshoot within +0.05 Hz and stabilizing time within 2 s after the disturbance
(6) System frequency overshoot after applying the designed PID-based virtual inertia control.

References

1. A. Visioli, *Practical PID Control* (Springer, London, UK, 2006)
2. Q.G. Wang, T.H. Lee, H.W. Fung, Q. Bi, Y. Zhang, PID tuning for improved performance. IEEE Trans. Control Syst. Technol. **7**(4), 457–465 (1999)
3. K.H. Ang, G. Chong, Y. Li, PID control system analysis, design, and technology. IEEE Trans. Control Syst. Technol. **13**(4), 559–576 (2005)
4. M.J. Neath, A.K. Swain, U.K. Madawala, D.J. Thrimawithana, An optimal PID controller for a bidirectional inductive power transfer system using multiobjective genetic algorithm. IEEE Trans. Power Electron. **29**(3), 1523–1531 (2014)
5. Y. Cheng-Ching, *Autotuning of PID Controllers* (Springer, London, UK, 2006)
6. J. Oravec, M. Bakošová, M. Trafczynski, A. Vasičkaninová, A. Mészáros, M. Markowski, Robust model predictive control and PID control of shell-and-tube heat exchangers. Energy **159**(15), 1–10 (2018)
7. I.D. Diaz-Rodriguez, S. Han, S.P. Bhattacharyya, *Analytical Design of PID Controllers* (Springer, Switzerland, 2019)
8. D. Rerkpreedapong, A. Hasanovic, A. Feliachi, Robust load frequency control using genetic algorithms and linear matrix inequalities. IEEE Trans. Power Syst. **18**(2), 855–861 (2003)
9. H. Bevrani, T. Hiyama, Robust decentralised PI based LFC design for time delay power systems. Energy Convers. Manag. **49**(2), 193–204 (2008)
10. J. Bai, X. Zhang, A new adaptive PI controller and its application in HVAC systems. Energy Convers. Manag. **48**(4), 1043–1054 (2007)
11. M. Andreasson, D.V. Dimarogonas, H. Sandberg, K.H. Johansson, Distributed PI-control with applications to power systems frequency control, in Proc. *American Control Conference*, 3183–3188 (2014)
12. C.S. Ali Nandar, Robust PI control of smart controllable load for frequency stabilization of microgrid power system. Renew. Energy **56**(1), 16–23 (2013)
13. H. Bevrani, Y. Mitani, K. Tsuji, Sequential design of decentralized load frequency controllers using μ synthesis and analysis. Energy Convers. Manag. **45**(6), 865–881 (2004)

14. H. Bevrani, T. Hiyama, H. Bevrani, Robust PID based power system stabiliser: design and real-time implementation. Int. J. Electr. Power Energy Syst. **33**(2), 179–188 (2011)
15. G. Magdy, E.A. Mohamed, G. Shabib, A.A. Elbaset, Y. Mitani, SMES based a new PID controller for frequency stability of a real hybrid power system considering high wind power penetration. IET Renew. Power Gener. **11**(12), 1304–1313 (2018)
16. P.K. Mohanty, B.K. Sahu, T.K. Pati, S. Panda, S.K. Kar, Design and analysis of fuzzy PID controller with derivative filter for AGC in multi-area interconnected power system. IET Gener. Transm. Distrib. **10**(15), 3764–3776 (2016)
17. W. Tan, Unified tuning of PID load frequency controller for power systems via IMC. IEEE Trans. Power Syst. **25**(1), 341–350 (2010)
18. B.K. Sahu, T.K. Pati, J.R. Nayak, S. Panda, S.K. Kar, A novel hybrid LUS-TLBO optimized fuzzy-PID controller for load frequency control of multi-source power system. Int. J. Electr. Power Energy Syst. **74**(1), 58–69 (2016)
19. W. Tan, Tuning of PID load frequency controller for power systems. Energy Convers. Manag. **50**(6), 1465–1472 (2009)
20. S. Sondhi, Y.V. Hote, Fractional order PID controller for load frequency control. Energy Convers. Manag. **85**(1), 343–353 (2014)
21. S.A. Taher, M. Hajiakbari Fini, S. Falahati Aliabadi, Fractional order PID controller design for LFC in electric power systems using imperialist competitive algorithm. Ain Shams Eng. J. **5**(1), 121–135 (2014)
22. G. Madgy, G. Shabib, A.A. Elbaset, Y. Mitani, *Renewable Power Systems Dynamic Security* (Springer, Switzerland, 2020)
23. T. Kerdphol, F.S. Rahman, M. Watanabe, Y. Mitani, D. Turschner, H.P. Beck, Enhanced virtual inertia control based on derivative technique to emulate simultaneous inertia and damping properties for microgrid frequency regulation. IEEE Access **7**(1), 14422–14433 (2019)
24. S. Skogestad, I. Postlethwaite, *Multivariable Feedback Control—Analysis and Design*, 2nd ed. (Wiley, USA, 2005)
25. K.S. Hong, U.H. Shah, Feedback control, in *Dynamics and Control of Industrial Cranes* (Springer, Singapore, 2019)
26. H. Bevrani, I. Francois, T. Ise, *Microgrid Dynamics and Control* (Wiley, Hoboken, NJ, USA, 2017)
27. Y. Zhang, Q.G. Wang, K.J. Astrom, Dominant pole placement for multi-loop control systems. Automatica **38**(7), 1213–1220 (2002)
28. R. Vilanova, IMC based Robust PID design: tuning guidelines and automatic tuning. J. Process Control **18**(1), 61–70 (2008)
29. H. Bevrani, *Robust Power System Frequency Control*, 2nd ed. (Springer, New York, USA, 2014)
30. I. Boiko, *Non-parametric Tuning of PID Controllers* (Springer-Verlag, London, UK, 2013)
31. MATLAB, Introduction: PID controller design *Matlab* (MathWorks, USA, 2017)

Chapter 6
Model Predictive Control for Virtual Inertia Synthesis

Abstract Nowadays, a new concept of modern power systems (i.e., smart/micro-grids), which includes various components such as smart meters, smart appliances, renewable energy sources (RESs), distributed generators (DGs), and controllable loads has increased attention worldwide due to its energy efficiency and environmental concerns. Such a modern system requires the employment of real-time application and intelligent control. To properly utilize the virtual inertia control regarding an intelligent ability in future predictions, the model predictive control (MPC) is necessary. The MPC has a fine performance in delivering fast dynamic response with robustness against disturbance and uncertainty, while keeping future control variables in account. Thus, it has been implemented in a wide range of industrial applications, including real-time measurement and control. In this chapter, the design of decentralized MPC-based virtual inertia control is introduced to emulate the suitable virtual inertia power, while predicting the future behavior or event regarding inertia control-based frequency regulation. The MPC controller applies a feedforward control technique to reject the disturbances from RES/DG and load penetration as well as system parameter uncertainty, ensuring rapid dynamic response with the robustness of system operation. The proposed MPC-based virtual inertia control is verified through a nonlinear control area system with high RESs/DGs penetration including the extended communication delay time.

Keywords Model predictive control (MPC) · Fast dynamic response · Frequency control · Inertia control · Nonlinear control · Parameter tuning · Predictive model · Real-time control · Robustness · Uncertainty · Virtual inertia constant · Virtual inertia synthesis

6.1 Introduction to Model Predictive Control

Model predictive control (MPC) is introduced in the late 70 s and has been developed until now. It is available in different names (e.g., dynamic linear programming, moving horizon control, dynamic matrix control, and rolling horizon planning). The idea of such control approaches is similar as they have the capability to forecast future events and compute the control actions with regards to the minimization

T. Kerdphol et al., *Virtual Inertia Synthesis and Control*, Power Systems,
https://doi.org/10.1007/978-3-030-57961-6_6

of an objective function. This technique creates linear controllers that practically provide a similar structure with suitable degrees of freedom. Up to the present, the applications of MPC have been widely used by the world industry and academics (e.g., robot control, cement industry, servo control, and steam generator control [1]. The remarkable performance of such applications ensures that the MPC's ability to highly obtain efficient control systems under a wide range of operational time. The major advantages of MPC are listed as follows:

- The designers require only a fundamental knowledge of control as the MPC concept is intuitive, and the tuning is quite simple.
- MPC is an open approach based on specific fundamentals that enable to predict future expansions.
- MPC is simply applied to govern a variety of processes from those with simple dynamic structures to advanced dynamic structures with long delay time.
- MPC deals with a solution at each sampling instant of a finite horizon optimal control issue with respect to the system state, dynamics, and constraints.
- MPC uses feedforward control in a usual style to compensate for considered disturbances.
- MPC is suitable for multivariable control issues working at different time scales.
- MPC can deal with real-time control in a natural mean, allowing the system to operate more closely to its limits.
- The MPC design process in dealing with constraints is easy and systematic.
- The obtained MPC controller is simple to apply a linear control law.

Consequently, the MPC is a powerful tool via stochastic formulation, allowing the controller the ability to optimize various conditions and scenarios at the same time. As a result, the distributed formulation optimizes the subsystems (e.g., components in micro/smart-grids) at the same time, and integrates into universal systems (e.g., power systems), thus determining the global objective function while respecting the subsystem objective functions [2].

Currently, a new concept of modern power systems called smart/micro-grids, which includes various components such as smart meters, smart appliances, renewable energy sources (RESs), distributed generators (DGs), and controllable loads has increased attention worldwide due to its energy efficiency and environmental concerns. The MPC is used to offer solutions to several issues found in modern power systems as follows:

- The MPC offers an optimal solution for system planning and operation in a synchronized mean in achieving the goals [3–5].
- The intermittency and disturbance of RESs, DGs, and controllable loads are designed as stochastic variables in the MPC control process; thus, the MPC control actions can overcome the uncertainty [6–10].
- MPC is applied when logical/binary variables are considered in the optimization process. This represents the cases of disconnection/connection of generating/load units (e.g., RESs/DGs, electric vehicles, energy storage units, and controllable loads) or the cases of different prices of energy for selling or purchasing [11–16].

- In cases of power systems with multiple agents, the power systems are distributed as micro-grids. Thus, MPC can address a complex problem in a distributed way, simply offering a distributed solution [17–20].
- During the abrupt changes in power systems (e.g., malfunctioning of a certain unit or contingency), MPC adjusts to the new circumstance by transforming its structure, allowing normal operation with degrees of freedom [21–27].

Focusing on frequency control, MPC has been widely utilized to solve several control problems as follows. In [6], the authors proposed a concept of MPC to solve load-frequency control problems. In [28], MPC is used to control the thermal power plant regarding frequency regulation. In [29], MPC is extended to deal with the operation and control of a nuclear power plant. The authors in [30] proposed the design of MPC for controlling multi-area power systems. Then, the integration of RESs/DGs was considered as a measured disturbance in the MPC process for controlling multi-area power systems [10]. The author in [31] proposed a decentralized MPC controller for secondary control of the system. In [32], the authors proposed the coordinated MPC control of plug-in electric vehicles (EVs) and wind farms for stabilizing power system frequency. Later, the charging/discharging states of plug-in EVs were considered in the MPC control process for system frequency stabilization [33].

Focusing on the virtual inertia synthesis with respect to frequency control, the modern system requires the employment of real-time or continuous monitoring and intelligent control due to the increasing size of intermittent RESs, DGs, and controllable loads. As a result, the RES/DG and load integration make the system more complex than ever. Regarding the previous chapter, the utilization of a PI/PID controller may require more resetting with longer computational time in response to high or continuous penetration of RESs/DGs and loads for stabilizing the system. Consequently, the PI/PID controller may face difficulty in applying the real-time or continuous application due to its long computational time. As the computation of the MPC control sequence minimizes an objective function, the MPC performance can deliver a quick dynamic response with robustness against parameter uncertainty and disturbance. By using a receding horizon technique in MPC, the horizon or control variable at each instant is relocated toward the future, which enables the employment of the first control signal of sequence to be determined at each step concerning the futurity. To utilize the virtual inertia control in the real-time or continuous application, the MPC can be properly respected as an applicable or intelligent control solution to predict the system control output at future time instants. In this chapter, the MPC application is extended to the main function of the virtual inertia control system, that is, the emulation of inertia power. Thus, the MPC controller is implemented in the tuning process of virtual inertia control to predict the suitable virtual inertia constant at future events regarding system frequency deviation. The MPC controller applies a feedforward control method to reject the disturbances from RES/DG and load penetration as well as system parameter uncertainty, ensuring rapid dynamic response with the robustness of operation. It is noted that this chapter is an extended version of the research presented in [11]. Finally, the proposed MPC-based virtual inertia control is verified through a nonlinear control area system under a wide range of parameter disturbance and uncertainty including the effect of communication delay.

6.2 Fundamental MPC Strategy

The MPC strategy is constructed in the discrete-time domain, and it allows the control variables to modify its values at discrete sampling instants. Compared to traditional linear control approaches, the MPC method can observe and handle the past, present, and future values of control variables. In fact, the MPC forecasts future errors and performs protective control actions, where the system will not be fallen on the vast errors, leading to robust system operation [1, 2]. The main operating strategy of MPC can be divided into three control actions as follows:

- **Prediction**: The MPC is operated based on online/real-time iterations or numerical optimization. This action involves the discrete-time model. The future values of the control input sequence (u) and state variable (y) are forecasted for a predicted horizon (N_p) using the feedback measurements and system model at the instant (t).
- **Optimization**: The forecasts have been examined by a cost function, which characterizes the desirable behavior or goals of a control system.
- **Receding horizon technique**: It consists of the processes of measuring new feedback variables, forecasting new behavior, and optimizing performance cost. This technique is repeated during each sampling interval.

Figure 6.1 describes a fundamental structure of MPC. It contains prediction and controller units [31]. The prediction unit includes the disturbance model and system model, which can effectively estimate future system actions based on its unmeasured/measured disturbances, control signal, and current output over the predicted horizon. The predicted output is driven by the control unit as the known parameters in an optimization issue, minimizing a goal function with respect to system

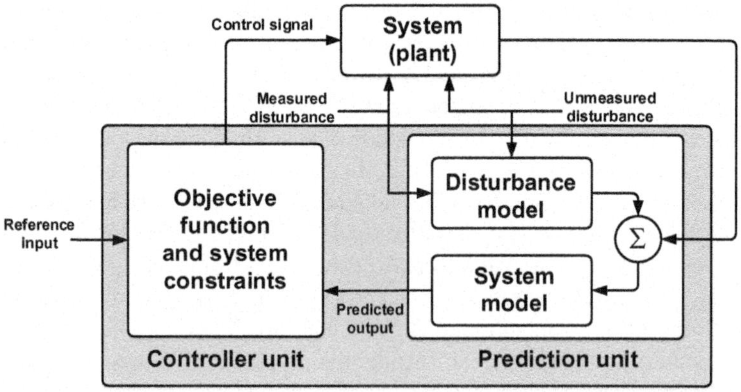

Fig. 6.1 Fundamental structure of MPC

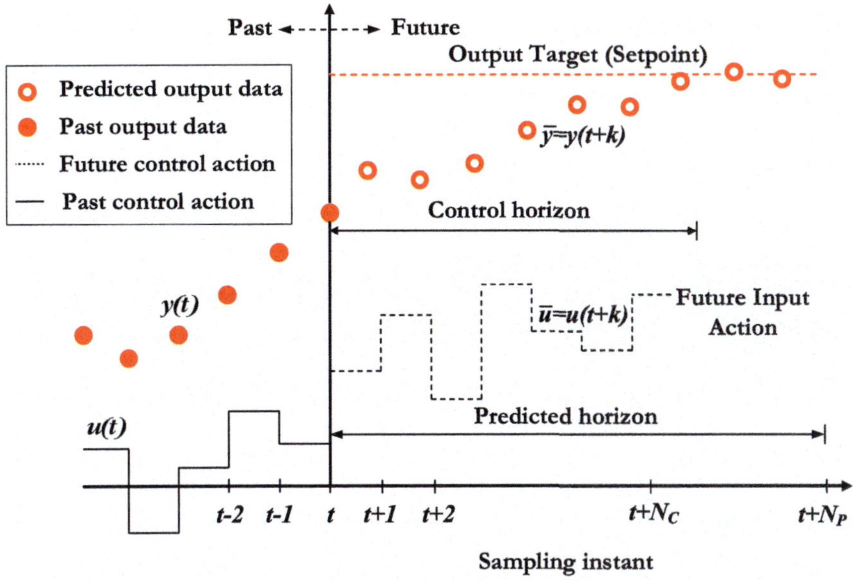

Fig. 6.2 Operating principle of MPC based on control and predicted horizons

limitations. After solving this issue, it results in an optimal control sequence over a control horizon. The first element of this sequence is added to the plant, and the entire process is repeated in the next sampling interval with the prediction horizon moved one sampling interval toward.

The MPC operating principle is described in Fig. 6.2. It is noted that the value of the variable at the instant $t + k$ computed at the current instant t, where $k = 1,\ldots,N_P$. The control process is briefly explained as follows:

1. The future (predicted) output ($\bar{y}$) for the predicted horizon is computed at each sampling instant t using the dynamic system model. The $\bar{y}$ depends on the actual values up to the instant t (current state and past inputs/outputs) and the future control signal ($\bar{u}$). The N_C and N_P are the control horizon and predicted horizon, respectively.
2. The $\bar{u}$ is computed by optimizing a specified constraint. This constraint tries to maintain the output nearby the reference (setpoint) trajectory. This process forms a quadratic function of errors between the predicted reference trajectory and the predicted output signal.
3. The control signal (u) is forwarded to the process, while the following control signals are discarded. This is because, at the next sampling instant,$y(t + 1)$ is recognized already via the feedback action. Then, the step 1 is repeated with this new value, and the whole process is updated. This process also applies the receding horizon concept regarding the future control and prediction.

The dynamic model of power systems is usually analyzed in the form of state-space equations, as discussed in Chaps. 3 and 4. To predict control problems, the

MPC-based state-space representation can be used to deal with multivariable systems as follows [1, 2, 31]:

$$x(t+1) = Ax(t) + Bu(t) \tag{6.1}$$

$$y(t) = Cx(t) \tag{6.2}$$

where *x(t)*, *y(t)*, and *u(t)* are the state vector.

To obtain offset-free control, the state-space is rewritten in the form of the control incremental $\Delta u(t)$ as an input as [1, 2, 31]:

$$\begin{bmatrix} x(t+1) \\ u(t) \end{bmatrix} = \begin{bmatrix} A & B \\ 0 & I \end{bmatrix} \begin{bmatrix} x(t) \\ u(t-1) \end{bmatrix} + \begin{bmatrix} B \\ I \end{bmatrix} \Delta u(t) \tag{6.3}$$

$$y(t) = \begin{bmatrix} C & 0 \end{bmatrix} \begin{bmatrix} x(t) \\ u(t-1) \end{bmatrix} \tag{6.4}$$

where

$$\Delta u(t) = u(t) - u(t-1) \tag{6.5}$$

By introducing a new state vector $\tilde{x}(t) = \begin{bmatrix} x(t) & u(t-1) \end{bmatrix}^T$, the general form of the state-space model can be obtained as [1, 2, 31]:

$$\tilde{x}(t+1) = R\tilde{X}(t) + L\Delta u(t) \tag{6.6}$$

$$y(t) = P\tilde{x}(t) \tag{6.7}$$

where *R*, *L*, and *P* are represented the functions of *A*, *B*, and *C*, respectively. It is noted that the incremental form (Δ) is used for MPC derivation.

Then, the MPC cost function is added to perform the minimization process. The main goal of the cost function is to make the future output in tracking a reference (setpoint) signal *w(t)* along the horizon. The general form of the cost function can be expressed as [1, 2]:

$$J(N_P, N_C) = \sum_{j=1}^{N_P} \delta(j)(\bar{y}(t+j) - w(t+j))^2 + \sum_{j=1}^{N_C} \lambda(j)(\Delta u(t+j-1))^2 \tag{6.8}$$

where $N_C \leq N_P$. The N_P sets the time instant limits, which is required for the output in tracking the reference. The coefficients $\lambda(j)$ and $\delta(j)$ are sequences which compute the relative weight of control effort and error along the horizon. In some cases, the state *x(t)* is added in the cost function instead of the future output. This makes the matrices C or P equals to the identity matrix (I).

Following the certain interval $N_C < N_P$, the control signals are constant. This means the $u(t+j)$ will not change after $j = N_C$ as [1, 2]:

$$\Delta u(t+j-1) = 0 \quad \text{for } j < N_C \tag{6.9}$$

According to the predictive ability of MPC, if the future development of the known reference $r(t+k)$ is recognized in priority, the MPC controller can respond to the change before occurring. Thus, the reference trajectory $w(t+k)$ is applied using a first-order filter for smoothing approximation as [1, 2]:

$$w(t) = y(t)w(t+k) = \alpha w(t+k-1) + r(t+k)(1-\alpha) \text{ for } k = 1 \ldots N_P \tag{6.10}$$

where α is the adjusted variable from zero to one. It is noted that the closer value to one represents the smoother approximation.

Alternatively, the general form of a predicted output with the objective function Eq. (6.8) can be defined as [1, 2]:

$$\bar{y}(t+j) = PR^j x(t) + \sum_{i=0}^{j-1} PR^{j-i-1} L\Delta u(t+i) \tag{6.11}$$

In this section, a simple MPC strategy is briefly explained. Interested readers can find the full-synthesis details, and more MPC based control techniques in [1, 2, 31].

6.3 MPC Disturbances

The MPC offers a remarkable ability to reject disturbances using the feedback function. The disturbances are measured or estimated in the dynamic model. Thus, the MPC controller manipulates the disturbance effects over the output. The disturbance effects $d(t)$ can be included to the MPC process. The general form of the MPC model, including disturbances, is expressed as [1, 2]:

$$x(t+1) = Ax(t) + Bu(t) + B_d d(t) \tag{6.12}$$

$$y(t) = Cx(t) \tag{6.13}$$

where B_d is the matrix representing the influence of disturbances on the states.

Then, the state vector Eqs. (6.6) and (6.7) are modified to include the disturbances $\tilde{X}(t) = \left[x(t)\ d(t)\right]^T$ as [1]:

$$\begin{bmatrix} \tilde{x}(t+1) \\ d(t+1) \end{bmatrix} = \begin{bmatrix} R & N_d \\ 0 & I \end{bmatrix} \begin{bmatrix} \tilde{x}(t) \\ d(t) \end{bmatrix} + \begin{bmatrix} L \\ 0 \end{bmatrix} \Delta u(t) \tag{6.14}$$

$$y(t) = \begin{bmatrix} P & 0 \end{bmatrix} \begin{bmatrix} \tilde{x}(t) \\ d(t) \end{bmatrix} \tag{6.15}$$

where the matrix N_d is B_d with m (a number of inputs) additional rows of zeros.

In this section, a general form of MPC disturbances is briefly explained. Interested readers can find several techniques of adding MPC disturbances in [1, 2, 31].

6.4 MPC Constraints

In real practice, a control system is subject to constraints (limitations). The constraints must be constructed in the MPC optimization process. The general forms of constraints with respect to the control input and output is defined as [1, 2]:

$$u_{min} \leq u(t) \leq u_{max} \forall t \tag{6.16}$$

$$\Delta u_{min} \leq \Delta u(t) \leq \Delta u_{max} \forall t \tag{6.17}$$

$$y_{min} \leq y(t) \leq y_{max} \forall t \tag{6.18}$$

It is noted that the constraints in the state *x(t)* are integrated using the same inequalities as applied for output constraints. This makes the matrices C in Eq. (6.2) or P in Eq. (6.7) equal to the identity matrix.

In this section, a general form of MPC constraints is briefly explained. Interested readers can find additional techniques for creating MPC constraints in [1, 2, 31].

6.5 MPC-Based Virtual Inertia Control

In this section, a decentralized MPC-based virtual inertia control is proposed for enhancing system frequency stability [11]. The controller is designed for predicting the suitable virtual inertia constant at future events regarding inertia control-based frequency regulation. This controller can effectively decrease the effects of RES/DG penetration, load disturbances, and system parameter uncertainties. The fast output sampling (FOS) is applied to the approximation of RES/DG and load changes, which is modeled as a measured disturbance for the proposed MPC controller [31]. To reject such effects, the proposed MPC controller has implemented a feed-forward control technique instead of feedback control, as shown in Fig. 6.3. When the effect of disturbances can be measured or estimated, the controller uses a feed-forward compensation to attenuate their impacts over the system output. Unlike feedback control,

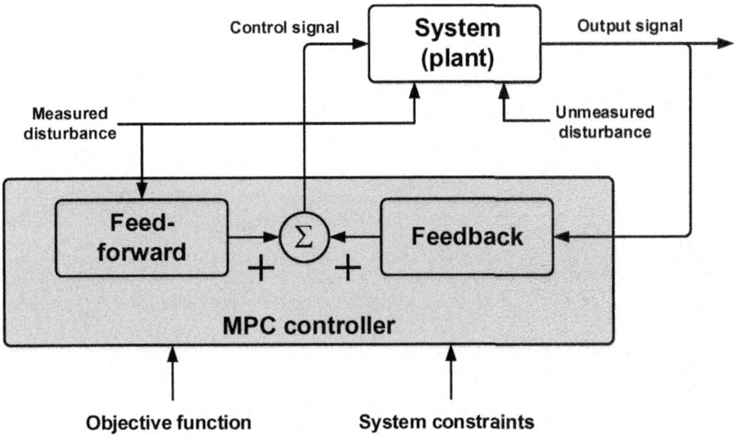

Fig. 6.3 MPC with feed-forward control

the feedforward control does not require to wait until the disturbance effect becomes obvious before performing corrected control actions. Consequently, the feed-forward control rejects the disturbance effect in the MPC controller more efficiently. By identifying a more precisely disturbance-output model, it results in a more efficient rejection of measured disturbances. Generally, there is difference between identified and exact models. Thus, the feed-forward control should be applied in combination with the feedback control to reduce the disturbance effects on the system response. The feed-forward control can be constructed in the MPC controller by adding the impacts of measured disturbances in the forecasts of future output.

Figure 6.4 shows the proposed control framework of the MPC-based virtual inertia control for enhancing system frequency stability. The MPC uses the system frequency deviations (Δf) and disturbances (ΔP_d) in RES/load changes to determine the suitable virtual inertia constant (K_{VI}) for generating the additional active power with required inertia and damping. In real practice, RES/DG and load changes in power systems are not practically measurable at every node or location. Thus, the RES/DG and load changes can be defined as the measured and unmeasured disturbances. Then, the FOS-based disturbance approximation technique is used to estimate the unmeasured disturbance. Later, the feed-forward control is applied to reject the effect of RES/DG and load disturbances.

The primary objective function of this MPC controller is to minimize system frequency oscillation and reject RES/DG and load disturbances. The calculated signal of the proposed MPC-based virtual inertia control can be computed as:

$$\Delta P_{VI}(t+1) = \sum_{j=1}^{N_P} \delta(j)\left(\Delta \overline{P_{VI}}(t+j) - w(t+j)\right)^2$$

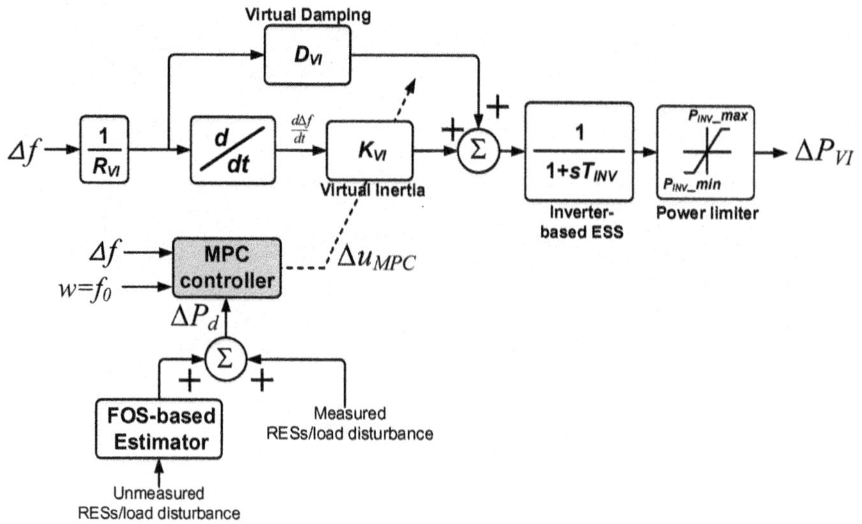

Fig. 6.4 Dynamic structure of the MPC-based virtual inertia control

$$+\sum_{j=1}^{N_C} \lambda(j)(\Delta u_{MPC}(t+j-1))^2 \tag{6.19}$$

subject to:

$$\Delta u_{MPC\_min} \leq \Delta u_{MPC} < \Delta u_{MPC\_max} \tag{6.20}$$

$$\Delta f_{min} \leq \Delta f < \Delta f_{max} \tag{6.21}$$

6.6 MATLAB-Based MPC

This section presents the implementation procedures for designing the MPC controller using MATLAB/Simulink® software [34]. The MPC controller block in Simulink® called "MPC control toolbox" offers an application with control functions for designing and simulating MPCs. The system (plant), disturbance models, constraints, horizons, and weights can be specified and designed in this toolbox via a function called "MPC Designer". The behavior of the MPC controller can be modified by changing weights and constraints at the processing time. By running closed-loop simulations, the performance of the MPC controller can be determined. This application is very useful for those who are not familiar with the C language. For embedded system implementation and rapid prototyping, the toolbox offers an automatic generation code of C language and IEC 61131-3 structured text generation.

In this section, the MPC control toolbox and associated objective functions are utilized to determine a suitable MPC control gain (u_{MPC}) for virtual inertia synthesis. The obtained MPC control gain represents the optimal virtual inertia constant (K_{VI}), which is required for evaluating the additional active power with required inertia and damping against changes in RES/DG and load disturbances. The design procedure for tuning a suitable MPC-based virtual inertia control is described as follows:

1. Open a 'Block Parameter' of the MPC controller (from 'Simulink Library Browser') and select a 'Design' tab for initialization and parameter configuration, see Fig. 6.5.
2. Then, a 'MPC Designer' window will appear as shown in Fig. 6.6. In the structure section, click a 'MPC Structure' tab and set the controller sample time. Next, choose a signal for the system input (e.g., measured/unmeasured disturbances or manipulated parameters). After that, click a 'Define and Linearize' tab for the controller initialization. By default, all system inputs are outlined as manipulated parameters and all system outputs are outlined as measured outputs.
3. The next process is to configurate a simulation scenario for testing the controller performance, see Fig. 6.7. Click a 'MPC Designer' tab, in the 'Edit Scenario' section, to test the MPC controller performance under a severe condition, in the 'Simulation duration', increase the simulation time to 20–30 s. Then, in the 'Reference Signals' dialog box, specify a 'Signal' as Gaussian. Later, in the 'Measured Disturbances' dialog box, specify a 'Signal' as Gaussian with the size

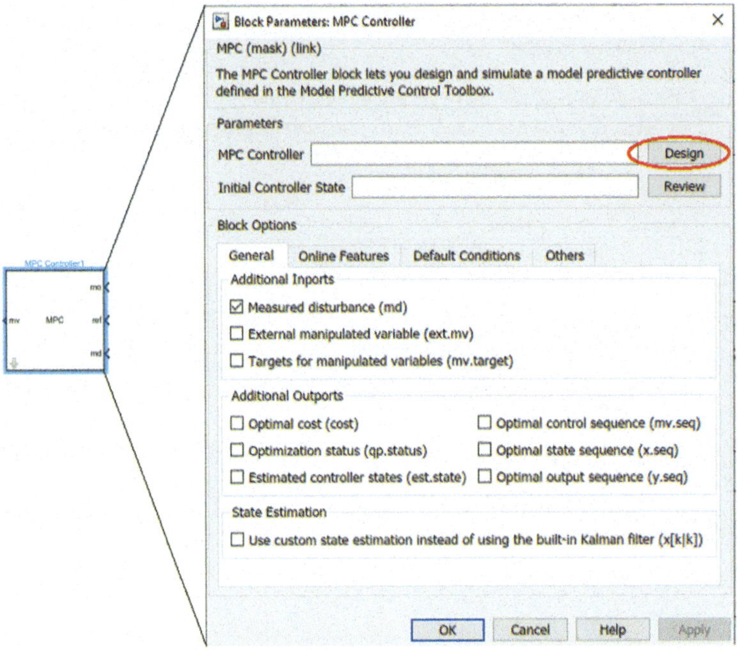

Fig. 6.5 Block parameter of the MPC controller

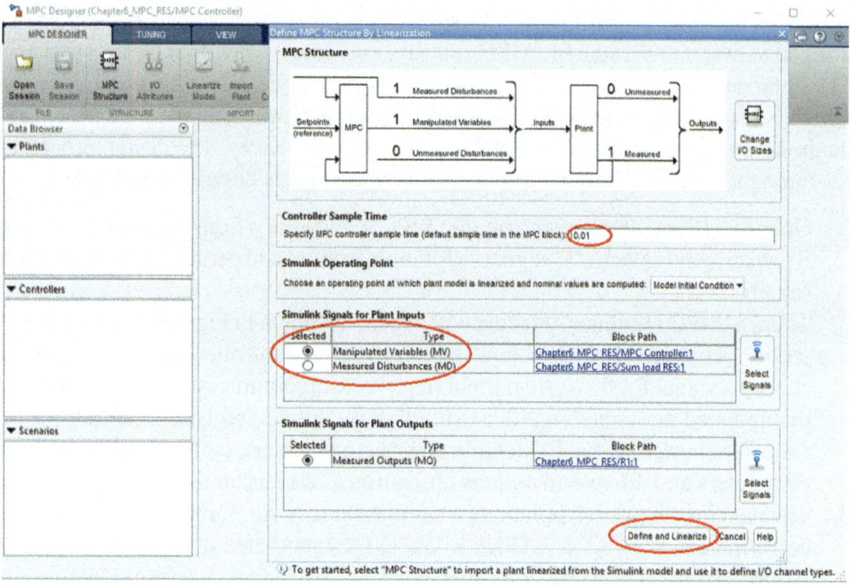

Fig. 6.6 'MPC Designer' window

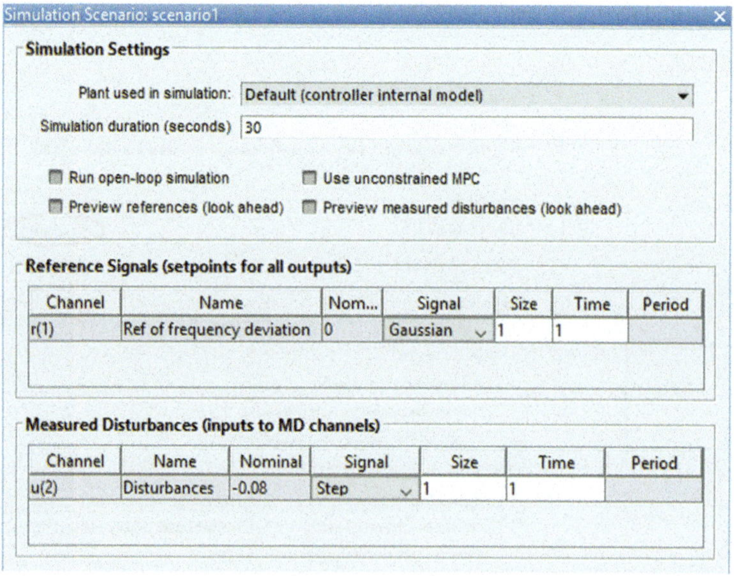

Fig. 6.7 Configuration of a simulation scenario

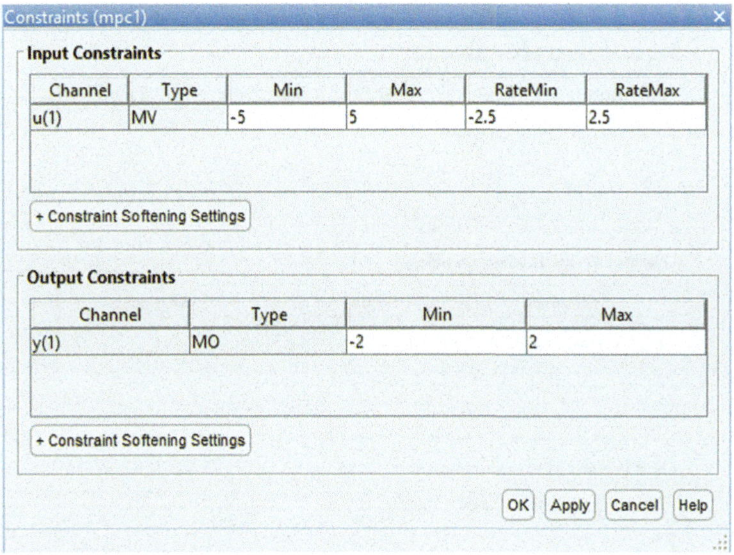

Fig. 6.8 Configuration of input/output constraints

of 1, and with a 'Time' of 1. In this study, it is noted that the measured disturbance is assigned to load units, and the unmeasured disturbance is assigned to RES/DG units.

4. To define input/output constraints, click a 'Performance Tuning' tab and then click a 'Constraints' tab, see Fig. 6.8. In the constraints dialog box, in the 'Input Constraints' section, enter the acceptable virtual inertia control gain (K_{VI}) upper and lower limits (in per-unit) in the 'Min' and 'Max' columns, respectively. In the 'Output Constraints' section, enter the acceptable frequency deviation upper and lower limits (in Hz) in the 'Min' and 'Max' columns, respectively.
5. To specify controller tuning weights, click a 'Performance Tuning' tab and then click a 'Tuning' tab, see Fig. 6.9. Increase the manipulated variables (MV) rate weight to 0.5. It is noted that increasing the MV rate weight penalizes large MV changes in the controller optimization cost function. For the 'Output Weights', keep the default values. This means all unmeasured outputs have zero weights.
6. To reduce system frequency oscillation, click a 'Performance Tuning' tab, see Fig. 6.10. Then, drag the closed-loop performance slider to the left until the system response (i.e., frequency deviation) has no overshoot. It is noted that moving the slider toward a left side simultaneously increases the manipulated parameter rate weight of the MPC controller and reduces the variable output weight, resulting in more robust control.
7. When the MPC performance is satisfied, click an 'Export Controller' tab to export the designed MPC controller into the associated power system model or MATLAB Workspace.

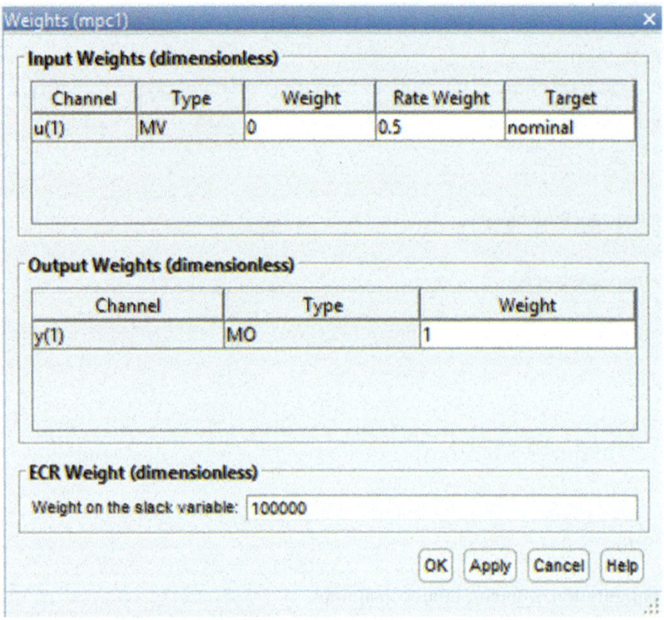

Fig. 6.9 Controller weight configuration

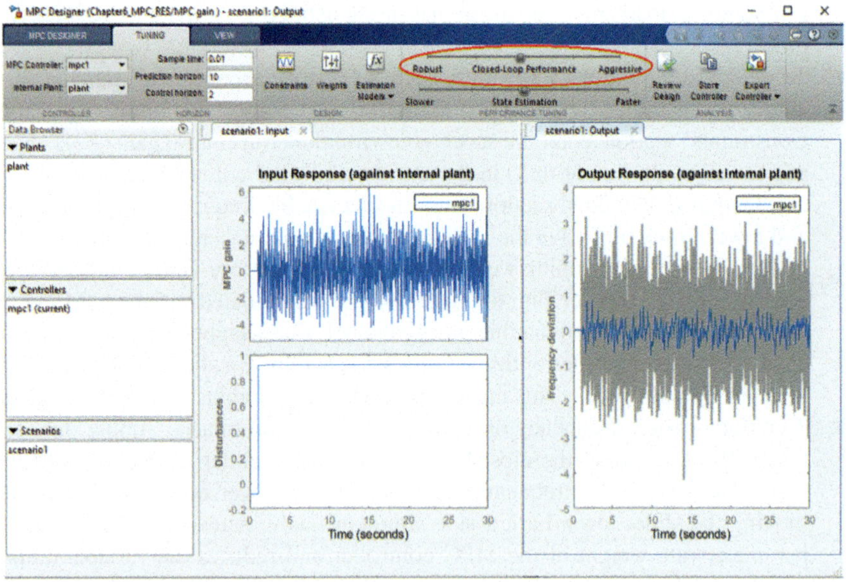

Fig. 6.10 MPC performance tuning

6.7 Simulation Results

The simulation results were carried out using MATLAB/Simulink® and tested through a single area power system in Fig. 3.9 with Table 3.1. Then, the effectiveness of the MPC-based virtual inertia control is verified under different scenarios of inertia and damping reductions, time delay, and high RES/load penetration. Then, the obtained results were compared with the PID-based virtual inertia control (obtained from Chap. 5) and conventional virtual inertia control (obtained from Chap. 3).

6.7.1 Efficacy of MPC-Based Virtual Inertia Control

In this section, the efficacy of the MPC-based virtual inertia controller is tested under normal operation; that is the high system inertia and damping property, see Table 3.2. The sudden load and RES/DG changes are considered as measured and unmeasured disturbances, respectively. The sudden load increase of 0.2 p.u. is applied at 2 s. Then, at 15 s, the sudden increase in solar power (0.1 p.u.) and wind power (0.1 p.u.) are simultaneously applied. The initial control parameters for the studied system are given in Table 3.1.

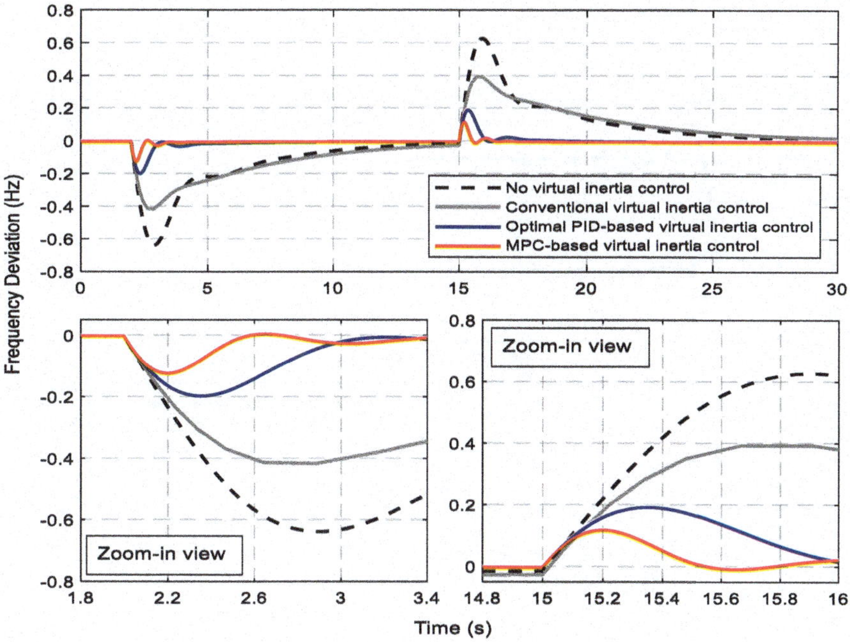

Fig. 6.11 Frequency response of the proposed MPC-based virtual inertia control

Figure 6.11 shows the frequency response of the system under normal operation. Due to the sudden load and RES disturbances, it yields the largest frequency drop and overshoot of ±0.65 Hz in the case of no virtual inertia control. By applying the conventional virtual inertia control (with fixed values in Table 3.1), the frequency drop and overshoot significantly reduce to ±0.40 Hz. The reduction is about 35% from the case of no virtual inertia control. By combining the optimal PID-based virtual inertia control (discussed in the previous chapter), the frequency drop and overshoot massively decrease to ±0.20 Hz due to its feedback control mechanism. Also, the stabilizing time enormously reduces. The reduction is about 50% from the case of conventional virtual inertia control. By applying the MPC-based virtual inertia control, it yields the lowest frequency drop and overshoot (about ±0.15 Hz) with the lowest stabilizing time due to its better ability in future prediction with regards to feedforward control for rejecting the disturbance. The reduction is about 25% from the case of PID-based virtual inertia control. It is obvious that the employment of the MPC results in the fastest dynamic response, which could allow the system in operating more closely to their limits toward the future prediction.

Figure 6.12 shows active power responses of the conventional generating unit and virtual inertia control unit. It is evident that the proposed MPC-based virtual inertia control provides the fastest power response with sufficient inertia (see the red line) compared with those cases of PID and conventional inertia control. Particularly, the MPC could generate sufficient power during the disturbances, leading to the lowest frequency drop and overshoot. Figure 6.13 shows the deviation of MPC control output (Δu_{MPC}) against the disturbances under normal operation.

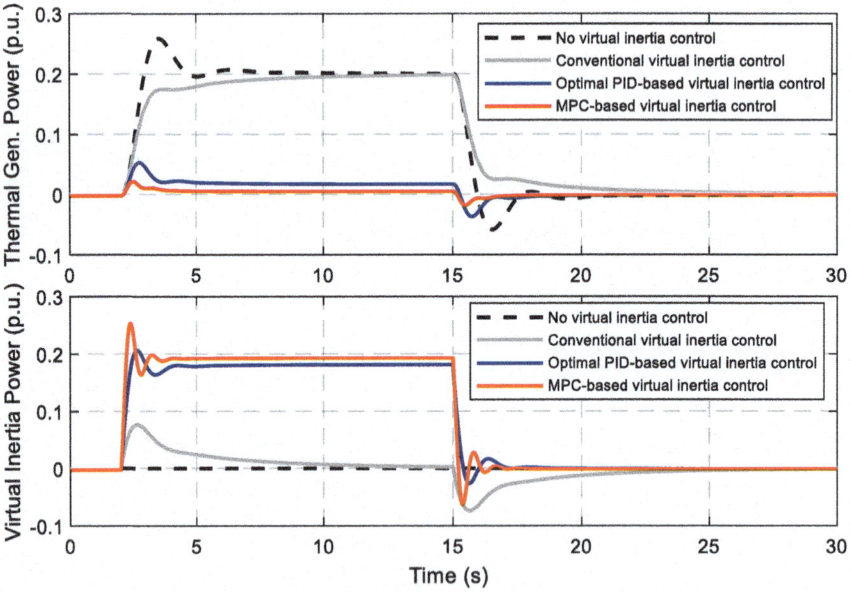

Fig. 6.12 Power responses (deviation) of the MPC-based virtual inertia control

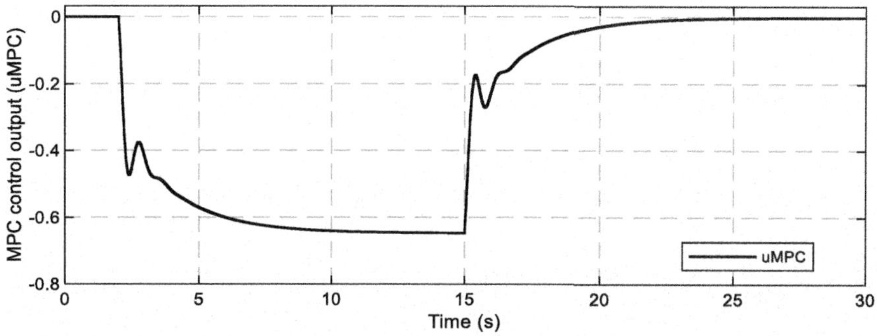

Fig. 6.13 Control output signal of MPC under normal operation

6.7.2 Robustness Against Inertia and Damping Reduction

In this section, the efficiency of the MPC-based virtual inertia controller is tested under off-normal operation regarding the vast reduction of system inertia and damping property (i.e., parameter uncertainty) driven by power electronics interfaces-based RESs/DGs penetration. The system inertia and damping property are reduced to 70% of its nominal values, as shown in Table 3.2. The sudden load increase of 0.2 p.u. is applied at 2 s. Then, at 15 s, the sudden increase in solar power (0.1 p.u.) and wind power (0.1 p.u.) are simultaneously applied.

Figure 6.14 shows the frequency response of the system under the degraded condition of system inertia and damping. As the system inertia and damping reduce, the system response is more oscillating with larger overshoot, resulting in longer stabilizing time. During the sudden disturbance injections, the largest frequency drop and overshoot of ±1.05 Hz occur in the case of no virtual inertia control. Compared with the normal operation, the frequency amplitude increases about 60% when the system inertia and damping drop to 70% of its nominal values. By adding the traditional virtual inertia control, the frequency drop and overshoot significantly reduces to ±0.52 Hz. The reduction is about 50% from the case of no virtual inertia control. By applying the optimal PID-based virtual inertia control, the frequency drop and overshoot slightly decrease to ±0.40 Hz due to its feedback control mechanism. In addition, the stabilizing time significantly reduces. The reduction is about 20% from the case of conventional virtual inertia control. By applying the MPC-based virtual inertia control, it generates the lowest frequency drop and overshoot (about ±0.25 Hz) with the fastest stabilizing time due to its outstanding ability in future prediction with regards to disturbance rejection. The reduction is about 38% from the case of PID-based virtual inertia control. Thus, under the degraded condition of system inertia and damping reductions, it is confirmed that the employment of the MPC could maintain the fastest dynamic response with stable operation against the parameter uncertainty, which indicates the system robustness.

Figure 6.15 shows the active power response of the conventional generating and virtual inertia control units under the degraded situation. It is obvious that the MPC-based virtual inertia control could offer the fastest power response with sufficient

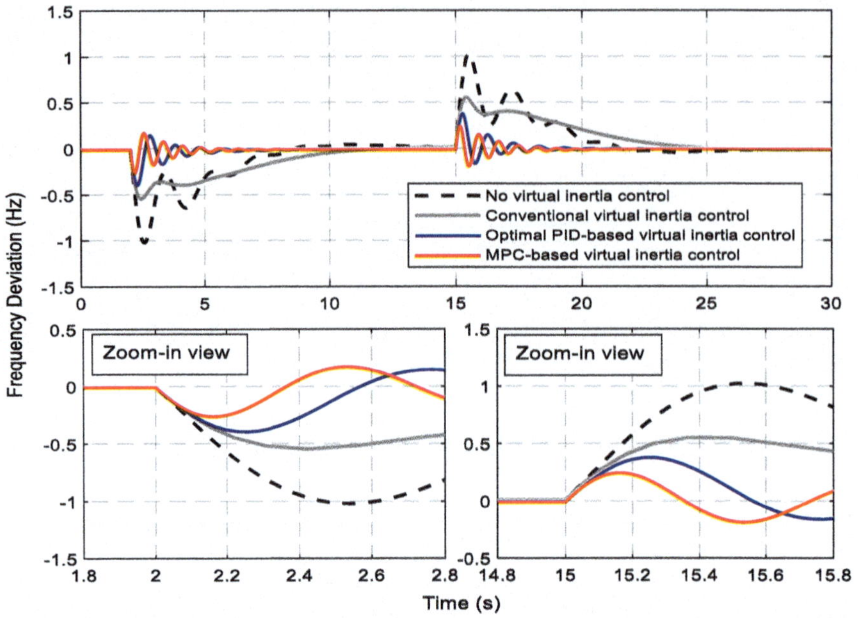

Fig. 6.14 Frequency response of the proposed MPC-based virtual inertia control under the effect of system inertia and damping reductions

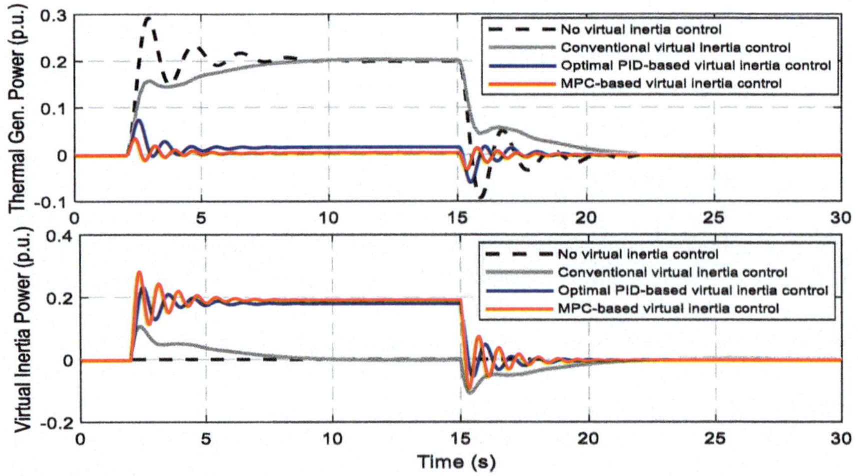

Fig. 6.15 Power responses (deviations) of the MPC-based virtual inertia control under the effect of system inertia and damping reductions

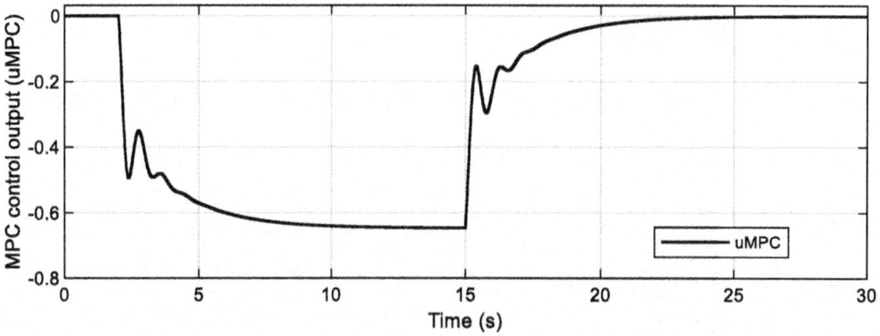

Fig. 6.16 Control output signal of MPC under the reductions of system inertia and damping

inertia and damping (see the red line) compared with those cases of PID and traditional inertia control. Under the degraded situation, it is confirmed that the MPC could generate the sufficient power over the parameter uncertainty, which leads to lower stress in the conventional generating unit. Figure 6.16 shows the deviation of MPC control output (Δu_{MPC}) against the inertia and damping variations.

6.7.3 Robustness Against Time Delay

In the real practice of a frequency control system, fast responses and changing devices of frequency are nearly unobservable according to several filters and delays related in the procedure [35]. Any signal filtering and processing create delays that must be analyzed. Practical filters on the secondary control (area control error: ACE) and tie-line measurements use about 1–2 s or more for the decision cycles and data acquisition of the frequency control system [31].

Currently, the communication delays in frequency control analysis have been received a more critical challenge according to the expanding and restructuring of functionality, complexity, physical setups of power systems. During the last decades, research works on frequency control analysis have ignored issues related to the communication network. Under such conventional communication links, it was considered as a valid assumption. Hence, the use of a communication infrastructure to assist ancillary services (e.g., secondary control or virtual inertia control) under the deregulated conditions is increasing concerns on the issues that may raise in the communication system.

Thus, in this section, the efficiency of the MPC-based virtual inertia control is tested via the effect of communication delay time. The time delay units have been added to the input of secondary control and virtual inertia control systems, as depicted in Fig. 6.17. The t_s is the communication delay time for secondary control, and t_{VI} is the communication delay time for virtual inertia control, where $t_s = 1$ s, and $t_{VI} = 0.5$ s. The rest of the initial control parameters are listed in Table 3.1. The system

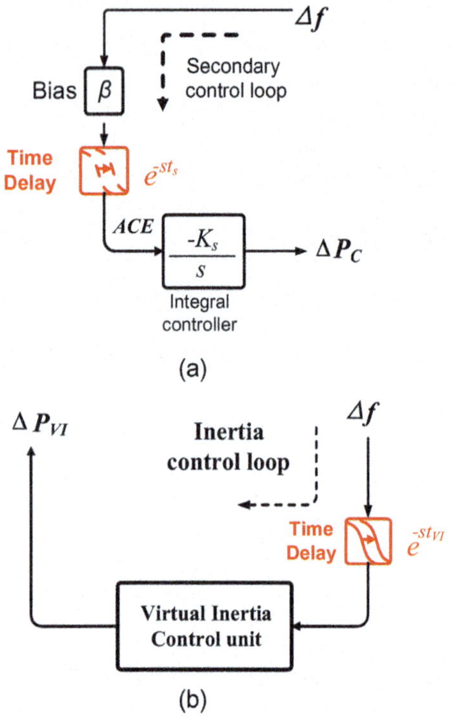

Fig. 6.17 Configuration of time delays in; **a** secondary control; **b** virtual inertia control

is operated under normal operation; that is the high system inertia and damping property (100%), see Table 3.2.

Figure 6.18 displays the frequency response of the system under the effect of time delay. It is obvious that the increasing time delay causes a longer stabilizing time and increases the frequency amplitude during the disturbances. The larger frequency drop and overshoot can be observed in the case of PID-based virtual inertia control. However, in the case of MPC-based virtual inertia control, it yields almost the similar behavior of frequency response compared with the case of MPC control without the time delay. This ensures that the MPC can provide remarkable performance in maintaining robust system stability under the increasing effect of communication delay time.

6.7.4 Robustness Against High Penetration of Renewables

In this section, the robustness of the MPC-based virtual inertia control is tested via high power penetration of wind, solar, industrial, and residential loads (see Fig. 6.19). To make the system more drastic, the communication delay units of $t_s = 1$ s and t_{VI} = 0.5 s have been added to this scenario. Moreover, the system is operated under the

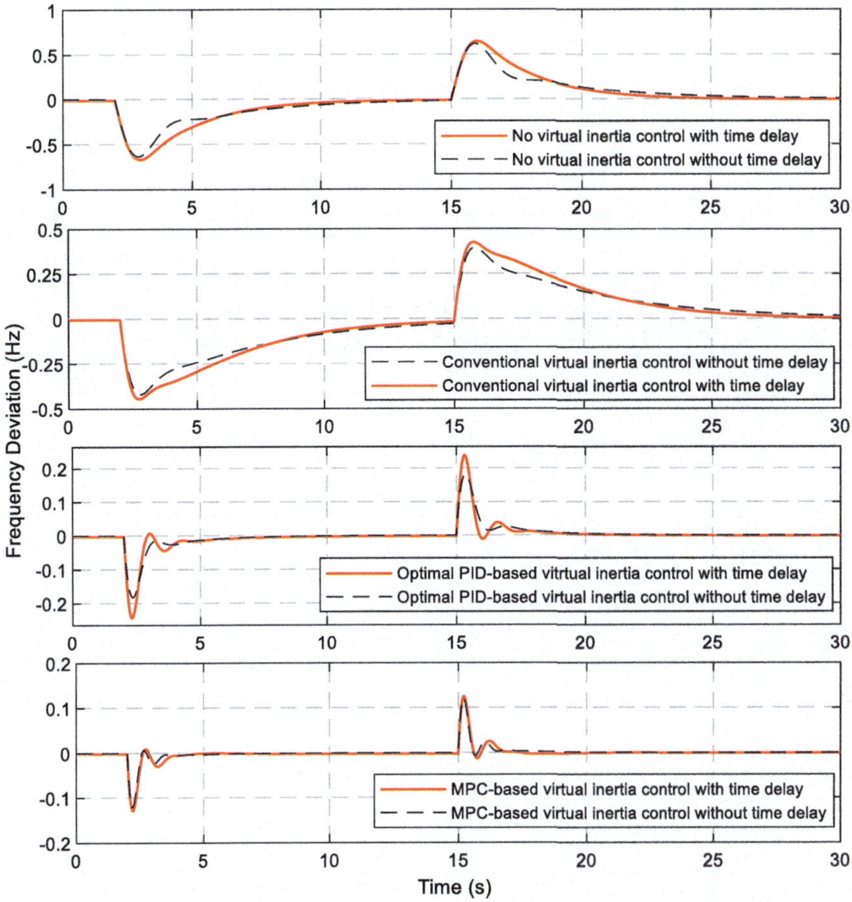

Fig. 6.18 Frequency response of the MPC-based virtual inertia control under the effect of time delay

severe condition of low system inertia and damping property (40% reduction from its nominal values), see Table 3.2. At the initial condition, the power is generated by the solar and thermal generating units and fed to the industrial and residential loads.

Figure 6.20 displays the frequency response of the system under the severe operating condition. At 200 s, the wind generating unit is connected to the system. It causes the large frequency overshoots of +0.6 Hz (in the case of no virtual inertia control), +0.3 Hz (in the case of conventional inertia control), +0.6 Hz (in the case of PID-based virtual inertia control), and +0.14 Hz (in the case of MPC-based virtual inertia control). Following the severe and continuous disturbances toward the nonlinearity, the PID-based virtual inertia control fails to stabilize the system frequency. It yields the larger frequency nadir and overshoot than the conventional virtual inertia control. In addition, the PID controller requires the resetting in respect to the applied disturbances. Clearly, the MPC could effectively arrest the disturbances due to its

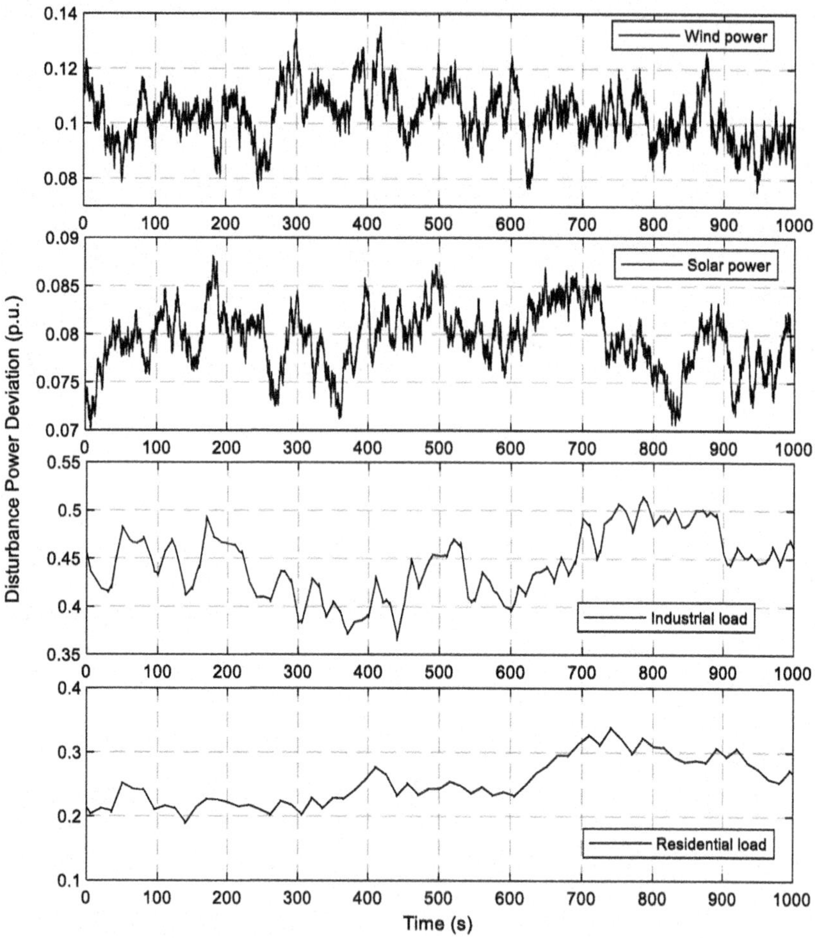

Fig. 6.19 High power disturbances in the presence of wind speed, solar irradiation, industrial and residential load consumption

ability in future prediction in combination with feedforward control in rejecting the disturbance, leading to the lowest frequency nadir and overshoot (see Fig. 6.21). At 800 s, the solar generating unit is disconnected from the system, causing the large disturbance. The frequency drops of −0.35 Hz (in the case of no virtual inertia control), −0.25 Hz (in the case of conventional inertia control), −0.32 Hz (in the case of PID-based virtual inertia control), and −0.12 Hz (in the case of MPC-based virtual inertia control) can be observed. Figure 6.22 shows the deviation of MPC control output (Δu_{MPC}) against the critical situation.

Evidently, during the severe and continuous disturbances, the proposed MPC can adjust to the new circumstance by changing its structure based on future prediction, allowing normal operation with degrees of freedom and robustness. Nevertheless,

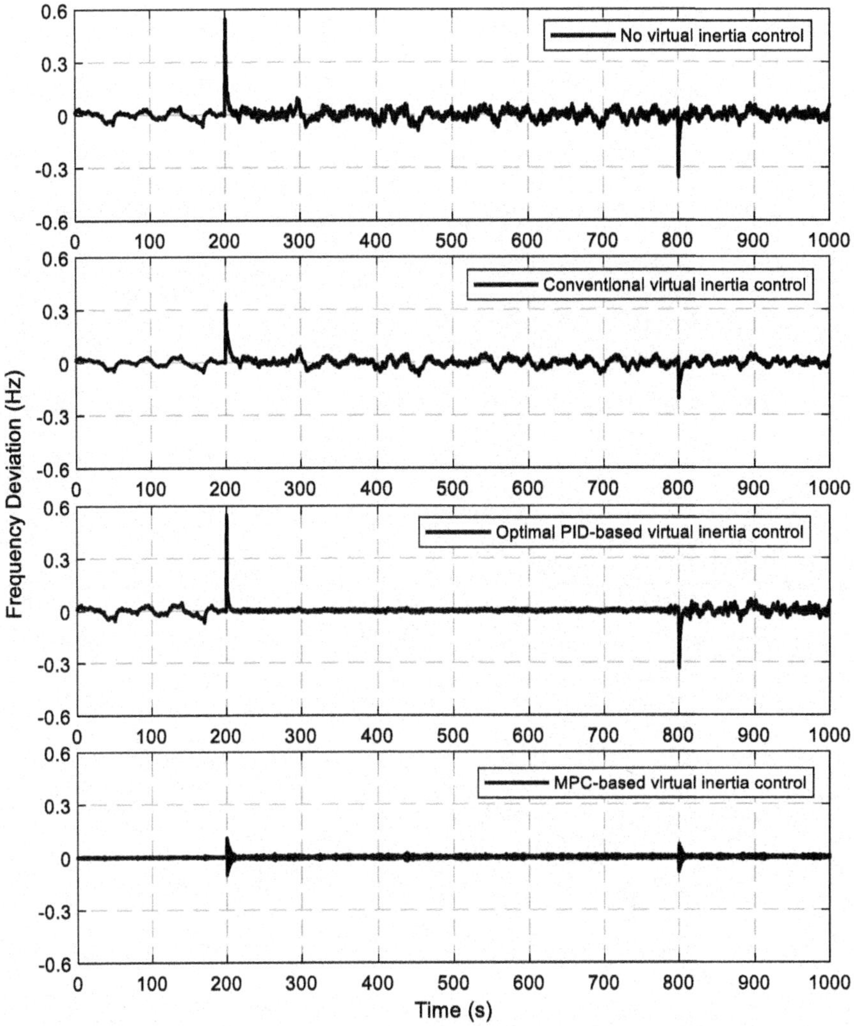

Fig. 6.20 Frequency response under severe disturbances

during normal/steady-state operation (no large disturbance event), the proposed MPC could maintain more stable frequency oscillation.

As a result, it is confirmed that the intermittency and disturbance of RESs, DGs, and controllable loads are designed as control variables in the MPC controller. Therefore, the proposed MPC-based virtual inertia control can overcome a wide range of disturbance and uncertainty with the robust operation. It is ensured that the employment of the proposed MPC could maintain the fastest dynamic response with the robust system operation, which is applicable for real-time utilization in smart/micro-grids.

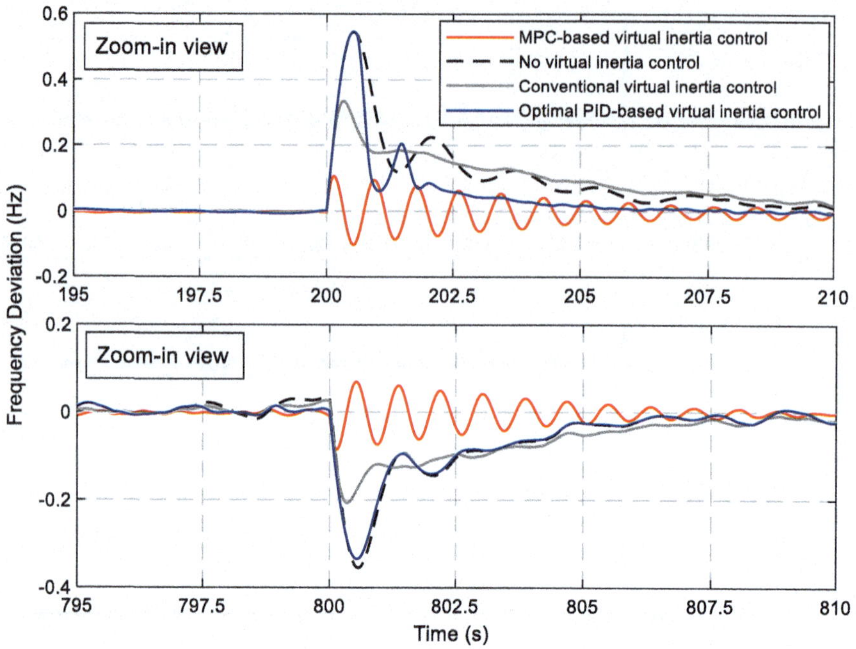

Fig. 6.21 Zoom-in views of frequency dip and drop under severe disturbances

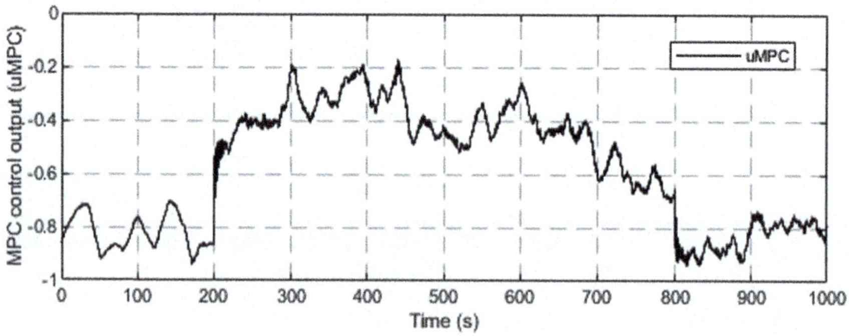

Fig. 6.22 Control output signal of MPC under the severe disturbances

6.8 Summary

In this chapter, the development of decentralized MPC-based virtual inertia control was introduced to emulate the suitable virtual inertia power, while forecasting the frequency events for real-time or continuous control applications in smart/micro-grids. The MPC controller used the future prediction ability in combination with a feedforward control strategy to reject the disturbances from RES/DG and load penetration as well as system parameter uncertainty, ensuring rapid dynamic response with the robustness of operation. The proposed MPC-based virtual inertia control was verified through a nonlinear control area system under a wide range of parameter variations, load conditions, high RESs penetration, and communication delays. The obtained results were compared with the PID-based virtual inertia control and traditional virtual inertia control systems.

References

1. C. Bordons, F. Garcis-Torres, M.A. Ridao, *Model Predictive Control of Microgrids* (Springer, Switzerland, 2020)
2. V. Yaramasu, B. Wu, *Model Predictive Control of Wind Energy Conversion Systems* (IEEE-Wiley, NJ, USA, 2016)
3. G. Serale, M. Fiorentini, A. Capozzoli, D. Bernardini, A. Bemporad, Model Predictive Control (MPC) for enhancing building and HVAC system energy efficiency: problem formulation, applications and opportunities. Energies **11**(3), 1–35 (2018)
4. T. Dragicevic, Model predictive control of power converters for robust and fast operation of AC microgrids. IEEE Trans. Power Electron. **33**(7), 6304–6317 (2018)
5. A.N. Venkat, I.A. Hiskens, J.B. Rawlings, S.J. Wright, Distributed MPC strategies with application to power system automatic generation control. IEEE Trans. Control Syst. Technol. **16**(6), 1192–1206 (2008)
6. A.M. Ersdal, L. Imsland, K. Uhlen, Model predictive load-frequency control. IEEE Trans. Power Syst. **31**(1), 777–785 (2016)
7. E. Perez, H. Beltran, N. Aparicio, P. Rodriguez, Predictive power control for PV plants with energy storage. IEEE Trans. Sustain. Energy **4**(2), 482–490 (2013)
8. Y. Zong, L. Mihet-Popa, D. Kullmann, A. Thavlov, O. Gehrke, H.W. Bindner, Model predictive controller for active demand side management with PV self-consumption in an intelligent building, in Proc. *IEEE PES Innovative Smart Grid Technologies Conference Europe*, 1–6 (2012)
9. G. Li, M.R. Belmont, Model predictive control of sea wave energy converters - Part II: the case of an array of devices. Renew. Energy **68**(1), 540–549 (2014)
10. T.H. Mohamed, J. Morel, H. Bevrani, T. Hiyama, Model predictive based load frequency control-design concerning wind turbines. Int. J. Electr. Power Energy Syst. **43**(1), 859–867 (2012)
11. T. Kerdphol, F.S. Rahman, Y. Mitani, K. Hongesombut, S. Küfeoğlu, Virtual inertia control-based model predictive control for microgrid frequency stabilization considering high renewable energy integration. Sustain **9**(5), 1–21 (2017)
12. H. Borhan, A. Vahidi, A.M. Phillips, M.L. Kuang, I.V. Kolmanovsky, S. Di Cairano, MPC-based energy management of a power-split hybrid electric vehicle. IEEE Trans. Control Syst. Technol. **20**(3), 593–603 (2012)
13. J. Ma, S.J. Qin, B. Li, T. Salsbury, Economic model predictive control for building energy systems, in Proc. *IEEE PES Innovative Smart Grid Technologies Conference Europe*, 1–6 (2011)
14. R. Halvgaard, N.K. Poulsen, H. Madsen, J.B. Jørgensen, Economic model predictive control for building climate control in a smart grid, in Proc. *IEEE PES Innovative Smart Grid Technologies*, 1-6 (2012)
15. L. Xie, M.D. Ilić, Model predictive economic/environmental dispatch of power systems with intermittent resources, in Proc. *IEEE PES General Meeting (IEEE PES GM)*, 1–6 (2009)

16. C. Finck, R. Li, W. Zeiler, Economic model predictive control for demand flexibility of a residential building. Energy **176**(1), 365–379 (2019)
17. R. Van Parys, G. Pipeleers, Distributed MPC for multi-vehicle systems moving in formation. Rob. Auton. Syst. **97**(1), 144–152 (2017)
18. R.R. Negenborn, B. De Schutter, H. Hellendoorn, Multi-agent model predictive control of transportation networks, in Proc. *IEEE International Conference on Networking, Sensing and Control*, 296–301 (2006)
19. S. Roshany-Yamchi, R.R. Negenborn, M. Cychowski, B. De Schutter, J. Connell, K. Delaney, Distributed model predictive control and estimation of large-scale multi-rate systems, in Proc. *IFAC Proceedings Volumes*, 416–422 (2011)
20. V. Javalera, B. Morcego, V. Puig, A multi-agent MPC architecture for distributed large scale systems, in Proc. *International Conference on Agents and Artificial Intelligence*, 1–9 (2010)
21. L. Jin, R. Kumar, N. Elia, Model predictive control-based real-time power system protection schemes. IEEE Trans. Power Syst. **25**(2), 988–998 (2010)
22. J.A. Martin, I.A. Hiskens, Corrective model-predictive control in large electric power systems. IEEE Trans. Power Syst. **32**(2), 1651–1662 (2017)
23. L. Jin, R. Kumar, N. Elia, Application of model predictive control in voltage stabilization, in Proc. *American Control Conference*, 5916–5921 (2007)
24. A. Bonfiglio et al., Improving power grids transient stability via model predictive control, in Proc. *Power Systems Computation Conference (PSCC 2014)*, 1–7 (2014)
25. I. Martinez Sanz, B. Chaudhuri, G. Strbac, Corrective control through HVDC links: a case study on GB equivalent system, in Proc. *IEEE PES General Meeting (IEEE PES GM)*, 1–5 (2013)
26. A. Bonfiglio et al., An MPC-Based approach for emergency control ensuring transient stability in power grids with steam plants. IEEE Trans. Ind. Electron. **66**(7), 5412–5422 (2019)
27. P. Xu, D. Song, W. Tang, C. Yang, S. Ma, "Coordinated frequency and voltage correction via model predictive control," in Proc. *International Conference on Power System Technology (POWERCON 2018)*, 1–6 (2019)
28. X. Kong, X. Liu, K.Y. Lee, Nonlinear multivariable hierarchical model predictive control for boiler-turbine system. Energy **93**(1), 309–322 (2015)
29. X. Liu, D. Jiang, K.Y. Lee, Quasi-min-max fuzzy MPC of UTSG water level based on off-line invariant set. IEEE Trans. Nucl. Sci. **62**(5), 2266–2272 (2015)
30. T.H. Mohamed, H. Bevrani, A.A. Hassan, T. Hiyama, Decentralized model predictive based load frequency control in an interconnected power system. Energy Convers. Manag. **52**(2), 1208–1214 (2011)
31. H. Bevrani, *Robust Power System Frequency Control*, 2nd ed. (Springer, New York, USA, 2014)
32. J. Pahasa, I. Ngamroo, Coordinated control of wind turbine blade pitch angle and PHEVs using MPCs for load frequency control of microgrid. IEEE Syst. J. **10**(1), 97–105 (2016)
33. J. Pahasa, I. Ngamroo, PHEVs bidirectional charging/discharging and SoC control for microgrid frequency stabilization using multiple MPC. IEEE Trans. Smart Grid **6**(2), 526–533 (2015)
34. L. Wang, *Model Predictive Control System Design and Implementation Using MATLAB®*. (Springer-Verlag, London, UK, 2009)
35. F. Milano, M. Anghel, "Impact of time delays on power system stability. IEEE Trans Circuits Syst. I Regul. Pap. **59**(4), 889–900 (2012)

Chapter 7
Fuzzy Logic Control for Virtual Inertia Synthesis

Abstract Currently, renewable energy sources (RESs) and distributed generators (DGs) are highly integrated into power systems regarding energy crisis, environmental concerns, and economic growth. The RESs/DGs penetration brings more complexity to the power system, since its systems are decentralized, and power outputs are intermittent or unpredictable against time-varying. Nevertheless, the RESs/DGs may not participate in stability regulation (e.g., frequency/voltage control), which causes the lack of inertia and damping property to the system, resulting in the weakening of grid stability. This situation can lead to system instability, cascading failures, and power blackouts. To deal with this problem, the system requires high-level (advanced) inertia controllers in tracking various levels of RESs/DGs penetration. Fuzzy logic control can be considered as one of the solution techniques due to the high reliability in nonlinear modeling with the fast processing time. In this chapter, a fuzzy logic technique is integrated into a virtual inertia control loop to enable the self-adaptive ability of virtual inertia constant against the different levels of RESs/DGs penetration regarding frequency control. As a result, the virtual inertia control unit can automatically adjust itself in emulating different amounts of inertia and damping responding the integrated levels of RESs/DGs at the specific time. At the beginning, the fundamental of fuzzy logic is discussed, and the recent achievements in fuzzy applications for frequency control problems are briefly reviewed. Then, a decentralized fuzzy controller in scheduling virtual inertia control constant is designed. Lastly, the effectiveness of the proposed control scheme is demonstrated through a nonlinear simulation under wide ranges of critical RESs/DGs penetration regarding system inertia and damping variations.

Keywords Fuzzy logic · Frequency control · Dynamic control · Inertia control · Nonlinear control · Parameter tuning · Uncertainty · Virtual inertia synthesis

T. Kerdphol et al., *Virtual Inertia Synthesis and Control*, Power Systems,
https://doi.org/10.1007/978-3-030-57961-6_7

7.1 Introduction to Fuzzy Logic Control

Fuzzy logic control has been credited with being a remarkable-accepted technique for designing controllers that can provide satisfactory performance under the presence of imprecision and uncertainty [1]. The fuzzy theory offers a special technique for less-skilled personal to create practical control techniques in a friendly way, which is close to human perception and thinking with a short amount of processing time. The fuzzy logic control can even create better performance than the PID controller. This is because the fuzzy control is a nonlinear controller and does not need a mathematical model of the control system. According to its nature of adaptability, the fuzzy logic can provide control actions in both linear and nonlinear systems [2]. In fact, most of the systems are nonlinear, but they are linearized for reducing complexity and offering simple understanding and so on. The linearization process could reduce the accuracy of systems, causing operational errors. These errors lead to the degraded performance of PID controllers, which required a mathematical model for the design process. For fuzzy controllers, the errors caused by linearization could not affect its performance as the fuzzy controllers do not require the mathematical model to generate the decision-making process. Subsequently, the fuzzy logic becomes an alternative control for nonlinear systems. The nonlinearity is compensated via fuzzy membership functions, rules, and decision-making processes. Choosing rules with reality allow the fuzzy system to deal with nonlinear systems better than conventional control techniques. Moreover, most of control systems have multiple inputs and variables which are needed to be designed and modeled, leading to complexity and time-consuming. The fuzzy rules can include nonlinear property so that the fuzzy system simplifies the application by combining multiple inputs with the fuzzy expression of "if-then" rule. This process is known as the implication, which performs the relation of fuzzy subsets used in the fuzzy rules. The fuzzy set is not different from the probability logic or Boolean logic. The difference of fuzzy is the outstanding ability to perform a more general theory of decision process (i.e., fuzzy process). The fuzzy process is special for approximating reasons (i.e., fuzzy reasoning), which uses the fuzzy set in combination with the fuzzy theory. The fuzzy reasoning is defined by fuzzy words that we use in our daily life, such as high, low, large, small, positive, negative, and so on. As a result, the fuzzy logic is able to control real-world systems (i.e., nonlinear systems) with minimal to no knowledge in mathematical models of the control process or plant.

Fuzzy logic control is widely integrated into several real-world applications. The first fuzzy logic controller was created by Mamdani and Assilian in 1975 for regulating a steam generator [3]. In 1976, the fuzzy logic was applied to a cement kiln controller [4]. In 1980s, fuzzy logic control applications were effectively implemented in several industries in Japan, including a water treatment system established by the Fuji electric company. In 1987, a fuzzy-based automatic train operation/control system was invented by the Hitachi company, and this system was practically implemented in the Sendai city's subway [5]. This implementation was motivated several

Japanese designers and engineers to utilize a wide range of fuzzy applications, leading to an era of a fuzzy boom in Japan [6]. In 1988, the Japanese Ministry of Internal Trade and Industry (MITI) had initiated large-scale research with collaboration between industries and universities. In 1990, the Panasonic company developed a washing machine-based fuzzy control and released a commercial campaign as a "fuzzy product". This campaign came out as a popular product for fuzzy technology [7]. Numerous home electronics companies tracked the Panasonic's way and developed fuzzy rice cooking, fuzzy vacuum cleaner, fuzzy camcorders for smoothing an image under hang shaking, fuzzy refrigerators, fuzzy camera for autofocus, and so on. As a result, the era of "fuzzy boom" has triggered a broad interest of fuzzy technology in the United States, Europe, Korea, and around the world. For example, several institutes and companies such as NASA [8] and Boeing [9] applied fuzzy logic for aviation and space applications. Until now, the fuzzy logic has been implemented in real-world applications, which is close to people lives in terms of small to large applications such as, smartphones, digital cameras, microwaves, televisions, refrigerators, washing machines, cars, elevators, vessels, aircrafts, heavy industry (cement, steel, petroleum), and so on [10].

Focusing on power system control aspects, the fuzzy logic control has been applied to several fields, such as bus voltage control, stability control, parameter estimation, power flow analysis, protection systems and so on [11–20]. The survey details are provided in [21–24]. Numerous studies are reported for fuzzy design-based frequency control in [25–37]. The fuzzy logic controllers are designed for secondary control in [27, 31, 33, 37, 38], energy storage control in [26, 33, 39–43], and RESs (e.g., wind and solar power utilization) in [15, 22, 44–48]. The applications of fuzzy-based frequency control are mainly classified into four types, as follows:

- Using fuzzy logic control as a dynamic controller
- Using fuzzy logic control in the form of a proportional-integral (PI) or proportional-integral-derivative (PID) controllers
- Using fuzzy logic control for other control aspects, such as economic dispatching
- Using fuzzy logic control as a primer for tuning gains of control systems.

Considering virtual inertia synthesis with respect to frequency control, several research works have used a fixed value of virtual inertia constant for countering to all integration levels of DGs/RESs [49–57]. The selection of virtual inertia constant is based on a tradeoff or graphical analysis-based the dynamic performance and response time of the system. This demands the extensive eigenvalue sensitivity analysis, which might result in unsuitable selection and errors, causing system instability. In addition to virtual inertia constant, system frequency and power deviate after RESs/DGs integration similar to synchronous machines, but the transient tolerance of power electronics-based RESs/DGs is much less than actual synchronous machines. Thus, implementing the fixed virtual inertia constant affects system frequency stability and performance under various integrating amounts of RESs/DGs. Larger frequency deviations after the disturbances might happen according to the fixed virtual inertia constant (i.e., unsuitable selection), weakening performance, stability, and resiliency of the system. To overcome such problem, a new idea of self-adaptive virtual inertia control is proposed by a few researchers in [58–61]. In [59], the authors proposed the virtual inertia control with an alternating moment of inertia to overcome power fluctuations of the system. In [61], the authors designed a self-tuning-based virtual inertia control technique to enhance frequency stability of the system. In [60], the adaptive virtual inertia control method concerning the impact of damping factor on frequency dynamics was proposed. These aforementioned methods are difficult to apply, and the high RESs/DGs effect has not been included or tested in the design process. The high levels of RESs/DGs penetration can trigger the self-adaptive inertia control techniques to be unstable to deliver an appropriate performance over a wide range of operation. Consequently, the inertia control systems might be inadequate and unstable after high-level integration of RESs/DGs, resulting in system instability. In the worst case, the inertia control systems can be interrupted due to fluctuations with high amplitude after RESs/DGs integration. This situation can lead to cascading failures and power blackouts.

To resolve the challenge, this chapter develops a decentralized fuzzy-based virtual inertia controller, which results in the effective system frequency stabilization against different levels of RESs/DGs integration. In this idea, the fuzzy logic control is applied as an automatic primer for tuning the suitable virtual inertia constant, which is the main function for inertia emulation. The proposed fuzzy system exploits the signals of RESs/DGs power injection and system frequency deviation to determine the fuzzy rules-based inertia emulation for enabling self-adaptive inertia response. Subsequently, the virtual inertia constant is automatically modified based on the measured input signals to prevent unsuitable selection and offer quick inertia response. Finally, the proposed fuzzy-based virtual inertia control is verified through a nonlinear control system under wide ranges of critical parameter uncertainties and high penetration of RESs/DGs and loads. It is noted that this chapter is an extended version of the research presented in [58].

7.2 Fundamental Fuzzy Logic

This section explains fundamental structures with definitions related to fuzzy logic control, which are required for basic understanding used in its design process. The important structures are described as follows.

7.2.1 Fuzzy Set

Fuzzy membership functions (inputs) are considered as a connection between the uncertain data and fuzzy system. The concept of fuzzy membership functions usually represents by a set, which contains a collection of intangible or tangible objects (e.g., shape, color). In classical logic, these sets have accurate boundaries. However, in fuzzy logic, this requirement can be compromised; thus, the fuzzy set boundaries

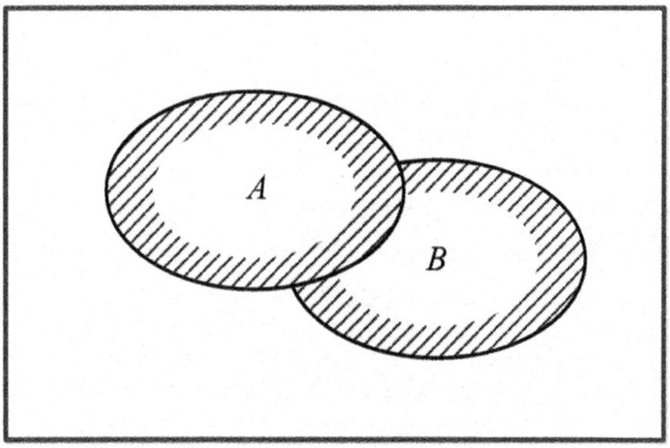

Fig. 7.1 Fuzzy intersection based on a logic diagram

are imprecise. Considering Fig. 7.1, the fuzzy sets *A* and *B* on Venn (logic) diagram can be seen as shading, which represents the boundary zone. The imprecise shaded zones expressing the set boundaries are the zones, where the partial membership $\mu$of the sets by elements is qualified. The μ is usually defined as $0 < \mu < 1$. Figure 7.2 displays the comparison between the classical logic and fuzzy logic, where elements are the members of a continuum. The element values are defined along the x-axis,

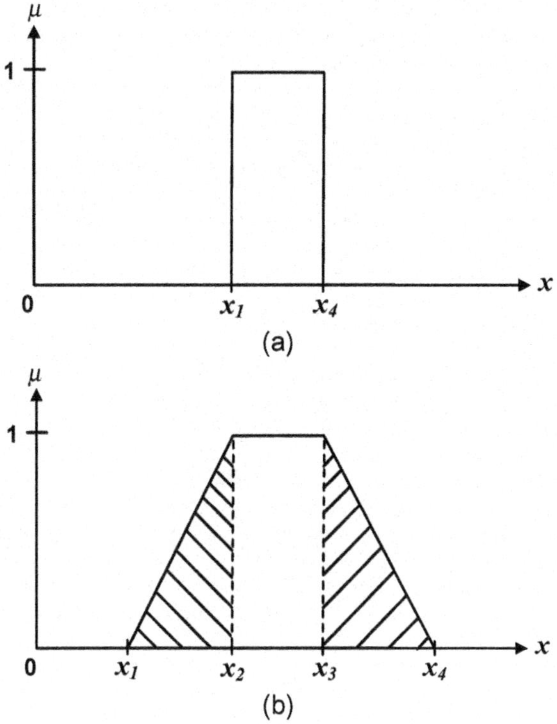

Fig. 7.2 Typical set membership patterns: **a** classical set; **b** fuzzy set

and the membership values of the elements are defined along the y-axis. In the case of classical logic, the membership is zero during $0 < x < x_1$ or $x < x_4$ and the membership is unity (i.e., one) during $x_1 < x < x_4$. In the case of fuzzy logic, the membership is zero during $0 < x < x_1$ or $x < x_4$. However, during $x_1 < x < x_2$, the membership can be increased from zero to unity. Similarly, during $x_3 < x < x_4$, the membership can be decreased from unity to zero. This ability could signify better ability of approximation [62].

7.2.2 Shapes of Fuzzy Set

The shapes of the fuzzy set for inputs are characterized by various forms (e.g., triangle, trapezium, bell, sigmoid, Cauchy, and Gaussian), which are subject to the constraint. The constraint defines that any element should not have more than one membership value of a specific set. The fuzziness in the universe of discourse (UD) depends on the structure of the shape. Due to the linearity structure, the trapezium and triangle are the most often used over other shapes. To obtain a desirable performance and perform operations on fuzzy sets easier, several shapes of membership functions are briefly described as follows [62].

Triangle membership function: This function is characterized by the combination of line equations, as shown in Fig. 7.3. The triangle function has a very sharp peak, which is very sensitive to variations in crisp variables. The triangle membership equation is represented as [24, 63]:

$$\mu_A(x) = \begin{cases} \frac{|x|-|x_1|}{|x_2|-|x_1|}, & \text{for } x_1 \leq x \leq x_2 \\ \frac{|x_3|-|x|}{|x_3|-|x_2|}, & \text{for } x_2 \leq x \leq x_3 \\ 0, & \text{otherwise} \end{cases} \tag{7.1}$$

Fig. 7.3 Triangle membership function

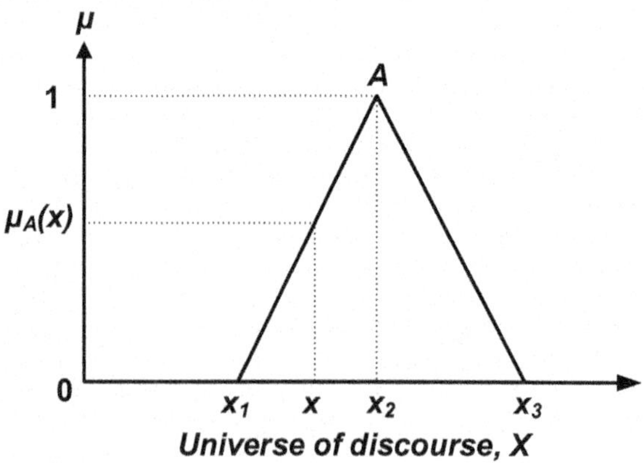

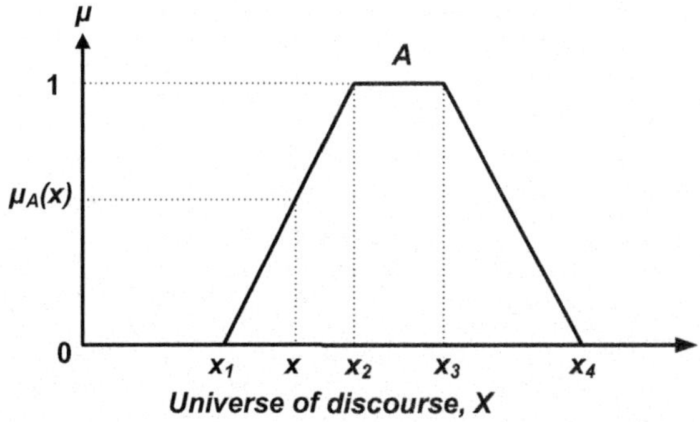

Fig. 7.4 Trapezium membership function

Trapezium membership function: This function is determined by the combination of line equations similar to the triangle function. The trapezium function has a flat top (see Fig. 7.4), which is not fuzzy, while the triangle has a sharp peak. The membership equation of the triangle is used in combination with a flat on the top. Then, the trapezium membership function is expressed as [24, 63]:

$$\mu_A(x) = \begin{cases} \frac{|x|-|x_1|}{|x_2|-|x_1|}, & \text{for} \quad x_1 \le x \le x_2 \\ 1, & \text{for} \quad x_2 \le x \le x_3 \\ \frac{|x_4|-|x|}{|x_4|-|x_3|}, & \text{for} \quad x_3 \le x \le x_4 \\ 0, & \text{otherwise} \end{cases} \tag{7.2}$$

Gaussian membership function: According to the high-sensitivity property in the triangle function, it is not preferred in some applications, which need a soft transition response to the variations in crisp variables. Thus, the Gaussian function, which has a soft peak, is created to reduce the sensitivity of the variable changes, see Fig. 7.5. The Gaussian membership equation is defined as [24, 63]:

$$\mu_A(x) = e^{-1/2\left(\frac{x-x_P}{w}\right)^2} \tag{7.3}$$

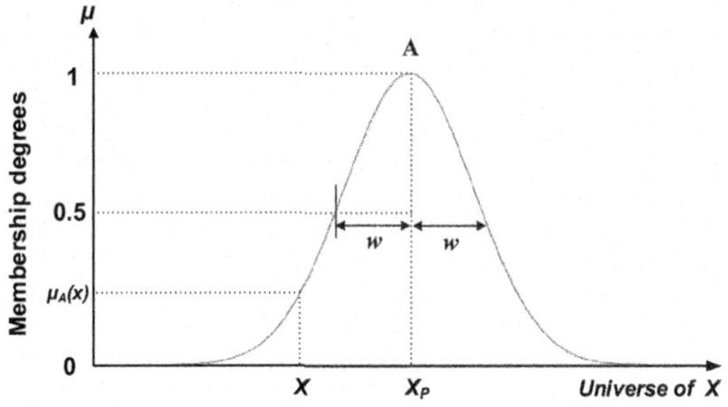

Fig. 7.5 Gaussian membership function

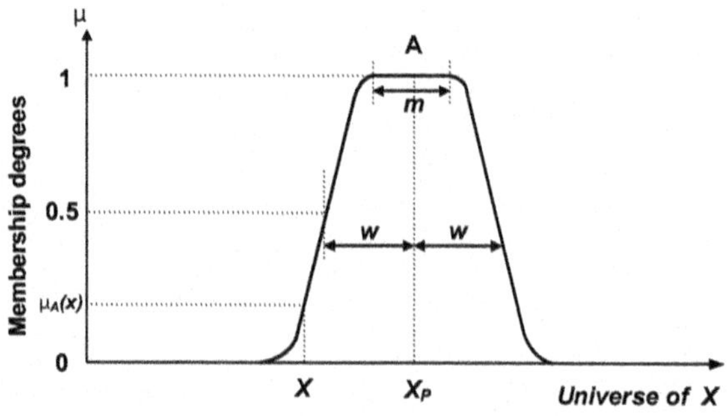

Fig. 7.6 Bell membership function

where x_p is the crisp number in the center of the membership function A with the maximum membership degree of one, x is the crisp variable expressed in the universe of X, and w is the distribution width of Gaussian membership function.

Bell membership function: This function is generated based on the Cauchy distribution used in probability theory. The bandwidth and flat top of the bell function can be adjusted by the w and m values, respectively, see Fig. 7.6. The adjustment of bandwidth and the flat top can increase the fuzziness. Applying a large value of m can change the function into a square crisp or perpendicular. Applying a small value of w can change the function into a more Cauchy curve. The bell membership equation is defined as [24, 63]:

$$\mu_A(x) = \frac{1}{1 + \left|\frac{x - x_p}{w}\right|^{2m}} \tag{7.4}$$

where m is the width of a flat top, and w is the distribution width.

Cauchy membership function: This function is a special form of the bell or Gaussian functions. The peak of this function is similar to the Gaussian, while it can be adjusted upon the width of a flat top (m) value, see Fig. 7.7. The w represents the distribution width. The Cauchy membership equation is defined as [24, 63]:

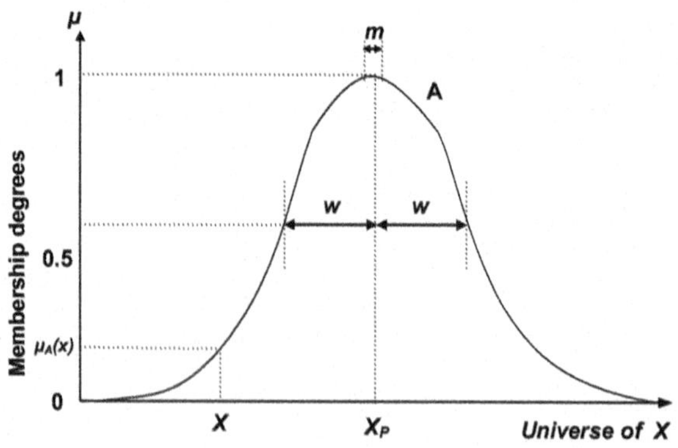

Fig. 7.7 Cauchy membership function

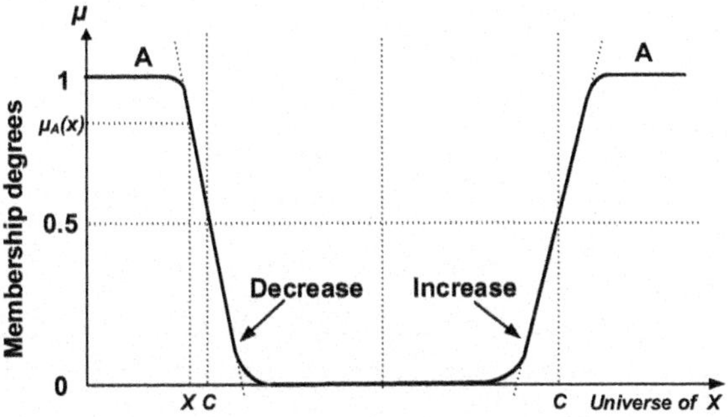

Fig. 7.8 Sigmoid membership function

$$\mu_A(x) = \frac{1}{1 + \left(\frac{x - x_P}{w}\right)^{2m}} \tag{7.5}$$

Sigmoid membership function: This function is designed for boundary sets used at upper and lower limits. It is also called the S-curve. The membership function tends to reduce in a sigmoid fuzzy subset for being used as a starting set at the lower boundary (see Fig. 7.8). On the contrary, the membership tends to increase for being used as a starting set at the upper boundary. The sigmoid function is often used in the applications, which is nonlinearity, such as the learning process of artificial neural networks. The sigmoid membership equation is expressed as [24, 63]:

$$\mu_A(x) = \frac{1}{1 + e^{-\alpha(x-c)}} \tag{7.6}$$

where α is the sign that represents the direction of the sigmoid function (i.e., positive means increasing and negative means decreasing). The α also evaluates the slope of the function. The range of the sigmoid function is adjusted by the values of α and c.

Figure 7.9 describes a set of complementary shape. The complementary equation can be defined as [24, 63]:

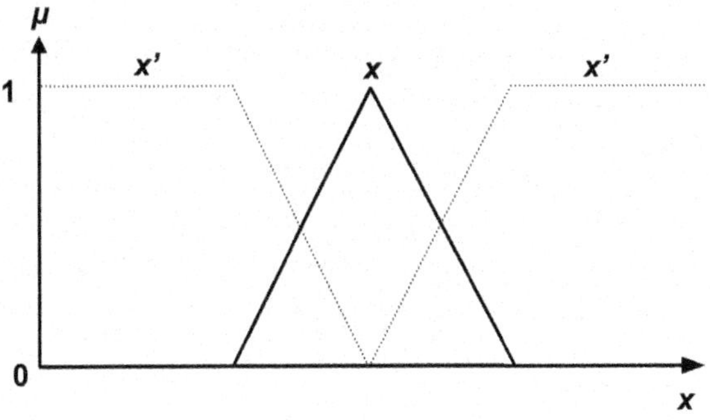

Fig. 7.9 Complementary of a fuzzy set

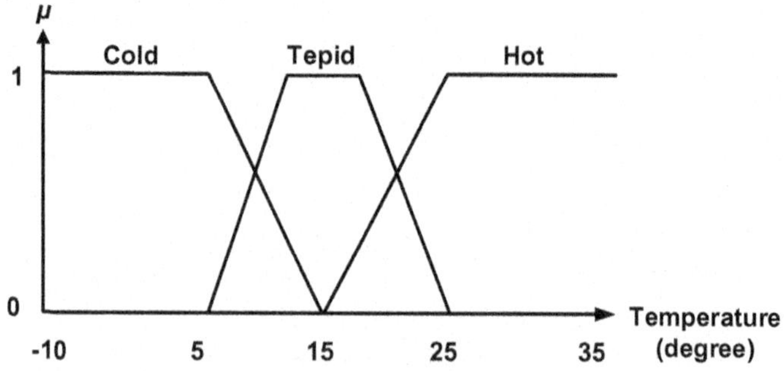

Fig. 7.10 Complementary of a fuzzy set for the UD of temperature

$$x' = 1 - x \tag{7.7}$$

where x is the fuzzy set.

For the shape of the UD, the elements of a universe are distributed over several sets. In fuzzy logic, the elements are members of more than one set, which have fractional membership values of sets. The sets of fuzzy logic are expressed over a UD of categories of specific attribution. Figure 7.10 shows an example of the UD of temperature, where the categories are changing degrees of "hot" [24, 63].

7.2.3 *Fuzzy Rule Base*

The fuzzy logic mechanism is created in the form of a rule base. Fuzzy rules are considered as an important tool for forming portions of knowledge in a fuzzy system. The rules are applied within a fuzzy system to evaluate the outputs based on the inputs. The most important rule is represented by the "if-then" statement as [10, 24, 63]:

$$\textbf{If } A, \textbf{ then } B. \tag{7.8}$$

where the "if" implies the antecedent, and the "then" implies the consequent. The antecedent expresses the fuzzy region in the input, whereas the consequent defines the output in the fuzzy region.

For example, if the brake is pushed, then the train will halt. This is the classical or crisp logic. Considering the fuzzy logic proposition, there are many pressure levels of the break, which may be implemented to halt the train. Also, there are several levels of halting from a sudden halt to gradual slowing down. Therefore, many categories of halting can be defined in fuzzy logic, and the classical logic cannot define such categories.

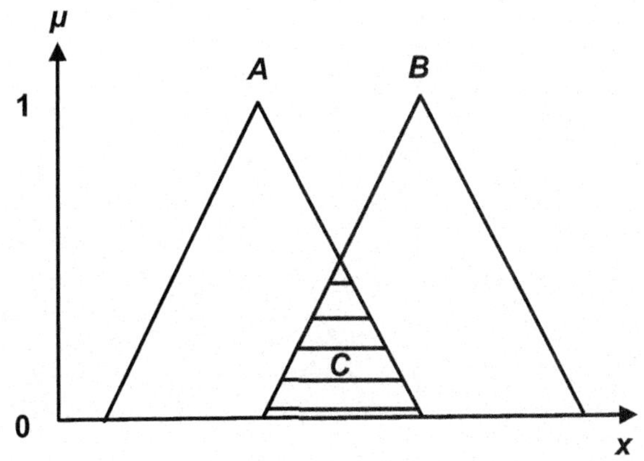

Fig. 7.11 Intersection of fuzzy sets

Each fuzzy rule expresses a fuzzy consequence between constraints and conclusion rules. Using fuzzy sets, object behavior represents the form of fuzzy relations. Such relations are combined with fuzzy expressions, which are linked using fuzzy logical operators. The important logical operators, which have been often used in fuzzy relations, are the intersection (and), union (or), and complement (not) [64].

For a complex control system, the compound rule base can be added and represented as:

$$\textbf{If } A \textbf{ and } B, \textbf{ then } C. \tag{7.9}$$

For example, if the brake (A) is pushed, and the pressure (B) is small, then the train (C) will gradually halt. This condition represents the intersection of fuzzy sets, as depicted in Fig. 7.11 [24, 63].

Alternatively, the "or" logic often uses in the compound rule base as:

$$\textbf{If } A \textbf{ or } B, \textbf{ then } C. \tag{7.10}$$

For example, if the brake (A) is pushed or the emergency light (B) is on, then the train (C) will halt. This condition represents the union of fuzzy sets, as shown in Fig. 7.12 [24, 63].

7.2.4 *Fuzzification*

This section offers an overview of the fuzzification process and inference mechanism. The fuzzification is the converting process, which transforms the crisp input values to the fuzzy values. This process is performed using the information of data or supervisor-based knowledge and learning. In addition, the crisp values can be obtained from quantitative measurements and transformed to linguistic values. This

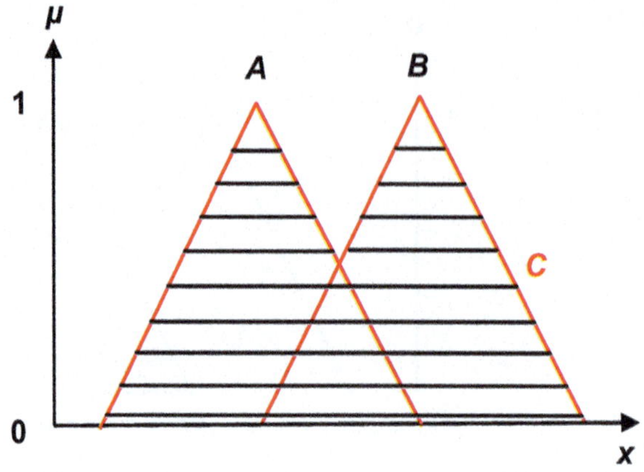

Fig. 7.12 Union of fuzzy sets

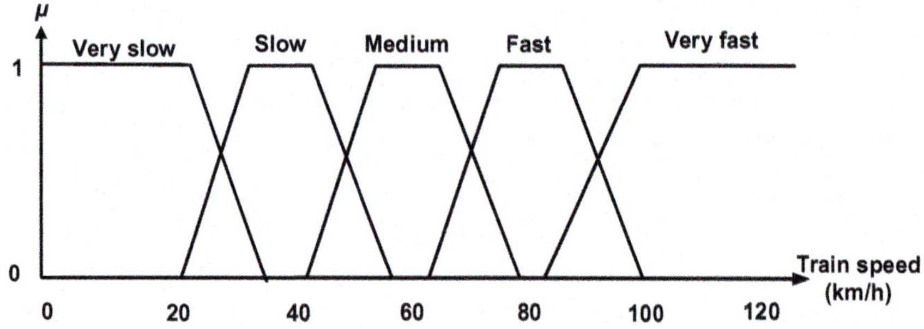

Fig. 7.13 Membership functions for fuzzification of a crisp train speed

process uses the membership functions to create the conversion. Figure 7.13 shows the membership functions for the fuzzification of a crisp train speed [10, 24, 63].

After the fuzzification process, the fuzzy rule bases are applied. These rules are presented in terms of "if-then" rules with the set of decisions when the input variables become certain fuzzy sets.

7.2.5 Fuzzy Inference System

The fuzzy inference system is an inference mechanism of formulating the mapping from a given input to an output using fuzzy logic [10, 24, 63]. This process interprets the values from an input vector and rules. Then, the process assigns the obtained values to an output vector [65]. The mapping offers a fundamental, in which the decisions can be created. The fuzzy inference mechanism involves membership functions and rule base. There are several types of fuzzy inference systems. The most commonly used types are the Sugeno fuzzy model and Mamdani fuzzy model. The difference is the consequents of their fuzzy rules, which produce different aggregation and defuzzification. A brief overview of each type is discussed as follows.

- **Mamdani Fuzzy Inference**
  This inference is the most often used fuzzy inference system. It was developed in 1975 by Ebrahim Mamdani. The process of this inference is performed in six steps as follows [66]:

  1. Determine the required fuzzy rules
  2. Convert the crisp inputs to fuzzy inputs using the membership function
  3. Aggregate the fuzzified inputs using the fuzzy rules
  4. Evaluate the rule consequent using the combination of rule strength
  5. Combine the consequents and obtain the output distribution
  6. Perform the defuzzification of output distribution.

  The process overview of Mamdani fuzzy inference is described in Fig. 7.14.

- **Sugeno Fuzzy Inference**
  This inference was proposed in 1985 by Takagi Sugeno. The model is similar to the Mamdani model, but the rule consequent is changed where the mathematical function is used for the input instead of the fuzzy set. The rule of the Sugeno inference model can be expressed as [66]:

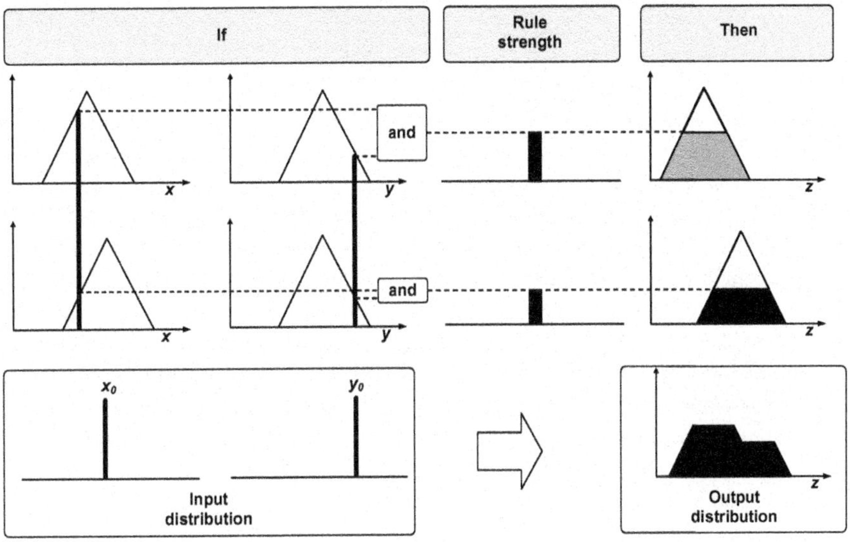

Fig. 7.14 Mamdani fuzzy inference process [66]

$$\text{If } x \text{ is } A \text{ and } y \text{ is } B \text{ then } z \text{ is } f(x, y) \tag{7.11}$$

where A and B are the fuzzy sets. x, y, and z are the linguistic variables. $f(x,y)$ is the mathematical function.

7.2.6 Defuzzification

Defuzzification is the procedure of transforming the degrees of membership of linguistic variables within their linguistic forms into crisp values. The results of the inference process are converted to crisp values. There are many methods of defuzzification available, such as adaptive integration, basic defuzzification distributions, bisector of area, center of gravity/area, center of sums, and weighted average [10, 24, 63].

In this section, the most popular defuzzification technique known as the center of the area, is presented. It determines the position of the center of the area/gravity of the subsets, which is also called a centroid method.

For a continuous membership function, the method is computed as [10, 24, 63]:

$$J = \frac{\int_{x=0}^{x=\infty} x \cdot \mu(x)dx}{\int_{x=0}^{x=\infty} \mu(x)dx} \tag{7.12}$$

For a discrete membership function, the method is computed as [10, 24, 63]:

$$J = \frac{\sum_{j=1}^{n} x_j \cdot \mu(x_j)}{\sum_{j=1}^{n} \mu(x_j)} \tag{7.13}$$

where n is the number of the fuzzy set for the discrete function, and $\mu(x)$ or $\mu(x_j)$ is the membership value for the point x or x_j, where x or x_j is the crisp value defined in the membership function corresponding to the center of the fuzzy set.

In summary, an overview of a basic scheme for a fuzzy control system is described in Fig. 7.15. The fuzzy system has four control units (i.e., fuzzification, rule base, inference mechanism, and defuzzification). At the begining (i.e., preprocessing), the scaling or normalizing processes can be applied to correct the input information. Crisp input information is transformed into fuzzy values for each input fuzzy set with its own fuzzification block. The inference evaluates how the fuzzy operation is conducted with the knowledge base and if-then rules. Then, the obtained data are altered to the crisp values using the defuzzification. Then, the postprocessing step can be applied to modify the output crisp values. Lastly, the scaled output as a control signal is implemented in the control system.

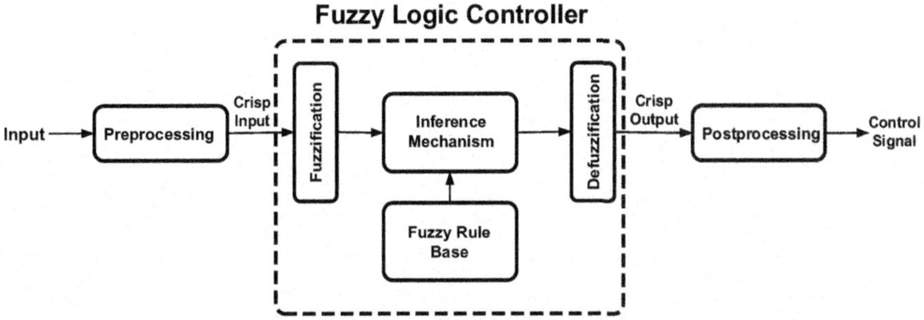

Fig. 7.15 General scheme for a fuzzy control system

7.3 Fuzzy-Based Virtual Inertia Synthesis

Recently, the connection of power electronics interfaces-based RESs/DGs is more decentralized, especially wind and solar systems. This situation results in more complicated power system than ever since the RESs/DGs may not be controllable or predictable. Thus, the availability and controllability of RESs/DGs differ from conventional power systems. Nevertheless, the high penetration of RESs/DGs begins to create a serious concern in frequency regulation of the power system [12–14]. The situation is significantly exacerbated in a single control area or an isolated system that has a small amount of inertia and damping property.

In addition to virtual inertia control, the high RESs/DGs penetration stimulates the inertia control system to be unstable to deliver an appropriate performance over a wide range of operations. Consequently, the systems might be inadequate and unstable after high-level integration of RESs/DGs, resulting in system instability. In the worst case, the inertia control systems can be interrupted due to fluctuations with the high amplitude after RESs/DGs integration. This situation can lead to cascading failures and power blackouts.

To resolve the challenge, this section suggests a decentralized fuzzy-based virtual inertia controller, which is demanded for the effective system frequency stabilization againt different levels of RESs/DGs toward high penetration. In this idea, the fuzzy logic control is applied as an automatic primer for tuning the suitable virtual inertia constant, which is the main function for inertia emulation. The dynamic structure of the proposed fuzzy-based virtual inertia controller is presented in Fig. 7.16. During frequency oscillations with low RES/DG levels, if the power system is functioned with the large virtual inertia constant, it results in the longer settling time in balancing the system frequency, causing slow damping of deviations. In this circumstance, the power system requires the small virtual inertia constant to regulate system frequency, delivering fast damping of deviations. On the other hand, during frequency oscillations with high RESs/DGs integration, large virtual inertia constant is extensively needed to arrest the significant frequency nadir/overshoot and diminish the deviations with high amplitude caused by the lack of system inertia and damping regarding high RESs/DGs penetration. Hence, if the virtual inertia control system can

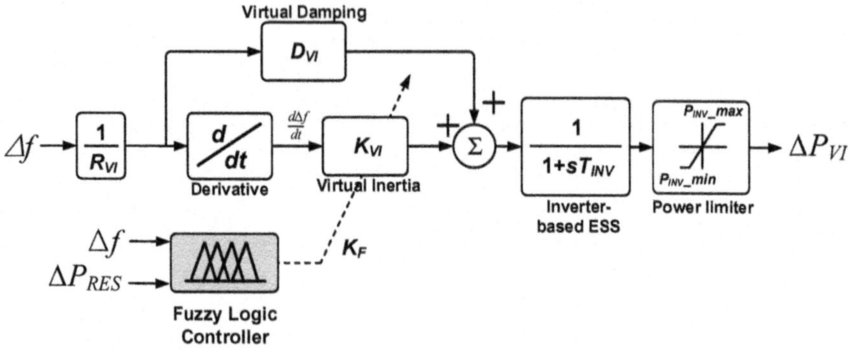

Fig. 7.16 Dynamic scheme for the fuzzy-based virtual inertia control

suitably adjust its control parameter to follow the different levels of RESs/DGs integration (i.e., disturbances), the optimal stability and performance of the power system can be reached, preventing system instability and power blackouts. To enable the self-adaptive virtual inertia control, the proposed fuzzy controller is used by combining the fuzzy rules and interface. Therefore, the virtual inertia constant is automatically modified by the fuzzy interface and knowledge base of RES/DG power changes and system frequency deviations.

The proposed fuzzy system contains four processing units; that is the fuzzification, inference mechanism, fuzzy rule base, and defuzzification, as shown in Fig. 7.17. The active power changes of RESs/DGs (ΔP_{RES}) and system frequency deviation (Δf) are used as the crisp inputs of the fuzzy controller. The output is a crisp (normalized) value of the virtual inertia constant (K_F). Thus, this fuzzy system contains two inputs and one output. Firstly, the scale factors (i.e., the preprocessing) is implemented to adjust the size of scale inputs. Later, the fuzzification process is performed to alter the actual inputs to the fuzzy values. The Mamdani inference model is used in the inference process. The *Input* 1 and *Input* 2 are formed as:

$$Input\ 1 = \Delta f \cdot K_1 \tag{7.14}$$

$$Input\ 2 = \Delta P_{RES} \cdot K_2 \tag{7.15}$$

where K_1 and K_2 are scale factors of the fuzzification system, where $K_1 = K_2 = 1$.

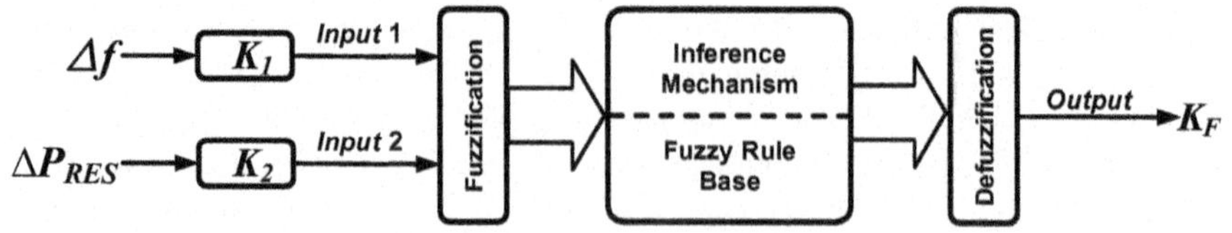

Fig. 7.17 Dynamic scheme for the fuzzy logic in scheduling virtual inertia constant

Next, the fuzzy values of *Input* 1 and *Input* 2 are progressed to the fuzzy inference mechanism with a fuzzy rule base. The fuzzy rules are the fundamental fuzzy operations for mapping the input signal to the output signal. To determine an optimal K_F value, the fuzzy rules are implemented by merging the input signals of Δf and ΔP_{RES}. From Table 7.1, fifteen fuzzy rules are used based on the practical experiences and knowledge of the virtual inertia control regarding RESs/DGs penetration and frequency deviation as follows:

1. When ΔP_{RES} and Δf are quite small, a zero value of K_F is applied to regulate the system frequency, distributing fast damping of deviations.
2. When ΔP_{RES} and Δf are quite large, a medium K_F value (in cases of large negative Δf) or very big K_F value (in cases of large positive Δf) should be applied to diminish the deviations with high amplitude driven by the lack of system inertia and damping regarding high RESs/DGs, preventing system instability and failures.
3. When Δf is quite small, and ΔP_{RES} is quite large, a medium K_F value (in cases of small negative Δf) or very big K_F value (in cases of small positive Δf) should be implemented to reduce the system frequency deviations with high amplitude driven by the lack of system inertia and damping regarding high RESs/DGs, preventing system instability and failures.
4. When Δf is quite large, and ΔP_{RES} is quite small, a zero value (in cases of large negative Δf) or medium value (in case of large positive Δf) of K_F should be applied to regulate the system frequency, delivering fast damping with low-frequency deviations.

By using Table 7.1, the fuzzy rules are expressed in terms of "if-then" expression as follows:

$$\textbf{If } Input\ 1\ is\ x\ \textbf{and } Input\ 2 \text{ is } y,\ \textbf{Then } J_F \text{ is } z.$$

where *x*, *y*, and *z* are the fuzzy set on the corresponding sets.

Numerous membership functions and fuzzy rules for the inputs and output of the fuzzy controller are proposed in Fig. 7.18 and Table 7.1. The input 1 has two trapezoidal and three triangular memberships, whereas the input 2 has one triangle and two trapezium memberships. The output of the fuzzy system has two trapezium and three triangular memberships. Due to the characteristics of simplified calculation and outstanding control performance, the triangle and trapezium membership functions

Table 7.1 Fuzzy rules of fuzzy-based virtual inertia control for linguistic variables

Control variable		Δf				
		NL	NS	ZO	PS	PL
ΔP_{RES}	L	ZO	ZO	ZO	ZO	M
	M	S	S	M	B	B
	H	M	M	M	VB	VB

have been chosen. In this study, the range of fuzzy variables are ΔP_{RES}= [−1, 1] p.u., and $\Delta f = [-0.5, 0.5]$ Hz. The output range of K_F is selected as [−1, 3], which is optimally evaluated based on the extensive eigenvalue sensitivity analysis in Chap. 3. The μ represents each membership grade. The inputs and output are separated into fuzzy subsets and expressed using linguistic variables. By defining linguistic variables, the linguistic words are defined by words that we use in our daily life with respect to the penetration levels and deviations as follows. PL is positive large, PS is positive small, ZO is zero, NS is negative small, NL is negative large, VB is very big, B is big, S is small, L is low, M is medium, and H is high.

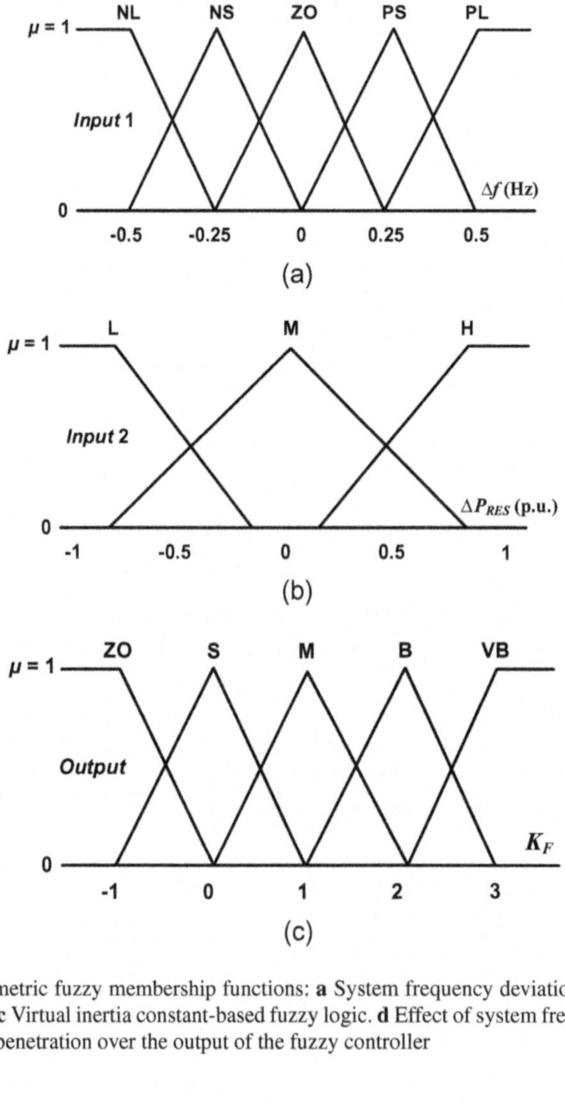

Fig. 7.18 Symmetric fuzzy membership functions: **a** System frequency deviations. **b** RESs/DGs power changes. **c** Virtual inertia constant-based fuzzy logic. **d** Effect of system frequency deviation and RESs/DGs penetration over the output of the fuzzy controller

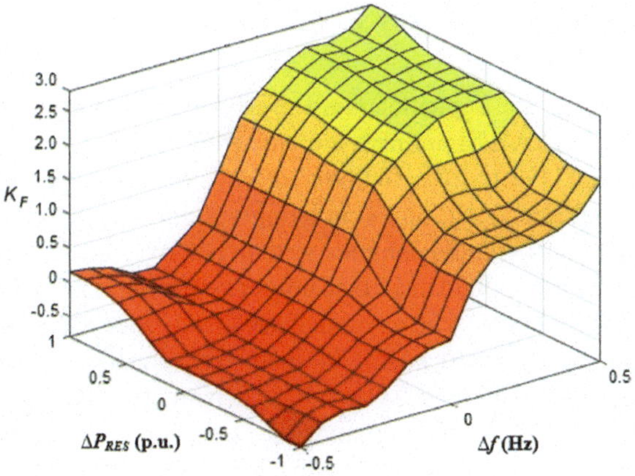

(d)

Fig. 7.18 (continued)

The fuzzy inference mechanism alters the rule base to the fuzzy linguistic output, as shown in Table 7.1. However, the linguistic output is unavailable for the signal of equalization control. To resolve the issue, the defuzzification process is demanded. In this study, the center of the area (centroid) technique is applied for defuzzification. Lastly, the obtained results from the fuzzy rules are forwarded to the defuzzification, and transformed to the crisp values as [67]:

$$K_F = \frac{\sum_{j=1}^{n} x_j \cdot \mu(x_j)}{\sum_{j=1}^{n} \mu(x_j)} \tag{7.16}$$

By implementing the proposed fuzzy controller, the virtual inertia constant is automatically modified to follow the variations in numerous integration levels of RESs/DGs (i.e., disturbances), allowing self-adaptive inertia response. The dynamic equation of the fuzzy-based virtual inertia control is obtained as:

$$\Delta P_{VI\_Fuzzy}(s) = \frac{K_F \cdot s + D_{VI}}{1 + sT_{INV}} \left(\frac{\Delta f(s)}{R_{VI}} \right) \tag{7.17}$$

Thus, the proposed fuzzy-based virtual inertia control is not only to offer sufficient inertia response with the self-adaptive ability but also to provide low-frequency deviation with fast damping, reducing high-frequency overshoots/drops driven by different penetration levels of RESs/DGs.

7.4 MATLAB-Based Fuzzy Logic Control

This section presents the implementation procedures for designing a fuzzy logic controller using MATLAB/Simulink® software [68]. A fuzzy logic controller block in Simulink® called "Fuzzy logic toolbox" offers an application with control functions for designing, analyzing, and simulating systems based on fuzzy logic. The toolbox lets engineers to model complex system behavior using simple logic rules. Afterward, the rules are applied in a fuzzy inference system. The control input/output data, membership functions with several shapes, inference systems, and rule base can be specified and designed in this toolbox via a function called "Fuzzy logic designer". Once a fuzzy inference system has been created, the evaluation and visualization can be modified. After completing the design process, engineers can use the fuzzy controller in Simulink and simulate the fuzzy systems within a comprehensive model of the whole dynamic system. Engineers can also generate the structured text for a fuzzy system applied in Simulink using the fuzzy logic controller block. This application is very useful for those who are not familiar with the C language. Alternatively, the deployment of a fuzzy system can be generated via C language either in MATLAB or Simulink. In this section, it is noted that the studied system is set based on Fig. 3.9 and Table 3.1 under the effect of 100% of system inertia and damping (see Table 3.2).

This section guides engineers and designers though the step-by-step of designing a fuzzy controller using MATLAB/Simulink. The proposed fuzzy-based virtual inertia controller has involved five steps as follows:

1. Understand the dynamics and characteristics of the control system. Next, determine the input/output control variables, states, and their variation ranges. Then, open a "fuzzy logic designer" function by typing "fuzzy" in the MATLAB command window. The function will appear as depicted in Fig. 7.19.
2. Define suitable fuzzy sets and membership functions. The input/output fuzzy set can be added by clicking a 'Edit' tab and select an 'Add variable' tab. Then, conduct the degree of membership functions for each input/output variable and complete the fuzzification process. The membership functions and its control ranges can be adjusted by double-clicking an 'input' or output section, see Fig. 7.20.
3. Select an appropriate inference mechanism by clicking a 'File' tab and 'New FIS' tab. Engineers can select either Mamdani inference model or Sugeno inference model. Next, create the rule base using the control rules, in which the system will operate under a desirable condition. By clicking an 'Edit' tab and 'Rule' tab, the fuzzy rules can be defined with linguistic variables, see Fig. 7.21.
4. Then, decide how the action will be examined by adjusting strengths (i.e., red line) to the rules, see Fig. 7.22. This process can be done by clicking a 'View' tab and 'Rule' tab.
5. Select a suitable defuzzification technique by clicking 'Defuzzification' tab as shown in Fig. 7.19. Then, perform the defuzzification and obtain the output as a control signal. Finally, implement the obtained fuzzy controller to the studied

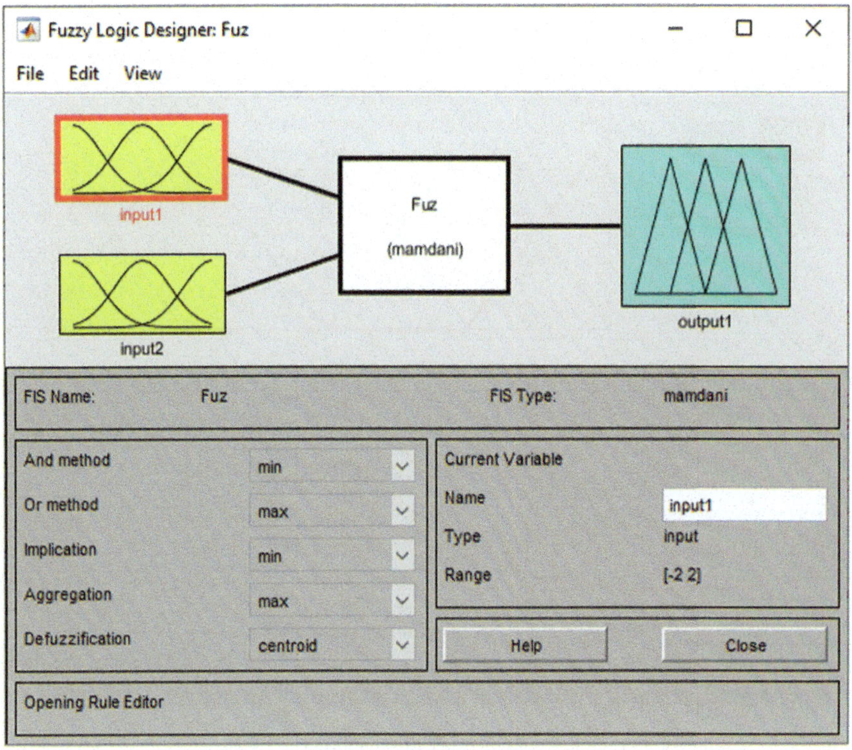

Fig. 7.19 Function Interface of fuzzy logic designer

power system-based Simulink model (see Fig. 3.9) through a "fuzzy logic control block" in 'Simulink Library Browser'. This transformation process can be done by clicking a 'File' tab, 'Export' tab and 'To Workspace' tap.

7.5 Simulation Results

The simulation results were carried out using MATLAB/Simulink® software and tested through a single area system in Fig. 3.9 with Table 3.1. Then, the effectiveness of the fuzzy-based virtual inertia control is verified under critical nonlinear simulations of contrastive RESs/DGs penetration, and mismatch parameters of primary and secondary control. Moreover, this section demonstrates the necessity of using the fuzzy-based virtual inertia control over the critical lack of system inertia and damping caused by high levels of RESs/DGs integration. Then, the obtained results were compared with the PID-based virtual inertia control (obtained from Chap. 5) and conventional virtual inertia control (obtained from Chap. 3). According to a practical frequency operating standard for a single area system used in today [69], the acceptable limit of frequency deviation is set as 49.5–50.5 Hz (±0.5 Hz) during

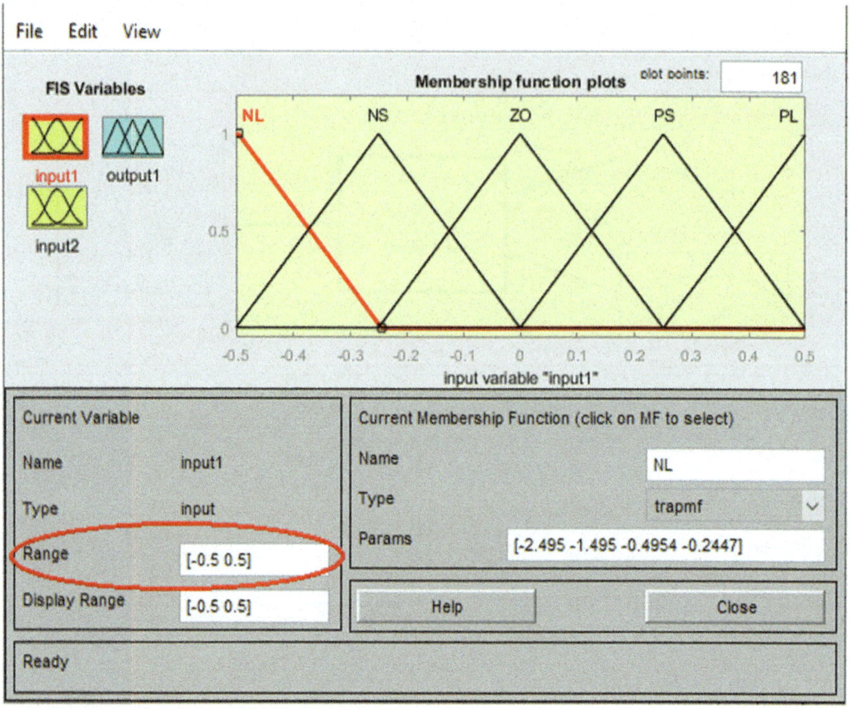

Fig. 7.20 Modification of input/output membership function

the generation/load events. The studied simulation applied a similar standard for frequency regulation of the system.

7.5.1 Effect of Low RESs Penetration

For the first scenario, the frequency response of the power system is verified under a normal operating condition; that is, high system inertia and damping (i.e., 30% drop from its nominal values, see Table 3.2) based on a small share of RESs/DGs penetration. The normal system operation is slightly interrupted by low power integration of wind and solar generations and high power consumption of industrial/residential loads, as shown in Fig. 7.23.

From Fig. 7.24, the sudden integration of wind-turbine power (about 0.1 p.u.) at $t = 200$ s causes the large frequency overshoot of 0.4 and 0.3 Hz in cases of no virtual inertia control and traditional control, respectively. By applying the PID-based virtual inertia control, the frequency overshoot is downsized to 0.28 Hz. Clearly, the frequency overshoot of the system is significantly reduced to 0.03 Hz in the case of the proposed control method due to the capability of self-adaptive virtual inertia constant.

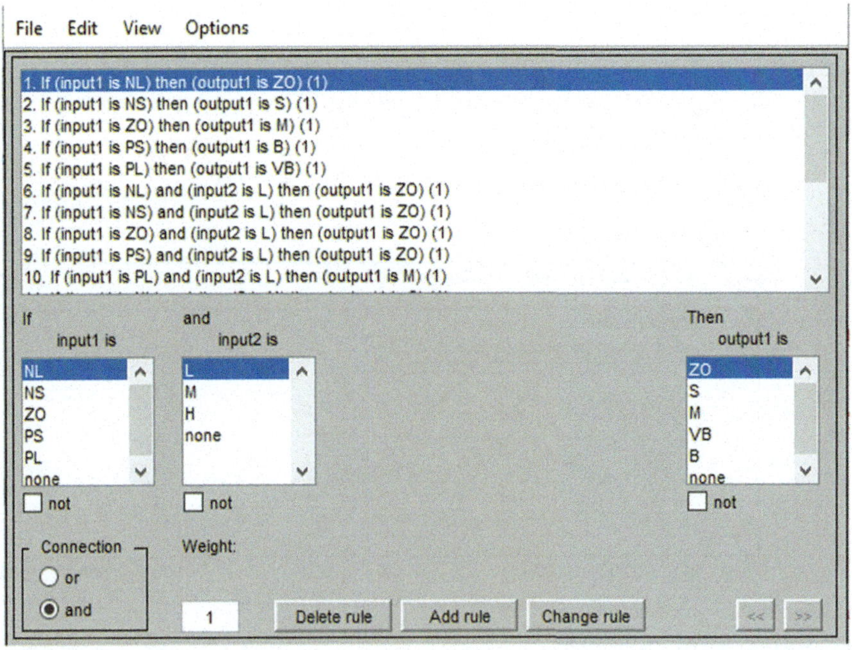

Fig. 7.21 Modification of fuzzy rules

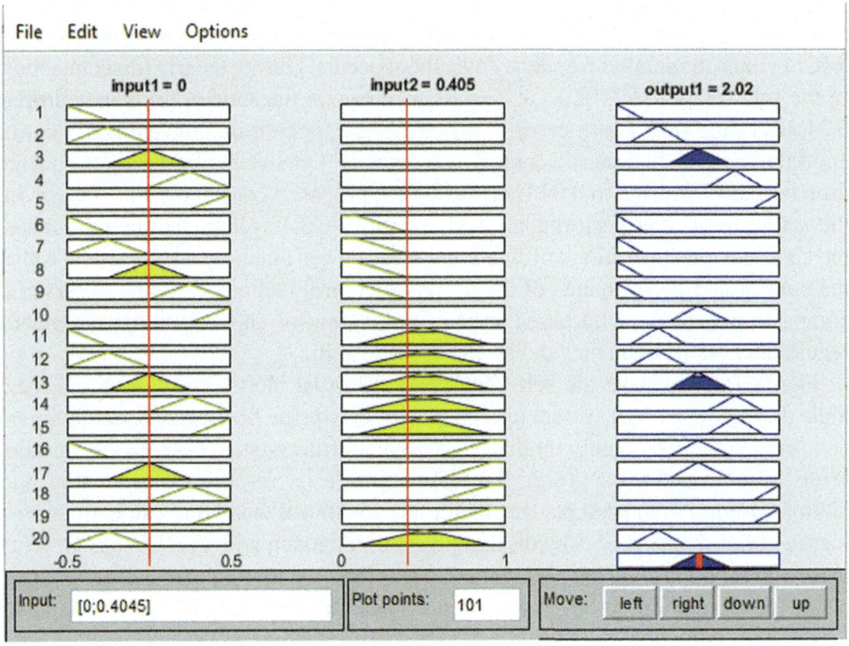

Fig. 7.22 Adjustment of action strength to fuzzy rules

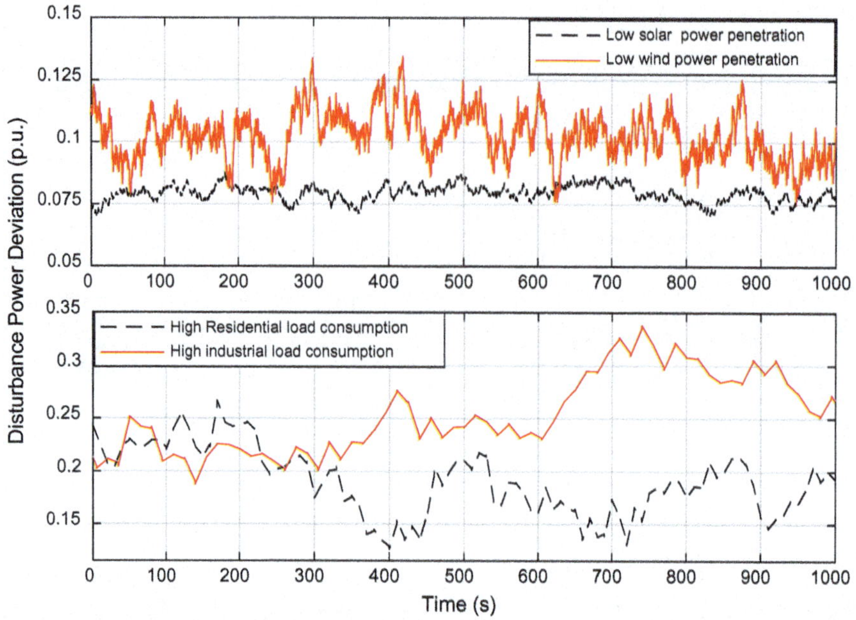

Fig. 7.23 Multiple disturbances in the presence of low wind speed, low solar irradiation, high industrial and residential load consumption

Evidently, the proposed control method can enhance the dynamic performance of the RoCoF; thus, the smaller frequency overshoot occurs. During the trip (disconnection) of the solar plant (0.075 p.u.) at $t = 800$ s, it causes the sudden frequency drop of 0.24, 0.21, and 0.19 Hz in cases of no virtual inertia control, conventional control, and PID-based virtual inertia control, respectively. Obviously, the system frequency drop is greatly arrested to 0.04 Hz by using the proposed control method, preserving the stable frequency performance and stability of the system. Ultimately, during the simultaneous operations of wind and solar power plants (e.g., $t = 300-700$ s), the outstanding performance of the proposed control technique can be observed in comparison with the PID-based virtual inertia control and traditional control for regulating system frequency deviations close to zero.

Figure 7.25 displays the self-adaptation of virtual inertia constant-based fuzzy logic during the normal system operation. Obviously, the fuzzy-based virtual inertia controller could adequately modify the virtual inertia constant to track the variation levels of wind and solar power plants, bringing faster response with smaller fluctuations. Figure 7.26 presents that the proposed control technique can extract more inertia power from the ESS by discharging in comparison with the PID-based virtual inertia control and conventional control. Therefore, the thermal generating unit (i.e., conventional generation-based synchronous machines) is less stressed and demands less power generation due to the proposed control method, as revealed in Fig. 7.27.

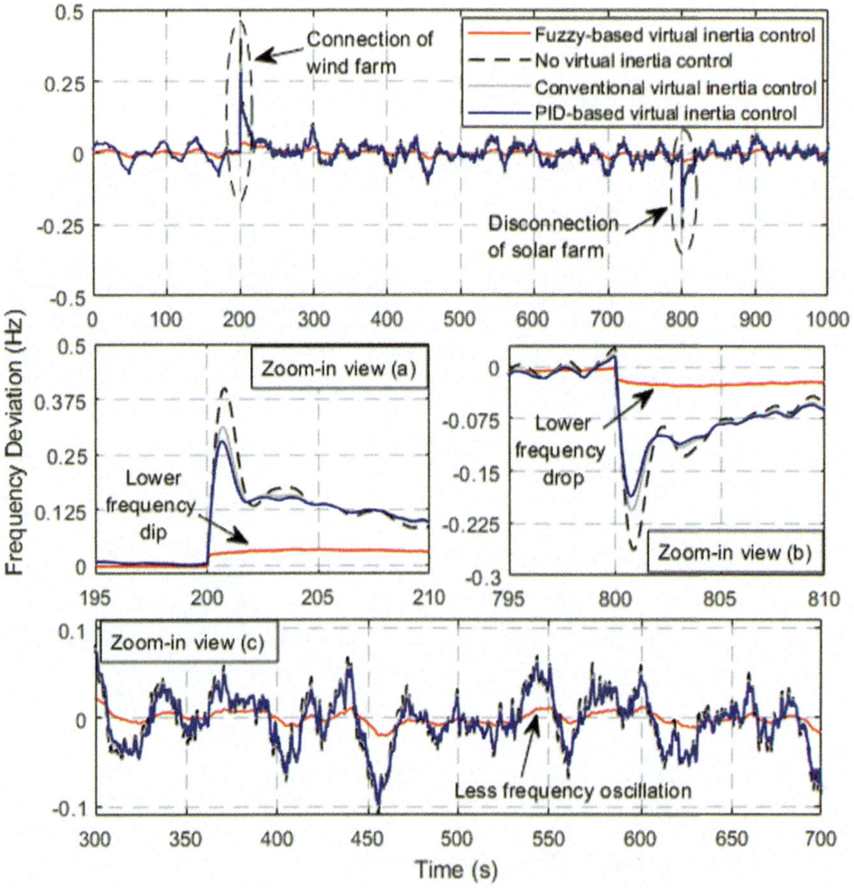

Fig. 7.24 Frequency response during low RESs/DGs integration regarding high system inertia and damping

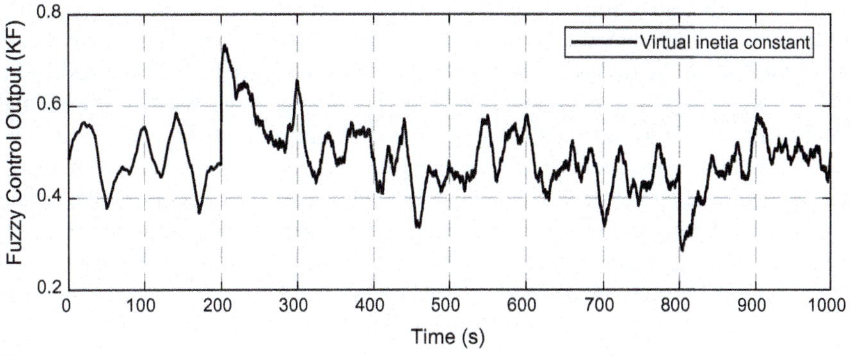

Fig. 7.25 Adaptation of virtual inertia constant using fuzzy control

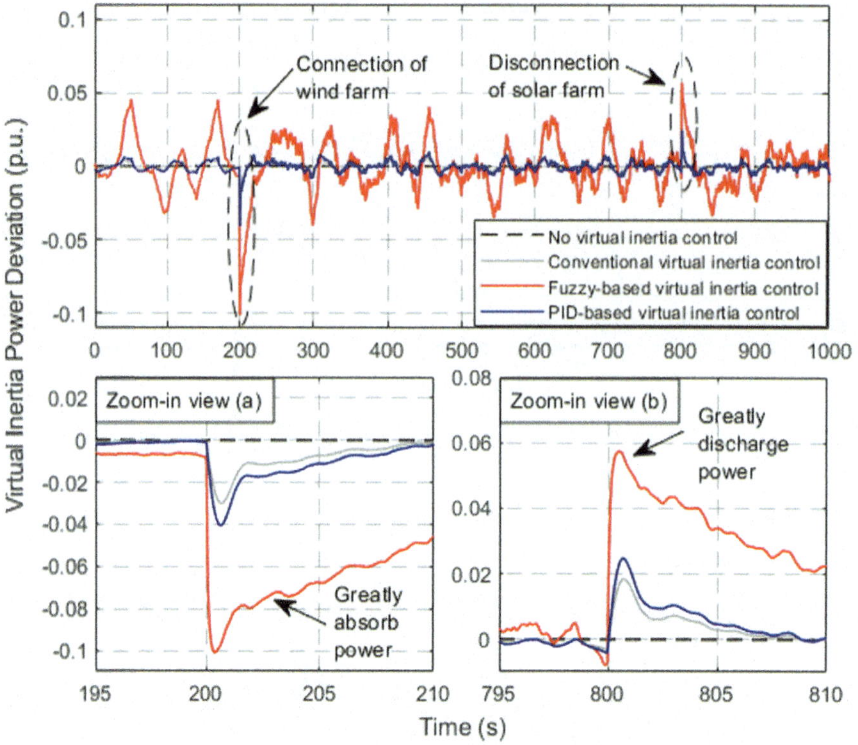

Fig. 7.26 Virtual inertia power emulation during low RESs/DGs penetration

7.5.2 *Effect of High RESs Penetration*

To perform the critical scenario, high amounts of solar and wind power are integrated into the system, as shown in Fig. 7.28. Then, the system frequency response is examined under the critical condition of low system inertia and damping (i.e., 70% drop from its nominal values, $H = 0.0249$, $D = 0.0408$). To perform more drastic operations from the frequency stability point of view, the industrial and residential load powers are consumed as low consumption (see Fig. 7.28), which creates a high load damping factor, causing the oscillations with high amplitudes.

During this severe scenario, the system frequency is more fluctuated than that of the previous scenario (see Fig. 7.29). Without the virtual inertia control, the system frequency severely fluctuates with the largest overshoots and stabilizing time. By applying the virtual inertia control methods, the system frequency can be alleviated. In the cases of the traditional and PID-based virtual inertia control, the transition states (at $t = 200$ and $t = 800$ s) cause the large transients with longer stabilizing time. The frequency transients also exceed the acceptable frequency operating standard limit of ± 0.5 Hz, which may cause unstable conditions of the system. This is because such control systems require certain time to perform the decision-making-based

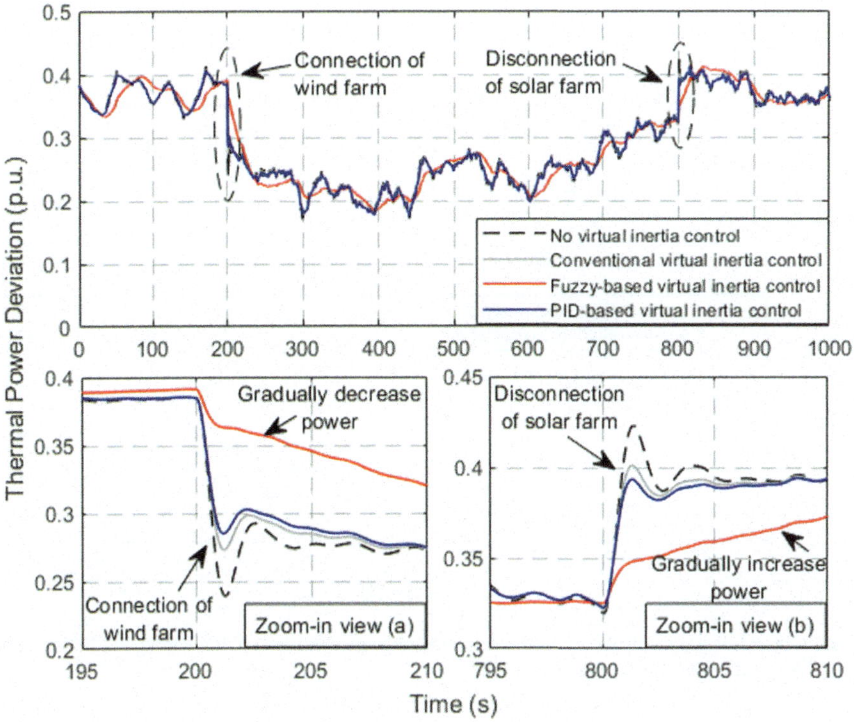

Fig. 7.27 Conventional power generation-based synchronous generator

virtual inertia constant for compensating nonlinearity behavior of high RESs/DGs penetration. On the contrary, the fuzzy controller requires less time for the decision-making, since the system can properly compensate the nonlinearity via the fuzzy membership functions and rule base. Choosing rules with reality allows the fuzzy system to deal with the nonlinear behavior better than conventional control techniques. Subsequently, the frequency transients are suddenly arrested, resulting in the lowest magnitude with less stabilizing time. Obviously, the more stabilizing effect during low system inertia and damping caused by high RESs/DGs can be observed in the case of fuzzy-based virtual inertia control. Particularly, the system frequency is effectively regulated within the allowable operating constraints, indicating the robust system operation.

Following the critical event, the self-adjustment of virtual inertia constant-based fuzzy logic during the high RESs/DGs integration is revealed in Fig. 7.30. The adjustment rate of virtual inertia constant obviously increases against high RESs/DGs penetration. Subsequently, the fuzzy-based virtual inertia controller is able to track the multiple variations of RES/DG penetration levels, leading to the fast damping of oscillations with lower frequency amplitudes.

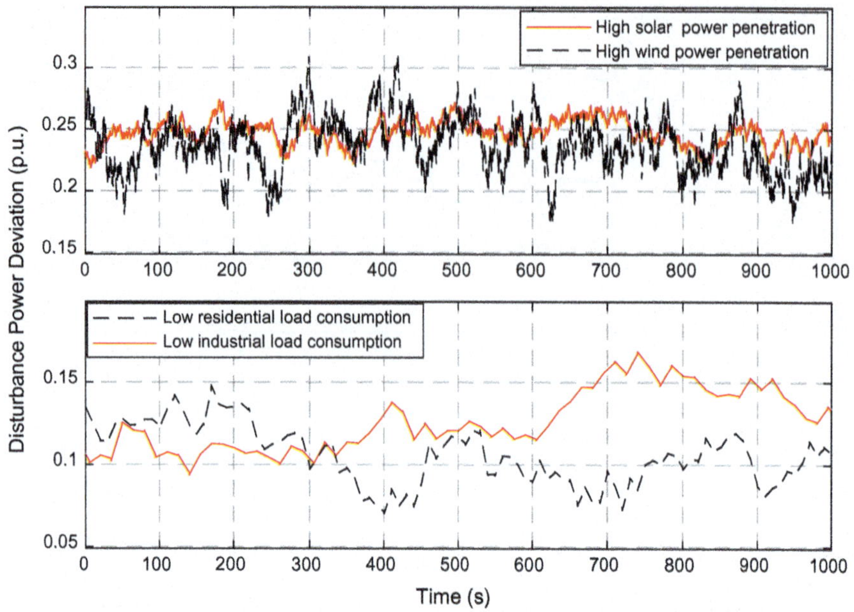

Fig. 7.28 Multiple disturbances in the presence of high wind speed, high solar irradiation, low industrial and residential load consumption

7.5.3 Mismatch Parameters of Primary/Secondary Control

In the real-world operation of power systems, it is probable to have a parameter imprecision approximated for designing the system. The system parameters may constantly change by time, seriously deteriorating the system dynamics and creating the system instability. Thus, it is necessary to analyze the performance and stability under the parameter changes, including primary and secondary controls. Previously, the generation-based synchronous generator provides the primary and secondary control services. In this section, the control parameters in those services are prone to the error changing by time. Thus, the virtual inertia control unit is additionally in charge of delivering primary and secondary frequency supports during the mismatch control parameters, while the main service is devoted to the inertia control support. To reveal the adaptive ability and robustness of the proposed fuzzy-based virtual inertia control, this case is integrated severe disturbances similar to Sect. 7.5.2, but the system parameters including primary and secondary controls are severely varied due to Table 7.2. To perform the drastic operation, the effect of communication delay time is considered in both secondary and inertia control units. The time delay units have been added to the secondary and inertia control systems (as depicted in Fig. 6.17), where the delay time is significantly increased. Moreover, the governor dead band and generation rate constraint (GRC) are also applied to the turbine-governor unit, as described in Fig. 2.28. The maximum value of the governor

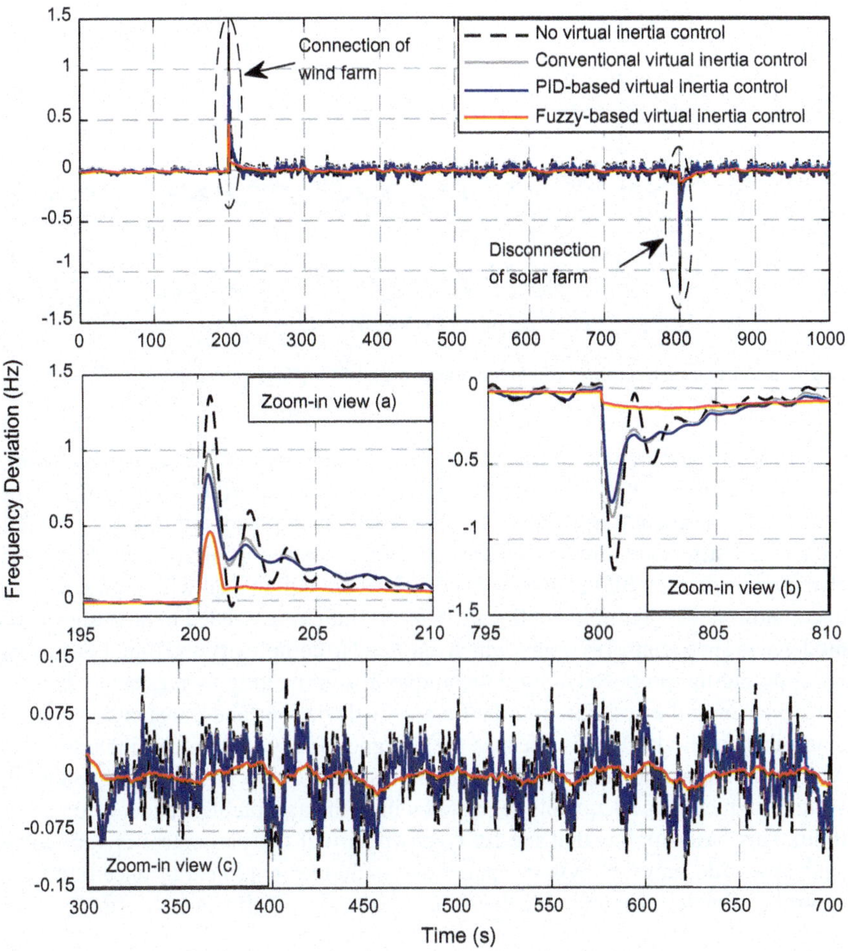

Fig. 7.29 Frequency response during high RESs/DGs integration regarding low system inertia and damping

dead band is set as 15% p.u./min. The GRC is set by applying the lower and upper constraints that limit the rate of valve/gate closing/opening speed ($V_L = -0.5$ and $V_U = 0.5$).

Following the severe operation, the critical frequency oscillations can be clearly observed (see Fig. 7.31). Without the virtual inertia control, it is impossible to regulate frequency stability, resulting in cascading failures and power blackouts. During the switching of wind and solar systems at $t = 200$ and $t = 700$ s, the traditional virtual inertia control with the fixed inertia constant cannot handle the integrated parameter perturbation, resulting in the larger frequency overshoot and drop about ± 1.6 Hz with longer settling time of about 10 s. This indicates the effect of slow damping of deviations. In addition, the traditional inertia control is not powerful enough to

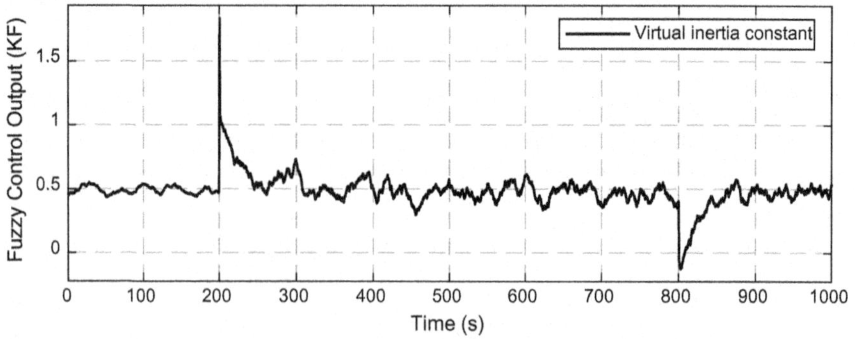

Fig. 7.30 Virtual inertia power emulation during high RESs/DGs penetration

deal with a wide range of multiple nonlinearity/uncertainties, significantly affecting the stable stability of system operation. By applying the PID-based virtual inertia control, the frequency overshoot and frequency drop during the transition slightly reduce and still exceed the acceptable frequency operating limit of ±0.5 Hz. This also implies that the PID controller is not robust enough to deal with a wide range of uncertainties. The multiple nonlinearity/uncertainties are needed to be designed and modeled in such control systems, which may lead to complexity and time-consuming. By applying the proposed control technique, it significantly increases the damping performance of the system; thus, the frequency deviations and transients are smaller than other methods. This is because the designed fuzzy system can include uncertainty properties and combine multiple inputs with a rule base. Consequently, the system frequency is adequately regulated within the acceptable frequency operating limit. This case implies that the fuzzy-based virtual inertia control is very robust against a wide range of system operations, including nonlinearity/uncertainties of control parameters, maintaining the stable frequency stability even in a more drastic scenario.

Figure 7.32 shows the self-adjustment of virtual inertia constant-based fuzzy logic. According to its nature of adaptability, the fuzzy controller can provide the control

Table. 7.2 Control parameters for mismatch of primary and secondary control schemes

Parameter	Value
Primary control	
Governor response time, T_g (s)	0.2 (+180%)
Turbine response time, T_t (s)	0.5 (+35%)
Secondary control	
Gain of secondary controller, K_s (s)	0.05 (−50%)
Communication delay time for secondary control, t_s (s)	2
Communication delay time for virtual inertia control, t_{VI} (s)	0.75

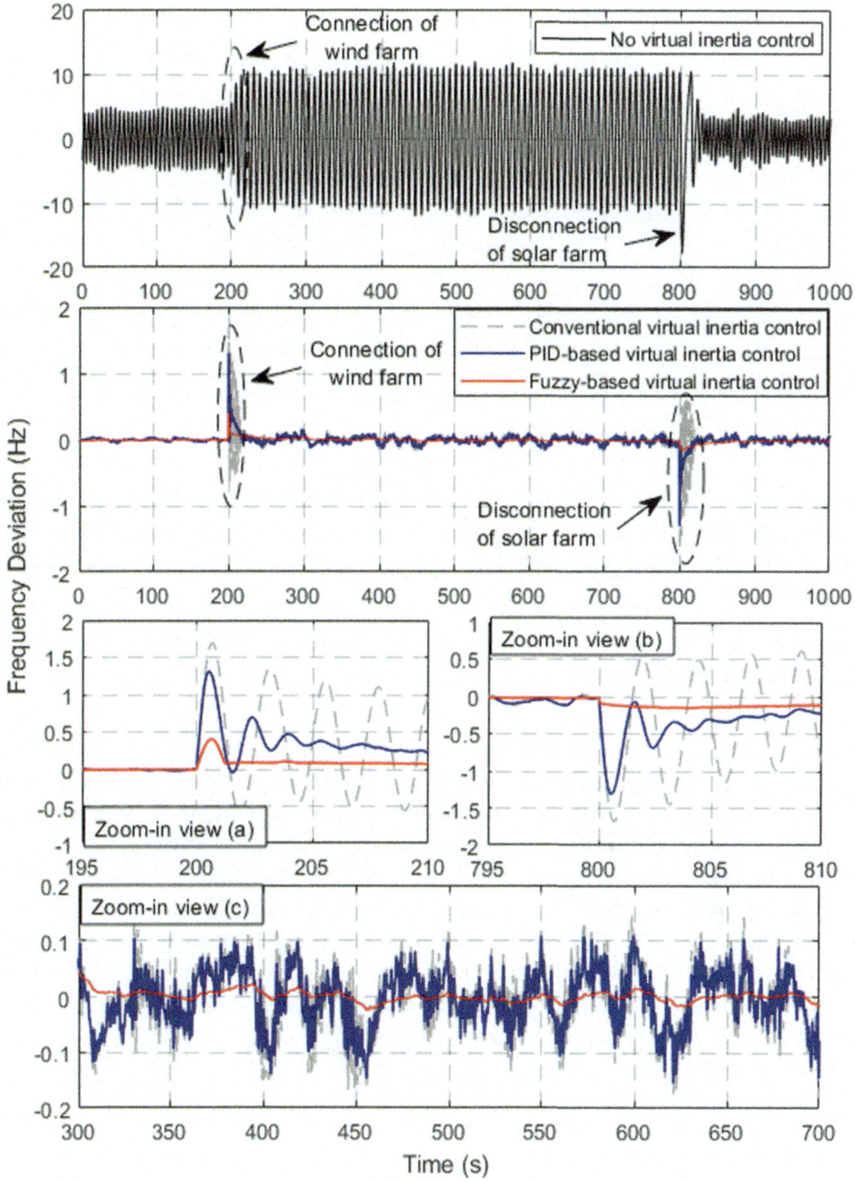

Fig. 7.31 Frequency response during mismatch parameters of primary and secondary control

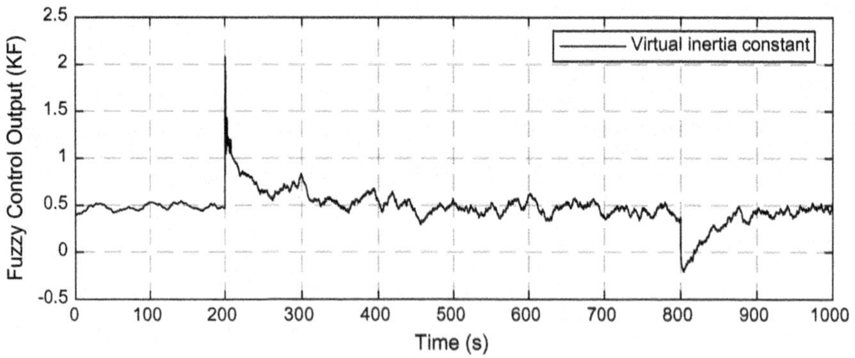

Fig. 7.32 Virtual inertia power emulation during mismatch parameters of primary and secondary control

action of K_F in the nonlinear system. As a result, the fuzzy-based virtual inertia controller is able to track the severe changes against the wide ranges of RESs/DGs operations, resulting in fast damping of system oscillations with lower amplitudes.

7.6 Summary

In this chapter, a fuzzy logic technique was applied to virtual inertia control to enable self-adaptive inertia mechanism, ensuring secure power system operation under high RESs/DGs penetration. The fundamental of fuzzy logic was discussed, and the recent achievements in the fuzzy applications for frequency control problems are briefly reviewed. Then, a decentralized fuzzy-based virtual inertia controller was designed to schedule the self-adaptive virtual inertia constant with a suitable value for responding the different levels of RESs/DGs penetration regarding frequency control. Lastly, the efficiency of the proposed control scheme is demonstrated through a nonlinear simulation under a wide range of critical operating conditions including the mismatch parameters of primary and secondary control services.

References

1. C.T. Lin, C.S.G. Lee, *Neural Fuzzy Systems* (Prentice-Hall, NJ, USA, 1996)
2. M. Sugeno, An introductory survey of fuzzy control. Inf. Sci. (Ny) **1**(36), 59–83 (1985)
3. E.H. Mamdani, Application of fuzzy algorithms for control of simple dynamic plant. Proc. Inst. Electr. Eng. **121**(1), 1585–1588 (1974)
4. L.P. Holmblad, J.J. Ostergaard, Control of a cement Kiln by fuzzy logic. Fuzzy Inf Decis Process. **1**(1), 389–399 (1982)

5. H. Oshima, S. Yasunobu, S. Ichi Sekino, Automatic train operation system based on predictive fuzzy control, in Proc. *International Workshop on Artificial Intelligence for Industrial Applications*, 485–489 (1988)
6. J. Yen, R. Langari, L.A. Zadeh, *Industrial Applications of Fuzzy Logic and Intelligent System* (IEEE Press, NJ, USA, 1995)
7. J. Yen, R. Langari, *Fuzzy Logic: Intelligence, Control and Information* (Prentice-Hall, NJ, USA)
8. M. Mouzé-Amady, E. Raufaste, H. Prade, J.P. Meyer, Fuzzy-TLX: Using fuzzy integrals for evaluating human mental workload with NASA-Task Load indeX in laboratory and field studies. Ergonomics **56**(5), 752–763 (2013)
9. L.L.L. McLauchlan, Fuzzy logic controlled landing of a Boeing 747, in Proc. *IEEE/RSJ International Conference on Intelligent Robots and Systems (IROS 2009)*, 5369–5375 (2009)
10. J.M. Mendel, H. Hagras, W.W. Tan, W.W. Melek, H. Ying, *Introduction To Type-2 Fuzzy Logic Control: Theory and Applications* (IEEE-Wiley, NJ, USA, 2014)
11. J.M. Mendel, *Electric Uncertain Rule-Based Fuzzy Systems* (Springer, Switzerland, 2018)
12. D.K. Chaturvedi, O.P. Malik, Neurofuzzy power system stabilizer. IEEE Trans. Energy Convers. **23**(3), 887–894 (2008)
13. P. Hoang, K. Tomsovic, Design and analysis of an adaptive fuzzy power system stabilizer. IEEE Trans. Energy Convers. **11**(2), 455–461 (1996)
14. I.H. Altas, J. Neyens, A fuzzy logic decision maker and controller for reducing load frequency oscillations in multi-area power systems, in Proc. *IEEE PES General Meeting (IEEE PES GM)*, 1–6 (2006)
15. T. Chaiyatham, I. Ngamroo, Improvement of power system transient stability by PV farm with fuzzy gain scheduling of PID controller. IEEE Syst. J. **11**(3), 1684–1691 (2017)
16. K.A. Varma, D.K. Mohanta, M.J.B. Reddy, Applications of type-2 fuzzy logic in power systems: a literature survey, in Proc. *International Conference on Environment and Electrical Engineering (EEEIC 2013)*, 1–6 (2013)
17. Y.A. Baysal, I.H. Altas, A fuzzy reasoning approach for optimal location and sizing of shunt capacitors in radial power systems, in Proc. *IEEE Energy Conversion Congress and Exposition*, 1–6 (2015)
18. M. Cortés-Carmona, R. Palma-Behnke, G. Jiménez-Estévez, Fuzzy arithmetic for the DC load flow. IEEE Trans. Power Syst. **25**(1), 206–214 (2010)
19. S. Kumar, D.K. Chaturvedi, Optimal power flow solution using fuzzy evolutionary and swarm optimization. Int. J. Electr. Power Energy Syst. **47**(1), 416–423 (2013)
20. A. Gupta, D.K. Jain, S. Dahiya, Management of power exchange between hybrid microgrids using intelligent control, in Proc. *IEEE International Conference on Power Systems (ICPS 2016)*, 1–6 (2016)
21. W.R. Anis Ibrahim, M.M. Morcos, Artificial intelligence and advanced mathematical tools for power quality applications: a survey. IEEE Trans. Power Deliv. **17**(2), 668–673 (2002)
22. L. Suganthi, S. Iniyan, A.A. Samuel, Applications of fuzzy logic in renewable energy systems—a review. Renew. Sustain. Energy Rev. **48**(1), 585–607 (2015)
23. M.E. El-Hawary, *Electric Power Applications of Fuzzy Systems* (IEEE Press, NJ, USA, 2010)
24. I.H. Altas, *Fuzzy Logic Control in Energy Systems with Design Applications in MatLab/Simulink* (The Institution of Engineering and Technology, UK, 2017)
25. K. Hongesombut, T. Kerdphol, Robust interline power flow controller with wind power source using phase-plane fuzzy logic control. Trans. Electr. Eng. Electron. Commun. **11**(2) (2013)
26. N. Tephiruk, W. Kanokbannakorn, T. Kerdphol, Y. Mitani, K. Hongesombut, Fuzzy logic control of a battery energy storage system for stability improvement in an islanded microgrid. Sustain. **10**(5), 1–16 (2018)
27. C.S. Rao, S.S. Nagaraju, P.S. Raju, Automatic generation control of TCPS based hydrothermal system under open market scenario: a fuzzy logic approach. Int. J. Electr. Power Energy Syst. **18**(3), 266–280 (2009)
28. A.P. Fathima, M.A. Khan, Design of a new market structure and robust controller for the frequency regulation service in the deregulated power system. Electr. Power Components Syst. **36**(1), 864–883 (2008)

29. P. Attaviriyanupap, H. Kita, E. Tanaka, J. Hasegawa, A fuzzy-optimization approach to dynamic economic dispatch considering uncertainties. IEEE Trans. Power Syst. **19**(3), 1299–1307 (2004)
30. C.F. Juang, C.F. Lu, Load-frequency control by hybrid evolutionary fuzzy PI controller. IEE Proc. Gener. Transm. Distrib. **153**(2), 196–204 (2006)
31. G.A. Chown, R.C. Hartman, Design and experience with a fuzzy logic controller for automatic generation control (AGC). IEEE Power Eng. Rev. **13**(3), 965–970 (1997)
32. A. Feliachi, D. Rerkpreedapong, NERC compliant load frequency control design using fuzzy rules. Electr. Power Syst. Res. **73**(2), 101–106 (2005)
33. A. Demiroren, E. Yesil, Automatic generation control with fuzzy logic controllers in the power system including SMES units. Int. J. Electr. Power Energy Syst. **26**(1), 291–305 (2004)
34. M.K. El-Sherbiny, G. El-Saady, A.M. Yousef, Efficient fuzzy logic load-frequency controller. Energy Convers. Manag. **43**(1), 1853–1863 (2002)
35. E. Yeşil, M. Güzelkaya, I. Eksin, Self tuning fuzzy PID type load and frequency controller. Energy Convers. Manag. **45**(1), 377–390 (2004)
36. E. Çam, I. Kocaarslan, A fuzzy gain scheduling PI controller application for an interconnected electrical power system. Electr. Power Syst. Res. **73**(3), 329–340 (2005)
37. Y.L. Karnavas, D.P. Papadopoulos, AGC for autonomous power system using combined intelligent techniques. Electr. Power Syst. Res. **62**(1), 225–239 (2002)
38. P. Subbaraj, K. Manickavasagam, Automatic generation control of multi-area power system using fuzzy logic controller. Eur. Trans. Electr. Power **18**(3), 266–280 (2008)
39. M.H. Ali, T. Murata, J. Tamura, Transient stability enhancement by fuzzy logic-controlled SMES considering coordination with optimal reclosing of circuit breakers. IEEE Trans. Power Syst. **23**(2), 631–640 (2008)
40. Q. Yaser, T. Kerdphol, Y. Mitani, Different optimization schemes for community based energy storage systems, in Proc. *International Conference on Electric Power and Energy Conversion Systems*, 1–5 (2015)
41. N.J. Schouten, M.A. Salman, N.A. Kheir, Fuzzy logic control for parallel hybrid vehicles. IEEE Trans. Control Syst. Technol. **10**(3), 460–468 (2002)
42. P.D. Brown, J.A. Peças Lopes, M.A. Matos, Optimization of pumped storage capacity in an isolated power system with large renewable penetration. IEEE Trans. Power Syst. **23**(2), 523–531 (2008)
43. A. A. Ferreira, J. A. Pomilio, G. Spiazzi, L. de Araujo Silva, Energy management fuzzy logic supervisory for electric vehicle power supplies system. IEEE Trans. Power Electron. **23**(1), 107–115 (2008)
44. N. Sa-ngawong, I. Ngamroo, Intelligent photovoltaic farms for robust frequency stabilization in multi-area interconnected power system based on PSO-based optimal Sugeno fuzzy logic control. Renew. Energy **74**, 555–567 (2015)
45. R.H. Liang, J.H. Liao, A fuzzy-optimization approach for generation scheduling with wind and solar energy systems. IEEE Trans. Power Syst. **22**(4), 1665–1674 (2007)
46. I.G. Damousis, M.C. Alexiadis, J.B. Theocharis, P.S. Dokopoulos, A fuzzy model for wind speed prediction and power generation in wind parks using spatial correlation. IEEE Trans. Energy Convers. **19**(2), 352–361 (2004)
47. Y.T. Chen, Y.C. Jhang, R.H. Liang, A fuzzy-logic based auto-scaling variable step-size MPPT method for PV systems. Sol. Energy **126**(1), 53–63 (2016)
48. T.L. Kottas, Y.S. Boutalis, A.D. Karlis, New maximum power point tracker for PV arrays using fuzzy controller in close cooperation with fuzzy cognitive networks. IEEE Trans. Energy Convers. **21**(3), 793–803 (2006)
49. Y. Hirase, K. Abe, K. Sugimoto, Y. Shindo, A grid-connected inverter with virtual synchronous generator model of algebraic type. Electr. Eng. Japan (English Transl. Denki Gakkai Ronbunshi) **132-B**(4), 10–21 (2013)
50. S. D'Arco, J.A. Suul, O.B. Fosso, A virtual synchronous machine implementation for distributed control of power converters in SmartGrids. Electr. Power Syst. Res. **122**(1), 180–197 (2015)

51. T. Shintai, Y. Miura, T. Ise, Oscillation damping of a distributed generator using a virtual synchronous generator. IEEE Trans. Power Deliv. **29**(2), 668–676 (2014)
52. L. Sigrist, I. Egido, E. Lobato Miguelez, L. Rouco, Sizing and controller setting of ultracapacitors for frequency stability enhancement of small isolated power systems. IEEE Trans. Power Syst. **30**(4), 2130–2138 (2015)
53. N. Soni, S. Doolla, M.C. Chandorkar, Improvement of transient response in microgrids using virtual inertia. IEEE Trans. Power Deliv. **28**(3), 1830–1838 (2013)
54. T. Kerdphol, F.S. Rahman, Y. Mitani, Virtual inertia control application to enhance frequency stability of interconnected power systems with high renewable energy penetration. Energies **11**(4), 981–997 (2018)
55. E. Rakhshani, D. Remon, A.M. Cantarellas, J.M. Garcia, P. Rodriguez, Virtual synchronous power strategy for multiple HVDC interconnections of multi-area AGC power systems. IEEE Trans. Power Syst. **32**(3), 1665–1677 (2017)
56. E. Rakhshani, P. Rodriguez, Inertia emulation in AC/DC interconnected power systems using derivative technique considering frequency measurement effects. IEEE Trans Power Syst **32**(5), 3338–3351 (2017)
57. P. Rodriguez, E. Rakhshani, A. Mir Cantarellas, D. Remon, Analysis of derivative control based virtual inertia in multi-area high-voltage direct current interconnected power systems. IET Gener. Transm. Distrib. **10**(6), 1458–1469 (2016)
58. T. Kerdphol, M. Watanabe, K. Hongesombut, Y. Mitani, Self-adaptive virtual inertia control-based fuzzy logic to improve frequency stability of microgrid with high renewable penetration. IEEE Access **7**(1), 76071–76083 (2019)
59. J. Alipoor, Y. Miura, T. Ise, Power system stabilization using virtual synchronous generator with alternating moment of inertia. IEEE J. Emerg. Sel. Top. Power Electron. **3**(2), 451–458 (2015)
60. D. Li, Q. Zhu, S. Lin, X.Y. Bian, A self-adaptive inertia and damping combination control of VSG to support frequency stability. IEEE Trans. Energy Convers. **32**(1), 397–398 (2017)
61. M. A. Torres L., L.A.C. Lopes, L.A. Moran T., J.R. Espinoza C, Self-tuning virtual synchronous machine: a control strategy for energy storage systems to support dynamic frequency control. IEEE Trans. Energy Convers. **29**(4), 833–840 (2014)
62. J. Harris, *Fuzzy Logic Application in Engineering Science* (Springer, Netherlands, 2006)
63. J. Espinosa, J. Vandewalle, V. Wertz, *Fuzzy Logic, Identification and Predictive Control* (Springer, London. UK., 2005)
64. A. Gunasekaran, *Agile Manufacturing: The 21st Century Competitive Strategy* (Elsevier, UK, 2001)
65. A. Soteris, *Kalogirou, Solar Energy Engineering* (Acedemic Press-Elsevier, USA., 2013)
66. H.D. Jude, D. Gupta, V.E. Balas, *Intelligent Data Analysis for Biomedical Applications.* (Acedemic Press-Elsevier, USA, 2019)
67. C. Wagner, H. Hagras, Toward general type-2 fuzzy logic systems based on zSlices. IEEE Trans. Fuzzy Syst. **18**(4), 637–660 (2012)
68. L. Wang, *Model Predictive Control System Design and Implementation Using MATLAB®* (Springer-Verlag, London, UK, 2009)
69. AEMC Reliability Panel, Stage One Final Determination: Review of the Frequency Operating Standard, Final Report, Australia, Sydney (2017)

Chapter 8
Synthesis of Robust Virtual Inertia Control

Abstract Regarding the previously elaborated inertia control techniques, they are not explicitly designed to deal with the effect of high uncertainty and disturbance. Thus, it is difficult to achieve a suitable trade-off between robust and nominal performances. In adaptive control techniques, the uncertainty formulation may not be appropriately included in the control design process. As a result, it is difficult to ensure the simultaneous robust performance and stability of the virtual inertia control in the presence of bounded modeling errors. Due to the capability of uncertainty formulation in its control synthesis, the robust control technique successfully resolves the concerned problem. Compared with the adaptive control theory, the robust control theory is static rather than adapting to measurements of variations. Subsequently, the robust controller is specially designed to operate assuming that certain varibles will be unknown but bounded. This chapter presents the application of robust uncertainty modeling theory for designing the $H\infty$ robust virtual inertia control system in the presence of high renewable energy sources (RESs)/distributed generators (DGs) penetration. Practical constraints and system uncertainties are appropriately considered during the robust synthesis process. The $H\infty$ robust control is used via a developed linear matrix inequalities (LMI) algorithm to reach an optimal solution between nominal and robust performances for design objectives. The robustness and performance of the $H\infty$-based virtual inertia controller are executed along with different sets of severe parametric uncertainty and external disturbance. The closed-loop system is verified through a nonlinear control system under the critical operating scenarios of uncertain control parameters with high RESs/DGs penetration.

Keywords $H\infty$ control · $H\infty$ norm · $H\infty$-SOF · Dynamic response · Frequency control · Inertia control · Nonlinear control · Performance index · Robust control · Uncertainty · Virtual inertia control · Virtual inertia synthesis

8.1 Introduction to Robust Virtual Inertia Control

Power systems are currently operated under serious conditions to provide a stable supply of electricity. Emerging various generations, increased competition, open transmission access, energy crisis, economic growth, and environmental concerns

T. Kerdphol et al., *Virtual Inertia Synthesis and Control*, Power Systems,
https://doi.org/10.1007/978-3-030-57961-6_8

are transforming the power system operations in new ways, which lead to major challenges for securing system operation [1]. Different ownership of generation, transmission, and distribution causes more complexity to power system control due to the reduction of coordination in system planning [2, 3]. As a result, the rising number of major blackouts has been reported as follows. In 2017, the northern Taiwan suffered from massive power blackout about 4 h due to an error from a power supply equipment [4]. A blast at the Northridge power plant caused a widespread blackout in San Fernando Valley, Los Angeles, for 13 h [5]. In 2018, a power blackout affected tens of millions of people in Brazil due to a failure of transmission lines [6]. An unexpected high temperature in a power plant caused a major power blackout in Azerbaijan for 8 h [7]. A power plant failure after the earthquake caused a power blackout in Hokkaido Island, Japan [8]. In 2019, a cascading failure from one power plant in Venezuela triggered the nation-wide blackouts for ten days [9]. A major power blackout affected 48 million people among Argentina, Uruguay, and Paraguay, and the cause remains unknown [10]. In Java, Indonesia, the problem in the power system caused a massive blackout in the western part of Java, including the capital city of Jakarta, for 20 h [11]. A failure in generators caused by lightning strike led to a major blackout in England and Wales, affecting nearly a million people [12]. With the increasing number of power blackouts, the power system operations need more awareness of all forms of control and stability issues.

Recently, the integrations of renewable energy sources (RESs) and distributed generators (DGs) have changed the power system generating infrastructures from a centralized to decentralized systems. Subsequently, it created a two-directional power transfer pattern, which results in new stability problems [13]. Moreover, intermittent behaviors of RESs/DGs could lead to the increased uncertainties and nonlinearity of the exploitation of transmission assets and corrected allocation of controls. In the worst situation, RESs/DGs may trip from the power grid, which could lead to system instability, cascading failures, and power blackouts [14, 15]. By interconnecting power systems, significant frequency deviations are considered as a major source of the cascading failures, and system collapses [16]. Thus, the violation of system frequency is reported as the major reason for power blackouts [17]. Operating power systems with RESs/DGs will be more complex than ever, due to the considerable levels toward high integration, and due to technical and economic limitations together with conventional requirements of system zsecurity and reliability.

Several organizations and countries have set their aggressive goals to integrate RESs/DGs towards high levels of RESs/DGs. For example, the United States has planned to install 80% of RESs/DGs by 2050. The California independent system operator (CAISO) was successfully integrated 50% of RESs/DGs (i.e., mainly solar) into the grid by 2013 [18]. In Europe, the European Union plans to achieve 20% and 55% of RESs/DGs by 2020 and 2050, respectively [19]. In 2050, Denmark plans to remove fossil energy, while Germany has promised to transform its electricity to 100% RESs/DGs (i.e., mainly wind and solar) [20]. According to the Australian energy market operator (AEMO)' s plan, 75% of RESs/DGs will be integrated into the grid by 2025 [21]. In Asia, China plans to achieve 80% of RESs/DGs by 2050. Japan also set a similar goal of installing 80% RESs/DGs by 2050. Numerous studies report that RESs/DGs could meet 100% of global energy demand by 2050 [22–24]. With increasing RESs/DGs toward high penetration levels, system inertia and damping are significantly reduced, which result in the weakening of system stability and resiliency. The high RESs/DGs penetration can cause virtual inertia control approaches to be unstable to provide a suitable performance over a wide range of system operations. Consequently, the inertia control systems might be inadequate and unstable after high-level integration of RESs/DGs, resulting in system instability. In the worst case, the inertia control systems may stop working due to fluctuations with high amplitude after RESs/DGs integration. This situation can lead to cascading failures and power blackouts. Hence, it is important to offer more robust inertia control techniques considering the bounded uncertainty to obtain a new tradeoff between system efficiency, security and dynamic robustness.

Regarding the previous inertia control techniques in Chaps. 5, 6, and 7, the significant uncertainty and behavior regarding high RESs/DGs penetration, system inertia, and damping reductions are neglected in their control synthesis. Additionally, the uncertainty formulation (e.g., structured uncertainty modeling) has not been considered and designed. Subsequently, it is difficult to acquire an appropriate trade-off between robust and nominal performances [3]. Moreover, an inaccuracy of estimated parameters for a designed virtual inertia control may occur. The virtual inertia control parameters may constantly change by time, significantly reducing system performance and stability. Hence, it is not easy to offer the simultaneous robust performance and stability during high RESs/DGs penetration. Due to the capability of uncertainty formulation in the robust control synthesis, this method could effectively solve the mentioned issues. The robust control is a method to design a controller that explicitly deals with the uncertainty. The uncertain parameters and disturbances are properly designed in the robust control process, while the certain variables are assumed to be unknown but bounded. The most important technique is known as the $H\infty$ (loop-shaping) control, which was developed by Keith Glover and Duncan McFarlane from Cambridge University. This technique could minimize the sensitivity of a control system against its frequency spectrum. Thus, it could guarantee that the system will not significantly deviate from the desirable trajectories when the disturbances occur in the system.

In the last decades, numerous research works on robust frequency control for power systems have been presented as follows. Robust frequency control techniques are designed for a single area system [25–27] and interconnected systems [28–30]. The $H\infty$ robust controller was developed for each generating unit of a power system in [31]. The $H\infty$ robust controller was built for power-sharing in both islanded and interconnected systems [32–34]. The $H\infty$ robust controller was established for secondary frequency control [25–27, 35, 36]. Then, the $H\infty$ robust controller was created for regulating energy storage systems [37–41] with regards to RESs penetration. Focusing on frequency regulation with regards to virtual inertia control, the primary emphasis is given on the improvement of virtual inertia control in the power electronics field (e.g., control and equipment scales), without analyzing the dynamic impact of high DGs/RESs integration. Thus, it is anticipated that integrating the robust virtual inertia control in a new environment can be more flexible and adaptive than traditional methods.

This chapter emphasizes on a new $H\infty$ robust control design for virtual inertia synthesis considering high RESs/DGs penetration. The proposed robust method is flexible enough to include uncertainties and disturbances in the system modeling process, significantly improving frequency stability of the system. The linear fractional transformation (LFT) process is added in the $H\infty$ control design, and the parametric perturbation is included as one block of bounded uncertainty. The developed linear matrix inequalities (LMI) procedure is used to determine an optimal solution between nominal and robust performances for the design objectives. Additionally, the comparative analysis between $H\infty$ and optimal PI-based virtual inertia control is performed. This chapter is an extended version of the research presented in [39]. The main contribution of this chapter over the existing virtual inertia control is listed as:

1. Considering the system inertia and damping as bounded sectors of uncertainties in the $H\infty$ control design process;
2. Considering the high penetration of RESs/DGs and loads as bounded sectors of disturbances in the $H\infty$ control design process;
3. Developing a robust $H\infty$-based virtual inertia controller to minimize frequency deviation of the system, guaranteeing stable system stability and resiliency against the severe disturbance and uncertainty.

8.2 $H\infty$ Robust Control Theory

In this section, a brief overview of the $H\infty$ (loop-shaping) robust control theory and design is described. The $H\infty$ robust control combines the conventional intuition of classical control techniques (e.g., Bode sensitivity analysis) with the $H\infty$ optimization techniques to perform the robust controllers whose the performance and stability maintain the bounded differences between the actual system (plant) in practice and the nominal system in design. The control design system expresses the desirable response and noise-suppression properties by weighting the transfer function of the system in the frequency domain, while the closed-loop-shaping is robustified via the optimization. From the control theorem point of view, a linear time-invariant system $G(s)$ including the state-space representation is expressed as [3, 39]:

$$\dot{x} = Ax + B_1 w + B_2 u \tag{8.1}$$

$$z = C_1 x + D_{12} u \tag{8.2}$$

$$y = C_2 x \tag{8.3}$$

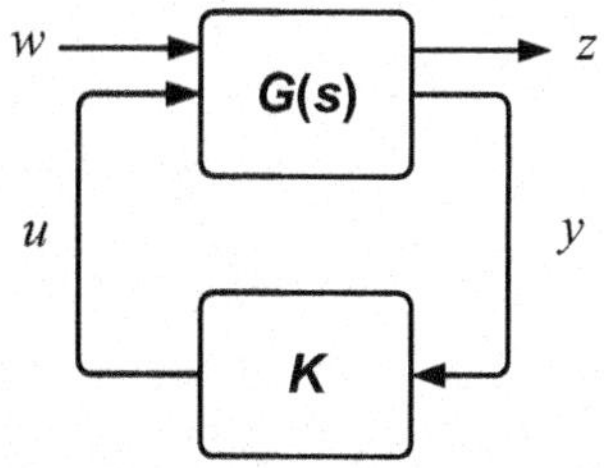

Fig. 8.1 The $H\infty$ robust control structure with a closed-loop system

where z is the controlled output vector, x is the state variable vector, w is the disturbance and other external input vectors, and y is the measured output vector.

The $H\infty$ robust control technique for $G(s)$ with the above state-space representation is formed to evaluate a given matrix K (i.e., static output feedback law $u = Ky$), see Fig. 8.1. To determine a suitable K matrix, two bounded constraints are expressed as:

1. The obtained closed-loop system must internally be stable.
2. The norm of $H\infty$ from w to z must be smaller than a specified positive number γ as:

$$\|T_{Zw}\|_\infty < \gamma \tag{8.4}$$

With the assumptions on system matrices, the following theory is applied to the $H\infty$ robust control issue.

Theory 8.1 The first theory is supposed that (A, B_2, C_2) is measurable and stabilizable. The proof of (8.5) is given in [42]. By solving matrix representation, the $H\infty$ dynamic controller is represented by the matrix K, if and only if there is a symmetric matrix $X > 0$ as [3, 39]:

$$\begin{bmatrix} A_{cl}^T X + X A_{cl} & X B_{cl} & C_{cl}^T \\ B_{cl}^T X & -\gamma I & D_{cl}^T \\ C_{cl} & D_{cl} & -\gamma I \end{bmatrix} < 0 \tag{8.5}$$

where,

$$A_{cl} = A + B_2 K C_2 \tag{8.6}$$

$$B_{cl} = B_1 \tag{8.7}$$

$$C_{cl} = C_1 + D_{12} K C_2 \tag{8.8}$$

$$D_{cl} = 0 \tag{8.9}$$

Then, Eq. (8.5) is re-established in the following form of matrix inequality as:

$$(\bar{A})^T \bar{X} + \bar{A}\bar{X} + K\bar{B}\bar{X}\bar{C} + (K\bar{B}\bar{X}\bar{C})\bar{X}^T < 0 \tag{8.10}$$

where,

$$\bar{A} = \begin{bmatrix} A & B_1 & 0 \\ 0 & -\gamma I/2 & 0 \\ C_1 & 0 & -\gamma I/2 \end{bmatrix} \tag{8.11}$$

$$\bar{B} = \begin{bmatrix} B_2 \\ 0 \\ D_{12} \end{bmatrix} \tag{8.12}$$

$$\bar{C} = \begin{bmatrix} C_2 \, 0 \, 0 \end{bmatrix} \tag{8.13}$$

$$\bar{X} = \begin{bmatrix} X \, 0 \, 0 \\ 0 \, I \, 0 \\ 0 \, 0 \, I \end{bmatrix} \tag{8.14}$$

Therefore, the $H\infty$ robust control-based problem is solved under the condition $X > 0$. Then, the matrix K is obtained by Eq. (8.10). For generalizing and solving a static output feedback stabilization problem of the system $(\bar{A}, \bar{B}, \bar{C})$, theory 8.2 is introduced.

Theory 8.2 This theory is supposed that (A, B, C) are defined in the forms of [3, 39]:

$$\dot{x} = Ax + Bu \tag{8.15}$$

$$y = Cx \tag{8.16}$$

By solving matrix representation, Eqs. (8.15) and (8.16) must be stabilized if and only if there is $X > 0$, $P > 0$, and the matrix K must be satisfied the quadratic matrix inequality [3, 39]:

$$\begin{bmatrix} XA + XA^T - PBB^T X - XBB^T P + PBB^T P & (B^T X + KC)^T \\ B^T X + KC & -I \end{bmatrix} < 0 \tag{8.17}$$

The proof of (8.17) is given in [42].

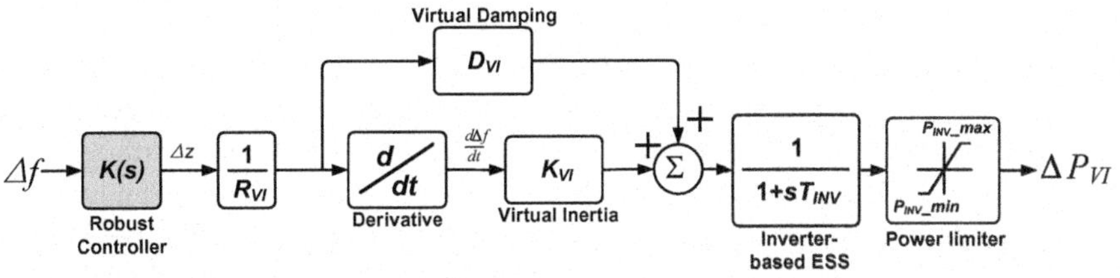

Fig. 8.2 The structure of $H\infty$ robust-based virtual inertia control

8.3 Design of $H\infty$ Robust Virtual Inertia Control

In this part, the control design of the $H\infty$ robust-based virtual inertia controller is constructed to emulate sufficient virtual inertia into the system against bounded disturbance and uncertainty (i.e., high RESs/DGs penetration and system inertia and damping degradations). Figure 8.2 shows the configuration of the proposed $H\infty$ robust-based virtual inertia control, where z is the control input produced by the robust controller.

To perform detailed analysis with respect to the $H\infty$ robust theory, the complete state-space representation of the system is important. In this study, high penetration levels of wind power (ΔP_W), solar radiation power (ΔP_{PV}), and load power (ΔP_L) are integrated as the disturbance signals. System damping (D) and system inertia (H) are integrated as the uncertain parameters. Then, the state-space model of the single area system with virtual inertia control is expressed in the form of Eqs. (8.1)–(8.3). State variables and coefficients of the state-space model, including the control output x, disturbance signal w, and control input u are given as Eqs. (3.15)–(3.17). The matrices (A, B_1, B_2) can be found in Eqs. (3.26)–(3.28). Finally, the complete state-space equation of the system, including primary control, secondary control, virtual inertia control and RESs/DGs penetration is presented as Eq. (3.25).

8.4 Modeling of Uncertainty and Disturbance

Power system characteristic naturally contains numerous phenomena of uncertainty/disturbance (e.g., continuous variations in system dynamics among generation-load and operating conditions). Thus, the uncertainty problems in system operation turn into an important feature for power system design and control. Considering a robust control research, several works have presented the designs of uncertainty modeling [35, 36, 43] in the systems. This is because the uncertainty (i.e., dynamic perturbation) represents the difference between actual and mathematical/nominal models.

By designing the $H\infty$-based virtual inertia control, the dynamic perturbations are modeled by lumping into a perturbation block of $\Delta(s)$. Then, the output multiplicative perturbation method is implemented for the uncertainty modeling of system inertia and damping. High power penetration levels of wind, solar, and load are considered as multiplicative disturbances. System inertia and damping are regarded as multiplicative uncertainties with ±50% change. These uncertainties have been extorted from the system as the structured uncertainty. Then, the uncertainty is modeled based on a basic configuration of the linear fractional transformation (LFT). Figure 8.3 shows the closed-loop model for the $H\infty$ control design procedure, involving the lumped uncertainty and weighting functions. $G(s)$ and $P(s)$ are the nominal system model without perturbation and nominal system with the actual dynamics (i.e., perturbed system/variable), respectively. $K(s)$ is the $H\infty$ robust controller. The weighting function $W_e(s)$ has the low-pass characteristic to examine the modeling error and robust stability. The purpose for applying a low-pass filter on the output y is due to no tracking needed at very high frequency for power system frequency control. The weighting function $W_u(s)$ has the high-pass characteristic to determine the weight on the control input u. In this work, the high penetration of the wind power change (ΔP_W), the solar irradiation power change (ΔP_{PV}), and the load power change (ΔP_L), are analyzed as the external disturbances. w is the disturbance signal, which is chosen to augment the robust stability and performance of the system. Δ_{in} and Δ_{out} are the uncertainty input and output. z is the desired performance signal. u is the control signal from the $H\infty$ controller. y is the measured output. To obtain suitable closed-loop performance and stability margin, the selections of the high-pass and low-pass components can be defined as:

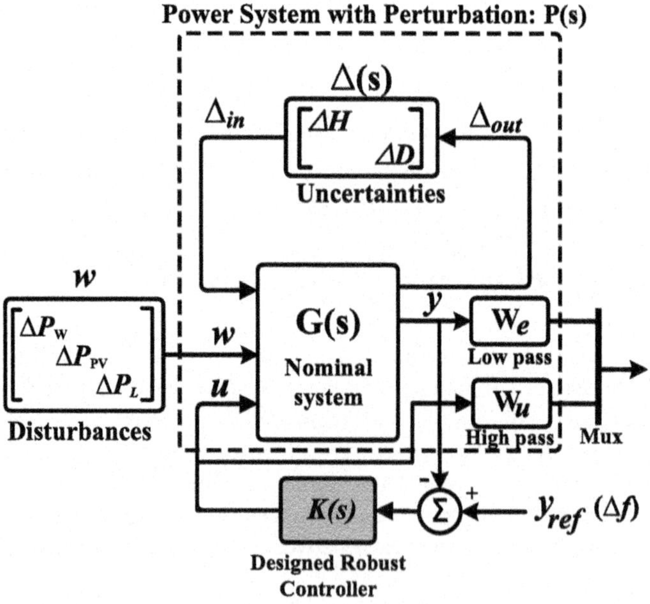

Fig. 8.3 The closed-loop system outline involving lumped multiplicative disturbances (high RESs penetration) and uncertainty (system inertia and damping variations)

$$W_e(s) = 30\frac{s+90}{s+0.01} \tag{8.18}$$

$$W_u(s) = 0.01\frac{0.5s+60}{0.0001s+20} \tag{8.19}$$

To suitably evaluate the weighting functions, the category and target of weighting functions (e.g., high or low pass filter) are initially computed. Then, we start by an initial weight, and continue the tuning procedure considering the closed-loop system performance based on Eq. (8.20) in a simulated environment until reaching the desirable performance. For more details, W_e and W_u are calculated by the appropriate technique provided in [3].

8.4.1 H∞ *Controller Design*

The $H\infty$ robust controller has an outstanding ability in handeling with nonlinear tracking issues by proposing a systematic structure for forming a robust nonlinear controller. This approach could properly determine a sufficient robust controller by minimizing the infinite (∞) norm of $E(G,K)$ as [3, 39]:

$$E(G, K)_{\infty} < 1 \tag{8.20}$$

Figure 8.4 shows the fundamental closed-loop LFT for the $H\infty$ control design. $E(G,K)$ is the transfer function matrix of the nominal closed-loop system from disturbance inputs to controlled outputs, which is represented by the transfer function T_{wz}. Considering Eq. (8.20), it is found that the optimization problem is not unique. Hence, the stabilizing K controller is optimally evaluated by the $H\infty$ norm. This means the $H\infty$ norm of (8.20) must be less than one.

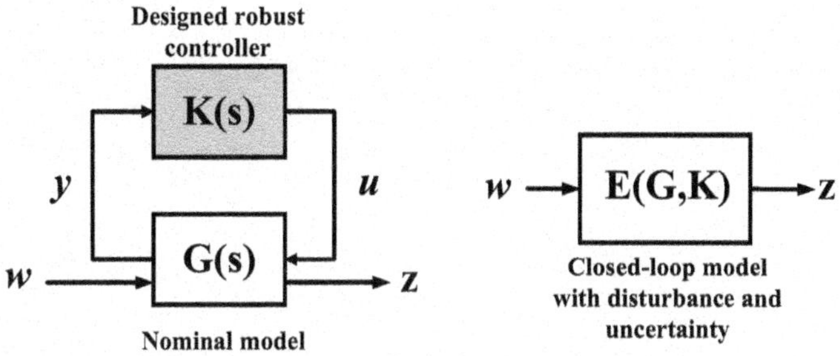

Fig. 8.4 The closed-loop system via $H\infty$ control framework

8.5 Closed-Loop Nominal Stability and Performance

The nominal performance and stability are fulfilled when the closed-loop system T_{wz} is internally stable for the calculated $K\infty$ controller. To determine the nominal performance and stability, the ∞-norm of the $K(s)$ function should be found less than a positive value. By implementing Eq. (8.21), the nominal stability and performance are evaluated, where W_e and W_u are the weighting functions [44, 45].

$$\left\| \begin{bmatrix} W_e(I+GK)^{-1} \\ W_u K(I+GK)^{-1} \end{bmatrix} \right\|_\infty < 1 \tag{8.21}$$

By employing the MATLAB® platform called *hinfsyn* from the robust control toolbox, the inequality from Eq. (8.21) is successfully solved. As a result, it is obvious that the ∞-norm inequality of (8.21) is fulfilled and always not go beyond one, see Fig. 8.5. Clearly, the closed-loop system effectively diminishes the disturbance influences (i.e., high integration levels of solar, wind, and load power). Then, the favorable performance is successfully obtained.

8.5.1 Closed-Loop Robust Stability and Performance

This section involves the modeling process of structured uncertainty. By considering the robust constraint, the robust performance and stability are fulfilled when the closed-loop system T_{wz} is internally stable for all possible plants $P = (I + \Delta(s))G(s)$, where Δ is the uncertainty block. To evaluate the robust stability and performance, the calculated $K\infty$ controller should achieve the performance condition for all possible plants involving the uncertainties as [3, 36]:

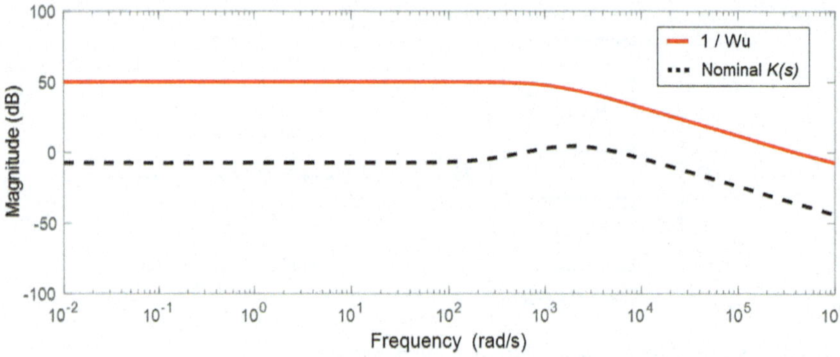

Fig. 8.5 $K(s)$ function of the nominal system

$$\left\| \begin{bmatrix} W_e(I+GP)^{-1} \\ W_u K(I+GP)^{-1} \end{bmatrix} \right\|_\infty < 1 \tag{8.22}$$

In this study, 50% perturbations of the system damping (D) and system inertia (H) are analyzed, as shown in Fig. 8.6. Obviously, the ∞-norm inequalities calculated by Eq. (8.22) are lower than one for all possible plants. In addition, the robust condition of $\|[W_u K(I + GP)^{-1}]\|_\infty < 1$ is satisfied, or $\|[K(I + GP)^{-1}]\|_\infty$ is less than $1/W_u$. Subsequently, the robust performance and stability are successfully found for the calculated K controller via the equation of ∞-norm inequality in Eq. (8.22).

Here, the design procedure of $K(s)$ controller is briefly described in the next step (see Fig. 8.7) as follows:

1. Calculate the state-space model for a studied control system.
2. Compute the constant weights and the optimal (guaranteed) robust performance index γ using the *hinfsyn* command in MATLAB based on a linear matrix inequality (LMI) toolbox to create the standard robust output controller for the studied system in Step 1. Next, set $\Delta\gamma = \Delta\gamma_0$ and let $\gamma = \gamma_0$ for an initial iteration, where γ_0 and $\Delta\gamma_0$ are positive values.
3. Solve the matrix x using the Riccati equation.
4. Minimize Eq. (8.5) subject to LMI constraints.

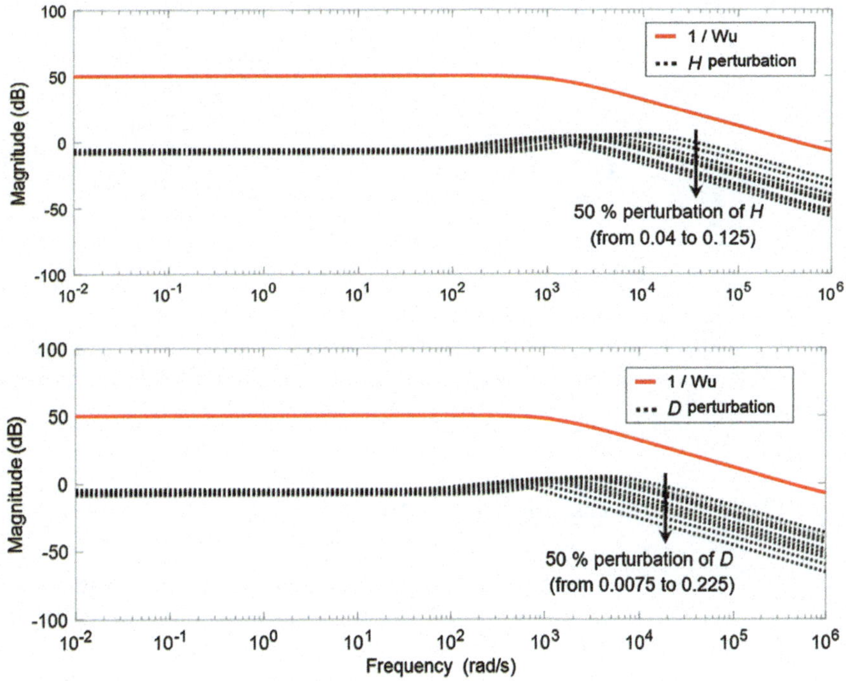

Fig. 8.6 $K(s)$ functions of the system over 50% perturbation

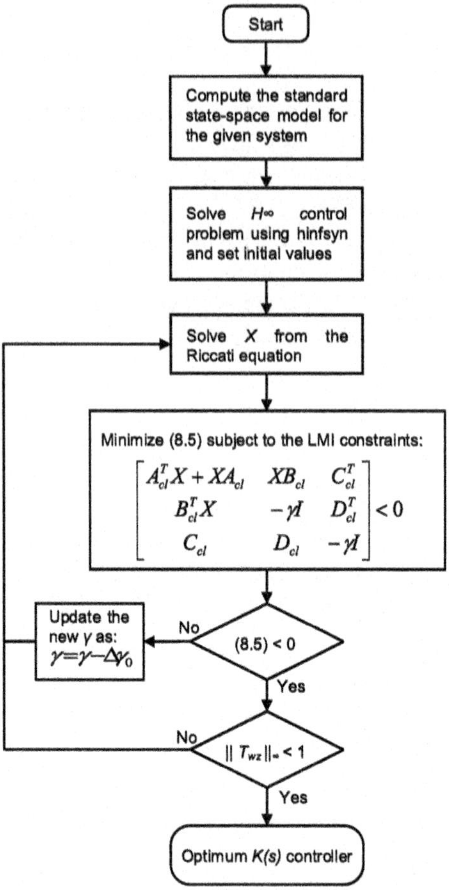

Fig. 8.7 The flowchart for determining an optimal $K(s)$ controller

5. Check if Eq. (8.5) is less than zero, proceed to Step 6. Else, compute the new γ for the following iteration and proceed to Step 3.
6. Check if $\|T_{wz}\|_\infty$ is less than one, then the obtained $K(s)$ indicates an optimal $H\infty$ controller such that the controller meets the constraint of Eq. (8.20). Otherwise, proceed to Step 3.

8.6 Order Reduction of $H\infty$ Controller

Most robust and optimum control methods yield complex state feedback with a high-order dynamic controller, which is not feasible for real-world implementation. By integrating the robust control approach, the order of the calculated controller is larger than or at least equal to the order of the studied plants. Consequently, the high-order of the designed $H\infty$ controller is considered as one of the primary issues of robust control methods, especially for high-order systems. To overcome such problem,

Table 8.1 Coefficients of the $H\infty$-based virtual inertia controller

a_6	274.3	b_6	2.188×10^4
a_5	1.856×10^4	b_5	2.783×10^6
a_4	1.975×10^5	b_4	8.049×10^7
a_3	3.837×10^5	b_3	6.682×10^8
a_2	3979	b_2	1.215×10^9
a_1	1.635	b_1	1.148×10^6
a_0	0.0001473	b_0	240.2

numerous approaches are presented for high-order reduction [46–48]. In this work, the Hankel-based order reduction technique [3, 46] is applied to diminish a high-order of the $H\infty$ controller for practical implementation.

Originally, the designed $H\infty$ controller has a high-order of 7th order. The calculated controller with the full-order can be obtained as [3, 25]:

$$K(s) = \frac{b_6 s^6 + b_5 s^5 + \ldots + b_1 s + b_0}{s^7 + a_6 s^6 + a_5 s^5 + \ldots + a_1 s + a_0} \tag{8.23}$$

The coefficients of the calculated $H\infty$ controller in the structure of (8.23) are displayed in Table 8.1. After integrating the Hankel technique, the controller order is suitably diminished to 4th order. Figure 8.8 shows the Bode analysis of the full-order (original) and reduced-order for the calculated $H\infty$ controller. It is confirmed that the high-order controller is declined to the low-order controller without performance reduction. Thus, the reduced-order transfer function of the robust controller is used as follow:

$$K(s) = \frac{2.189 \times 10^4 s^3 + 2.508 \times 10^6 s^2 + 4.846 \times 10^7 s + 3.177 \times 10^4}{s^4 + 261.8 s^3 + 1.525 \times 10^4 s^2 + 154.4 s + 0.01941} \tag{8.24}$$

8.7 Simulation Results

The single area system from Fig. 3.9 and Table 3.1 is used as the studied system. Simulation results under high RESs/DGs penetration (i.e., disturbance), and severe parameter variations (i.e., uncertainty) are performed by the MATLAB®/Simulink software. To determine the efficiency of the $H\infty$ robust-based virtual inertia controller (reduced-order), it is compared with the optimal PI-based virtual inertia controller. The optimal PI parameters are computed using the internal model control (IMC) technique for tuning PI/PID controller described in Chap. 5. The optimal parameters of K_P and K_I are found as 3.581 and 22.114, respectively. Five contrastive case studies are performed to examine the efficiency and robustness of the proposed system.

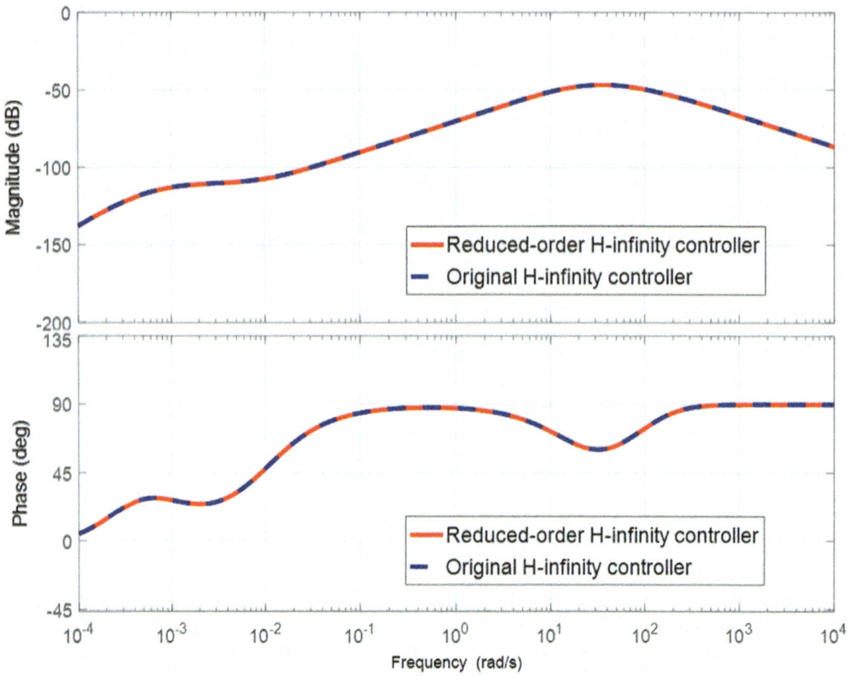

Fig. 8.8 Comparison between original (dashed) and reduced-order (solid) $H\infty$ controllers

8.7.1 Effect of Abrupt Change

Scenario 1: The efficiency of the proposed $H\infty$ robust-based virtual inertia control technique is verified by encountering the sudden load change; that is 10% step change of load power under normal operation (i.e., 100% of system inertia and damping). After the sudden load change at 1 s, it can be seen the system with virtual inertia controllers could successfully augment frequency stability and reduce transient excursion compared with the system without the virtual inertia controller, see Fig. 8.9a. Clearly, by applying the $H\infty$-based virtual inertia controller, the frequency stability and performance of the system are significantly improved. For a clearer comparison, a zoom-in view of the system frequency response is depicted in Fig. 8.9b. Under the effect of sudden disturbance, the $H\infty$-based virtual inertia control produces the outstanding stability in terms of transient curtailment, tracking property, disturbance rejection, and zero steady-state error compared with the optimal PI-based virtual inertia control. In addition, the optimal PI-based virtual inertia control produces larger frequency oscillation and overshoot under the same operating condition.

Scenario 2: This case studies the effect of a similar 10% step load change, while the system inertia and damping are exacerbated to the low operating condition (i.e.,

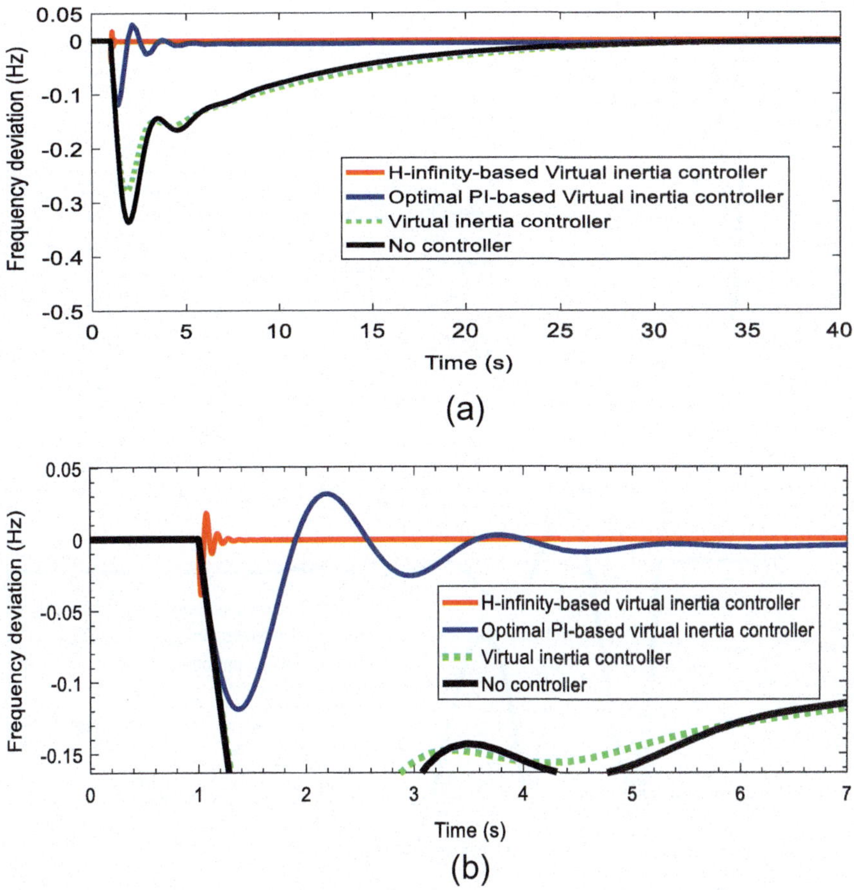

Fig. 8.9 Frequency oscillation: **a** Under high system inertia and damping, **b** Comparison between $H\infty$ and optimal PI controllers under the normal condition of high system inertia and damping

50% reduction from its nominal values). The impact of the system frequency over system inertia and damping reduction is clearly explained in this case. Figure 8.10 shows system frequency under the presence of low system inertia and damping. It is observed that the system frequency is more oscillating, and a lower frequency drop could be seen during this degradation event. Focusing on the transient curtailment, tracking property, disturbance rejection, and error minimization, the system with $H\infty$–based virtual inertia controller could achieve the favorable performance and stability than those of other comparison methods. Under the lack of system inertia and damping, the designed $H\infty$ controller is more robust in dealing with the abrupt disturbance and follows the system operating point better than the optimal PI-based virtual inertia controller.

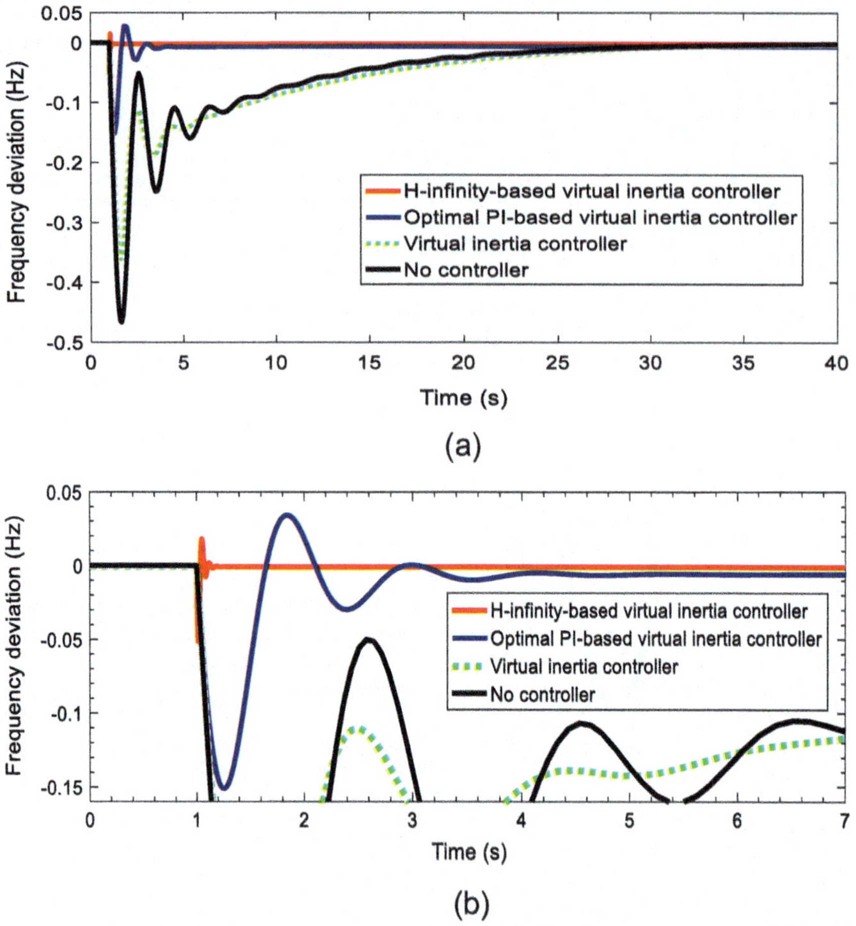

Fig. 8.10 Frequency oscillation: **a** Under 50% reduction of system inertia and damping, **b** Comparison between $H\infty$ and optimal PI controllers under of 50% reduction in system inertia and damping

8.7.2 High Penetration of RESs and Loads

The scenarios 3, 4, and 5 are conducted by applying different penetration levels of RESs/DGs and loads (see Fig. 8.11) with the contrastive operating conditions in Table 8.2. These critical operations demonstrate the dynamic effects of actual (real-world) system operations. The robustness of the proposed $H\infty$ robust controller is verified against high RES/load penetration and system inertia and damping variations. Moreover, the effect of high RESs penetration on overall system frequency characteristics can be evidently observed from these real-world scenarios.

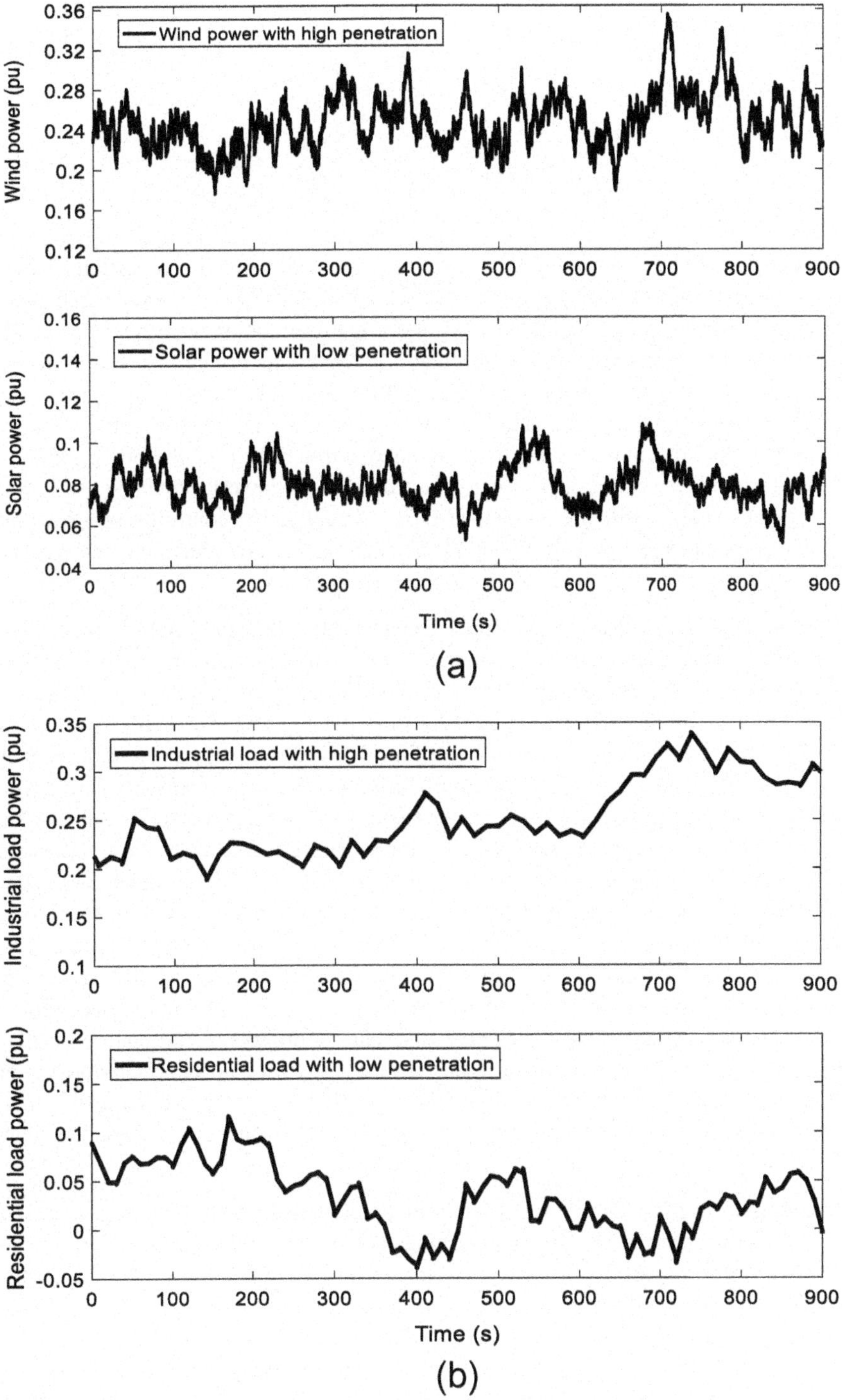

Fig. 8.11 Power deviation pattern: **a** Wind and solar power, **b** Industrial and residential loads

Table 8.2 Multiple operating conditions of the real-world power system

Disturbance source	Starting time (s)	Stopping time (s)
Wind farm	450	–
Solar farm	Initial	–
Residential load	Initial	700
Industrial load	200	–

Scenario 3: The system frequency response is tested under the condition of high system inertia and damping (i.e., 95% of its nominal values). Figure 8.12 displays system frequency response under high RES/load integration with respect to high system inertia and damping environment. Without the virtual inertia controller, it yields the largest frequency transient about ±1.1 Hz during the contingency (at 200 s). By applying the traditional virtual inertia controller, it could improve frequency stability, and the transient is arrested within ±0.7 Hz. Obviously, the $H\infty$-based virtual inertia controller effectively preserves the lowest frequency oscillation within ±0.05 Hz, while the optimal PI-based virtual inertia controller produces the larger frequency oscillation of about ±0.25 Hz. It is observed that the greatest result can be found from the $H\infty$-based virtual inertia controller.

Scenario 4: To perform a more drastic scenario, the studied system is functioned under the partial reduction of system inertia and damping (i.e., 55% reduction from its nominal values). Following the reduction in system inertia and damping together with high RESs/DGs integration, the system frequency response is more oscillating with higher transient and deviation, see Fig. 8.13. It is clear that the reduction of system inertia and damping is particularly affecting the system stability and performance. Without virtual inertia control, the system frequency response turns into instability. Without the proper control scheme, it may cause cascading failures and power blackouts after the connection of the industrial load. During the connection such load, it results in a large frequency transient of −1.2 Hz in the case of the traditional virtual inertia controller. To stabilize the system, it may require the abrupt operation of under-frequency load shedding (UFLS). The load connection at 200 s results in the lower frequency transient of −0.4 Hz in the case of the optimal PI-based virtual inertia controller. The load connection also causes the lowest frequency transient of −0.03 Hz in the case of the $H\infty$-based virtual inertia controller. Obviously, the proposed $H\infty$ controller effectively holds the oscillation within the frequency operating range of ±0.05 Hz according to the real-world frequency operating standard of AEMO.

Scenario 5: To make a critical test scenario, the governor and turbine time constants of the thermal generation (i.e., synchronous generator) are increased by 90% of its nominal values, creating the mismatch (uncertain) parameter effect. This condition could occur during the offline change of the actual turbine and governor, while the controller maintains its nominal values. This means that the thermal power plant becomes unstable, leading to the malfunctioning of primary and secondary control services. Nevertheless, the system is operated under the severe environment of low

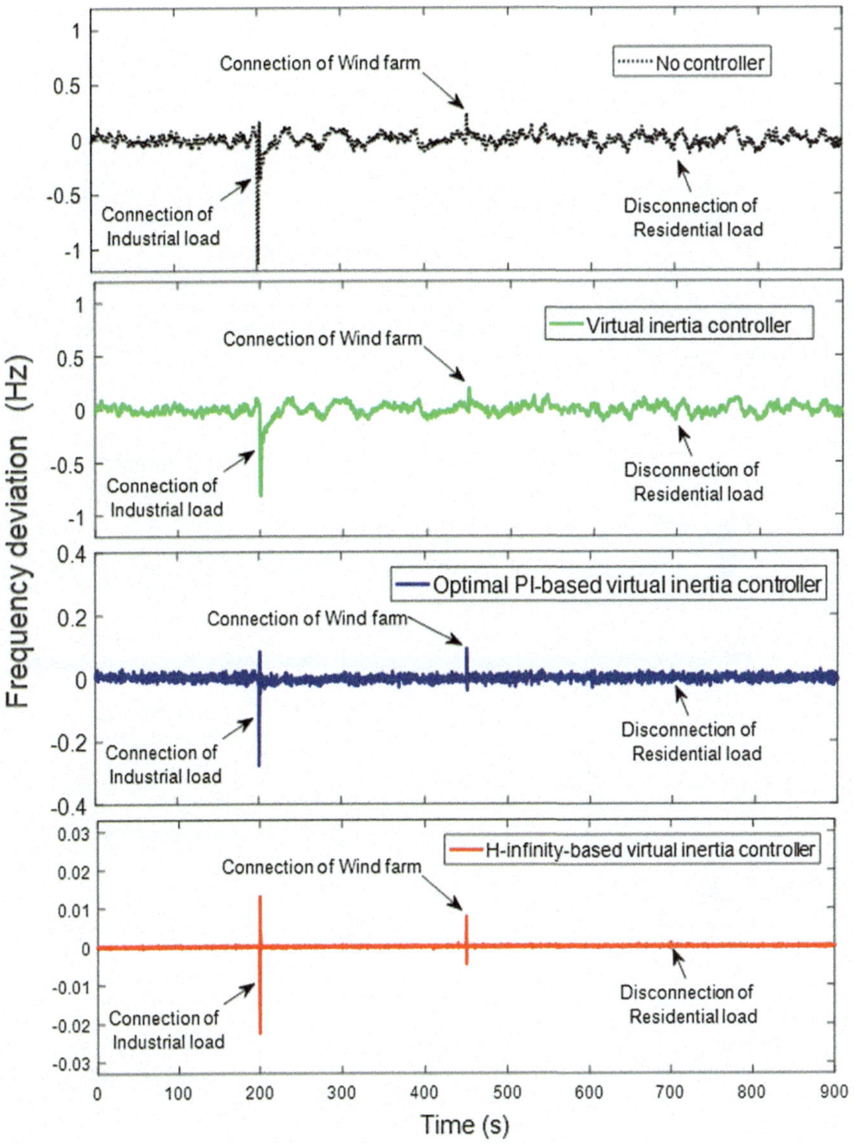

Fig. 8.12 Frequency response of the system under high RESs/DGs penetration with respect to high system inertia and damping

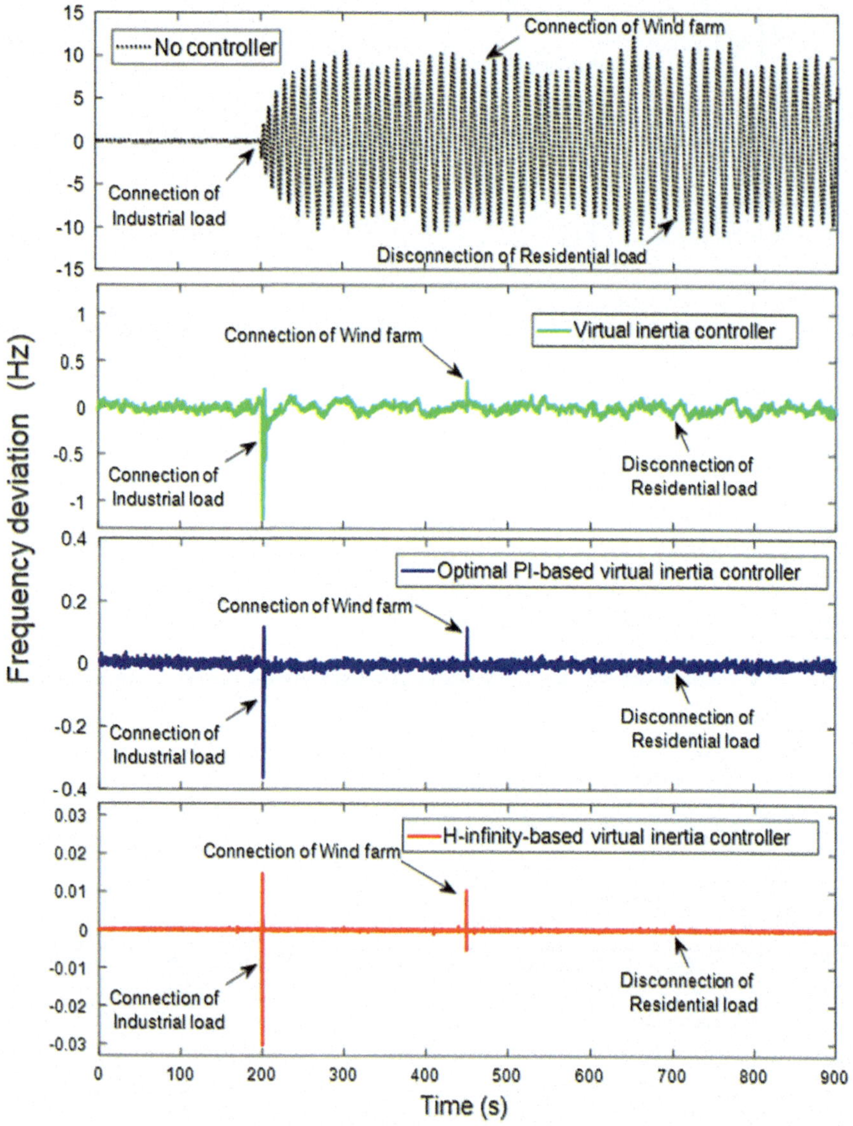

Fig. 8.13 Frequency response of the system under high RESs/DGs penetration with respect to the partial (medium) level of system inertia and damping reduction

system inertia and damping (i.e., 85% reduction from its nominal value). Figure 8.14 shows the system frequency response under the severe operating conditions. It is obvious that the magnitude of the frequency drop greatly increases due to the effect of low system inertia and damping. The lower inertia and damping result in a larger frequency drop. Without the virtual inertia controller, the system completely becomes

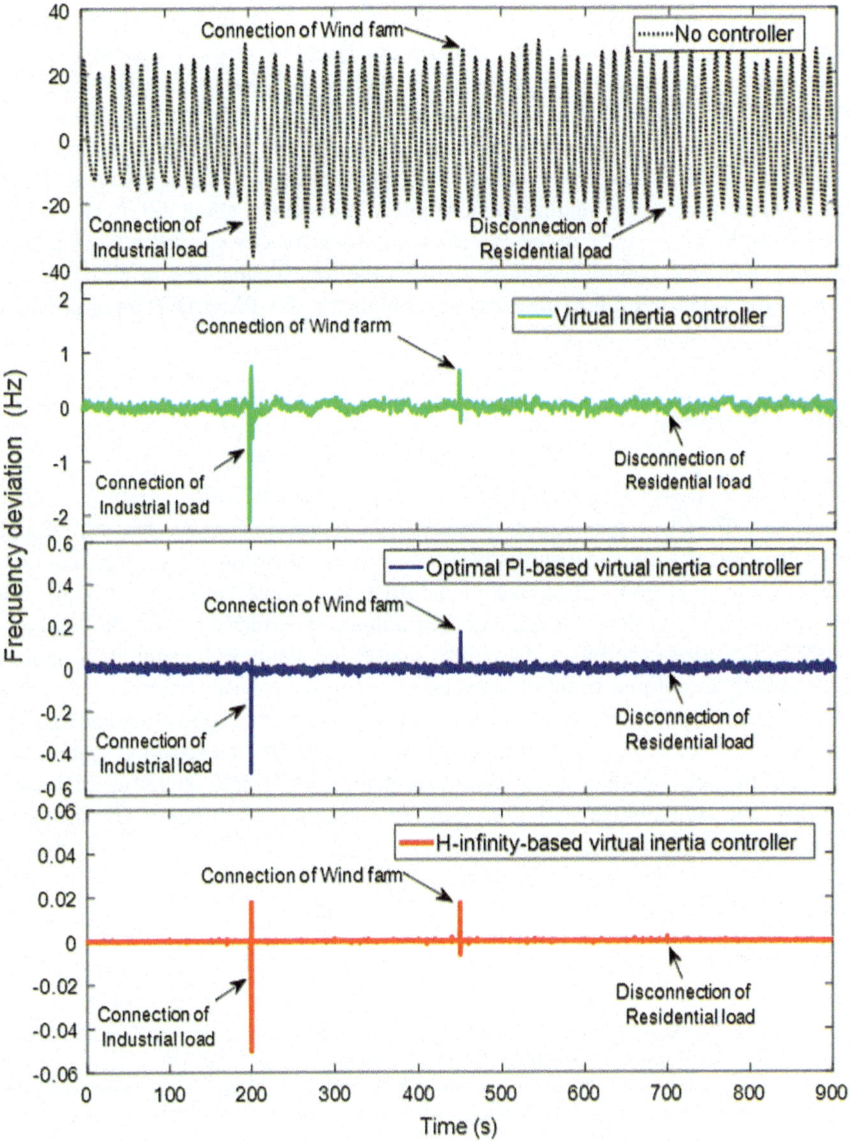

Fig. 8.14 Frequency response of the system under high RESs/DGs penetration with respect to low system inertia and damping

unstable, causing the system collapse and power blackout. By using the traditional and optimal PI-based virtual inertia controllers, larger frequency transient and deviation occur, causing a higher range of frequency oscillation. This situation may result in a trip of frequency relays and under-frequency load shedding. Clearly, the $H\infty$-based virtual inertia controller successfully prevails the integrated uncertainty due to its ability in uncertainty modeling. Subsequently, the frequency deviation quickly reaches to zero, leading to the smallest transient of about ± 0.05 Hz (within the desirable range), even in the extreme operation. As a result, it is confirmed that the proposed $H\infty$ control technique delivers the robust capability in virtual inertia control for today and future power systems with massive RESs/DGs integration.

8.8 Summary

This chapter proposed a new $H\infty$ robust design for virtual inertia synthesis considering high RESs/DGs penetration. The proposed robust method was feasible enough to include uncertainty and disturbance in the system modeling, improving frequency stability of the system. Then, the linear fractional transformation technique was added in the $H\infty$ control design, and the parametric perturbation was outlined as a block of bounded uncertainty. The linear matrix inequalities algorithm was applied to approach an optimal solution between nominal and robust performances for the design objectives. The controller robustness and performance were evaluated through parametric uncertainty and external disturbances. The $H\infty$ closed-loop system was verified through a nonlinear power system under the critical operating scenarios. Besides, the comparative study between $H\infty$ and optimal PI-based virtual inertia control was performed.

References

1. X. Wang, F. Blaabjerg, Harmonic stability in power electronic-based power systems: concept, modeling, and analysis. IEEE Trans. Smart Grid **10**(3), 2858–2870 (2019)
2. A. Anvari-Moghaddam, A. Rahimi-Kian, M.S. Mirian, J.M. Guerrero, A multi-agent based energy management solution for integrated buildings and microgrid system. Appl. Energy **203**(1), 41–56 (2017)
3. H. Bevrani, *Robust Power System Frequency Control,* 2nd ed. (Springer, New York, USA, 2014)
4. H. Hui, Y. Ding, K. Luan, and D. Xu, Analysis of '8•15' blackout in Taiwan and the improvement method of contingency reserve capacity through direct load control, in Proc. *IEEE PES General Meeting (IEEE PES GM),* 1–6 (2018)
5. CBS NEWS, Power fully restored after widespread Los Angeles outage, *CBS Interactive Inc.,* 09-Jul-2017
6. Y. Liu, Analysis of brazilian blackout on March 21st, 2018 and revelations to security for human grid, in Proc. *International Conference on Intelligent Green Building and Smart Grid,* 422–426 (2019)

7. AP News, Azerbaijan hit by massive blackout, worst in decades, *The Associated Press*, 04-Jul-2018
8. K. Asano, T. Iwata, Source rupture process of the 2018 Hokkaido Eastern Iburi earthquake deduced from strong-motion data considering seismic wave propagation in three-dimensional velocity structure. Earth Planets Sp. **71**(1), 1–17 (2019)
9. BBC News, Venezuela blackout: Power cuts plunge country into darkness, *BBC News*, 23-Jul-2019
10. CNN, Massive failure' leaves Argentina, Paraguay and Uruguay with no power, *Cable News Network*, 16-Jul-2019
11. BBC News, Indonesia blackout: Huge outage hits Jakarta and surrounding area, *BBC News*, 05-Aug-2019
12. BBC News, UK power cut: National Grid promises to learn lessons from blackout, *BBC News*, 10-Aug-2019
13. J.O. Petinrin, M. Shaabanb, Impact of renewable generation on voltage control in distribution systems. Renew. Sustain. Energy Rev. **65**(1), 770–783 (2016)
14. A.J. Veldhuis, M. Leach, A. Yang, The impact of increased decentralised generation on the reliability of an existing electricity network. Appl. Energy **215**(1), 479–502 (2018)
15. G. Andersson et al., Causes of the 2003 major grid blackouts in North America Europe, and recommended means to improve system dynamic performance. IEEE Trans. Power Syst. **20**(4), 1922–1928 (2005)
16. Y.V. Makarov, V.I. Reshetov, V.A. Stroev, N.I. Voropai, Blackout prevention in the United States, Europe, and Russia, Proc. of IEEE, **93**(11), 1942–1955 (2005)
17. UCTE, "Final report of the investigation commitee on the 28 september 2003 blackout in Italy," 2004
18. I. Green, CAISO experience with impact of high penetration of renewable resources on short-term voltage stability, in Proc. *IEEE PES General Meeting (IEEE PES GM)*,1–6 (2015)
19. W. Zappa, M. Junginger, M. van den Broek, Is a 100% renewable European power system feasible by 2050? Appl. Energy **233–234**(1), 1027–1050 (2019)
20. K. Hansen, B.V. Mathiesen, I.R. Skov, Full energy system transition towards 100% renewable energy in Germany in 2050. Renew. Sustain. Energy Rev. **102**(1), 1–13 (2019)
21. B.S. Howard, N.E. Hamilton, M. Diesendorf, T. Wiedmann, Modeling the carbon budget of the Australian electricity sector's transition to renewable energy. Renew. Energy **125**(1), 712–728 (2018)
22. X.J. Yang, H. Hu, T. Tan, J. Li, China's renewable energy goals by 2050. Environ. Dev. **20**(1), 86–90 (2016)
23. M.A. Delucchi, M.Z. Jacobson, Providing all global energy with wind, water, and solar power, Part II: reliability, system and transmission costs, and policies. Energy Policy **39**(3), 1170–1190 (2011)
24. M.Z. Jacobson, M.A. Delucchi, Providing all global energy with wind, water, and solar power, Part I: technologies, energy resources, quantities and areas of infrastructure, and materials. Energy Policy **39**(3), 1154–1169 (2011)
25. A. Fathi, Q. Shafiee, H. Bevrani, Robust frequency control of microgrids using an extended virtual synchronous generator. IEEE Trans. Power Syst. **33**(6), 6289–6297 (2018)
26. H. Bevrani, T. Hiyama, Robust decentralised PI based LFC design for time delay power systems. Energy Convers. Manag. **49**(2), 193–204 (2008)
27. H. Bevrani, T. Hiyama, On load-frequency regulation with time delays: design and real-time implementation. IEEE Trans. Energy Convers. **24**(1), 292–300 (2009)
28. H. Bevrani, K. Tsuji, Y. Mitani, On robust load-frequency regulation in a restructured power system. IEEJ Trans. Power Energy **124**(2), 190–198 (2004)
29. A. Haddadi, B. Boulet, A. Yazdani, G. Joos, A μ-based approach to small-signal stability analysis of an interconnected distributed energy resource unit and load. IEEE Trans. Power Deliv. **30**(4), 1715–1726 (2015)
30. N. Sa-ngawong, I. Ngamroo, Intelligent photovoltaic farms for robust frequency stabilization in multi-area interconnected power system based on PSO-based optimal Sugeno fuzzy logic control. Renew. Energy **74**, 555–567 (2015)

31. Y. Han, P.M. Young, A. Jain, D. Zimmerle, Robust control for microgrid frequency deviation reduction with attached storage system. IEEE Trans. Smart Grid **6**(2), 557–565 (2015)
32. M.J. Hossain, H.R. Pota, M.A. Mahmud, M. Aldeen, Robust control for power sharing in microgrids with low-inertia wind and PV generators. IEEE Trans. Sustain. Energy **6**(3), 1067–1077 (2015)
33. M. Babazadeh, H. Karimi, A robust two-degree-of-freedom control strategy for an islanded microgrid. IEEE Trans. Power Deliv. **28**(3), 1339–1347 (2013)
34. D.J. Lee, L. Wang, Small-signal stability analysis of an autonomous hybrid renewable energy power generation/energy storage system part I: time-domain simulations. IEEE Trans. Energy Convers. **23**(1), 311–320 (2008)
35. H. Bevrani, Y. Mitani, K. Tsuji, Robust decentralized AGC in a restructured power system. Energy Convers. Manag. **45**(15–16), 2297–2312 (2004)
36. D. Rerkpreedapong, A. Hasanovic, A. Feliachi, Robust load frequency control using genetic algorithms and linear matrix inequalities. IEEE Trans. Power Syst. **18**(2), 855–861 (2003)
37. T. Kerdphol, F.S. Rahman, M. Watanabe, Y. Mitani, Robust virtual inertia control of a low inertia microgrid considering frequency measurement effects. IEEE Access **7**(1), 57550–57560 (2019)
38. N. Tephiruk, K. Hongesombut, and T. Kerdphol, Robust control of combined optimized resistive FCL and ECS for power system transient stability improvement, in Proc. *International Electrical Engineering Congress (iEECON 2014),* 1–4 (2014)
39. T. Kerdphol, F.S. Rahman, Y. Mitani, M. Watanabe, S. Kufeoglu, Robust virtual inertia control of an islanded microgrid considering high penetration of renewable energy. IEEE Access **6**(1), 625–636 (2017)
40. R. Jiang, J. Wang, Y. Guan, Robust unit commitment with wind power and pumped storage hydro. IEEE Trans. Power Syst. **27**(2), 800–810 (2012)
41. R.A. Jabr, S. Karaki, J.A. Korbane, Robust Multi-Period OPF with storage and renewables. IEEE Trans. Power Syst. **30**(5), 2790–2799 (2015)
42. P. Gahinet, P. Apkarian, A linear matrix inequality approach to H∞ control. Int. J. Robust Nonlinear Control **4**(1), 421–448 (1994)
43. A. Packard, J. Doyle, The complex structured singular value. Automatica **29**(1), 71–109 (1993)
44. S. Vachirasricirikul, I. Ngamroo, Robust controller design of heat pump and plug-in hybrid electric vehicle for frequency control in a smart microgrid based on specified-structure mixed H2/H∞ control technique. Appl. Energy **88**(11), 3860–3868 (2011)
45. X. Li, Y. J. Song, S. Bin Han, Frequency control in micro-grid power system combined with electrolyzer system and fuzzy PI controller. *J. Power Sources* **18**(1), 468–475 (2008)
46. D.W. Gu, P.H. Petkov, M.M. Konstantinov, *Robust control design with MATLAB®* (Springer-Verlag, London, UK, 2013)
47. P. Apkarian, D. Noll, H.D. Tuan, Fixed-order H∞ control design via a partially augmented Lagrangian method. Int. J. Robust Nonlinear Control **13**(12), 1137–1148 (2003)
48. M.P.S. Gryning, Q. Wu, M. Blanke, H.H. Niemann, K.P.H. Andersen, Wind turbine inverter robust loop-shaping control subject to grid interaction effects. IEEE Trans. Sustain. Energy **7**(1), 41–50 (2016)

Chapter 9
Optimization of Virtual Inertia Control Considering System Frequency Protection Scheme

Abstract To maintain frequency stability of power system subject to large contingency resulting in large generation-load imbalance, the frequency protection scheme such as underfrequency load shedding (UFLS) scheme is usually applied in the power system. However, with the high penetration of renewable generation, the system inertia will significantly decrease. Moreover, due to the variable nature of RESs, the system inertia will also vary depending on the actual RESs penetration. Therefore, in this condition, the existing frequency protection scheme might be insufficient to protect the system in the case of large contingency. While the reconfiguration of the frequency protection scheme could solve this issue, it is not practical from the system operation point-of-view to always modify its setting following the penetration level of RESs. In this chapter, a new method to select proper virtual inertia constant by considering the system frequency protection scheme is presented. The particle swarm optimization (PSO), a well-known optimization technique, is implemented for obtaining proper virtual inertia constant. Using the proposed method, optimal virtual inertia support could be achieved for different system inertia conditions without the necessity to reconfigure the existing frequency protection scheme.

Keywords Frequency control · Frequency protection scheme · Optimization · Particle swarm optimization (PSO) · Underfrequency load shedding (UFLS) · Virtual inertia constant · Virtual inertia control

9.1 Introduction

From the previous chapters, various techniques have been developed to obtain the desired performance of the virtual inertia control from the perspective of its control system. In addition to a proper design of virtual inertia control in terms of its control technique, one of the important aspects in designing virtual inertia control is to properly select the value for its control parameters, including virtual inertia constant and virtual damping coefficient. Improper value of virtual inertia control parameters may result in insufficient dynamic frequency support from the virtual inertia control. In particular, the virtual inertia constant has a large influence on dynamic frequency

T. Kerdphol et al., *Virtual Inertia Synthesis and Control*, Power Systems,
https://doi.org/10.1007/978-3-030-57961-6_9

support from virtual inertia control. By using higher virtual inertia constant, the frequency deviation during system transient could be suppressed. The frequency nadir subject to a particular disturbance will also be lower compared with using lower virtual inertia constant. However, these benefits come in exchange for more oscillations in the system [1] and higher energy consumption from the energy storage system to supply the required energy for the virtual inertia control [2]. Therefore, selecting the appropriate values of the virtual inertia control parameter would be important to obtain the optimal trade-off between energy requirement and the improvement in system frequency deviation.

Regarding the techniques for the optimization of virtual inertia control parameters, several methods have been proposed in [3–6]. However, the methods have several disadvantages, such as the use of qualitative analysis and the dependence of the optimal value to the selected upper and lower boundary [3], the requirement for extensive calculation [4–6], and long calculation time [5]. Moreover, the methods have not addressed the issue of the optimization of the virtual inertia control in regards to the stability of the system subject to a large disturbance.

One of the frequency stability issues in the power system caused by the reduction in system overall inertia is regarding the compatibility of the existing frequency protection scheme for the application in low inertia conditions. In low inertia conditions, the frequency-based protection (e.g. underfrequency load shedding (UFLS), anti-islanding protection) will potentially mal-operate more often, leading to more frequent interruption of customer loads [7]. When a large disturbance occurs under severe system inertia condition, the mal-operation might also lead to system frequency instability. While the modification of the frequency protection setting based on system inertia condition might solve this issue, from the system operation point-of-view, it is impractical to modify the setting of the existing frequency protection scheme depends on the penetration level of RESs.

This issue related to the impact of low system inertia on the overall performance of the existing frequency protection scheme is important. Therefore, in this chapter, a method to select a proper virtual inertia constant value while also retaining the pertinence of the existing frequency protection scheme is proposed. Particle swarm optimization (PSO) [8], as one of the available optimization techniques, will be utilized to obtain a proper virtual inertia constant value so that the level of frequency nadir subject to a major contingency in normal high inertia condition could be retained at the same level in low inertia condition. Thus, the frequency protection setting in normal high inertia conditions would also be adequate in low inertia conditions without changing its settings. To determine the proper virtual inertia constant, frequency nadir is used as the only index, leading to a simple optimization process with reduced calculation burden and time. This chapter is an extended version of the research presented in [9].

9.2 Particle Swarm Optimization

In this section, a brief overview of the particle swarm optimization is described. Particle swarm optimization (PSO) introduced by Kennedy and Eberhart [8] is a robust and simple technique for solving the optimization problem and has been used for various applications. It has several advantages, such as efficient computation, easy implementation, stable convergence characteristic [10], and also suitable for online optimization due to its simplicity [11].

PSO is formulated based on the behavior of a group/swarm of birds in finding the location of food [8]. The illustration of PSO is shown in Fig. 9.1. Initially, each bird flies at a different location in a certain area (search space). To find the food location, each bird will move based on the closest position it achieves to the food location and the position of the bird in the group with the closest distance to the food location. The birds will move until arriving at the food location.

In the PSO algorithm, the birds are defined as particles. Each particle has its own position and velocity in the search space. Using the PSO algorithm, the position of each particle will be updated based on the specified objective function until an optimal solution is obtained (i.e. the birds will move until arriving at the food location). The particle position update is realized by using the value of $pbest_j$ (i.e. the closest position achieved by a particular j bird) and $gbest$ (i.e. the position of the bird in the group with the closest distance to the food location). The obtained optimal solution is determined by the value of $gbest$ after a certain number of iteration or when a certain stopping condition is fulfilled.

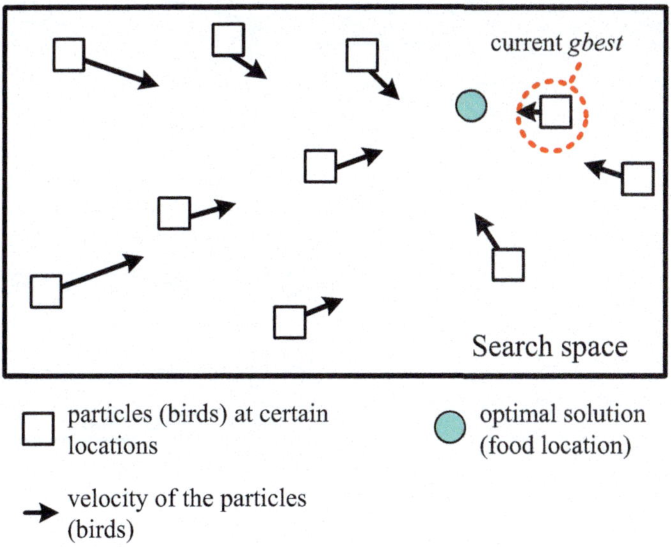

Fig. 9.1 The illustration of particle swarm optimization (PSO)

The general flowchart of the PSO algorithm is shown in Fig. 9.2. In the flowchart, the number of particles used in PSO algorithm is designated as N_p. Particle position (x_j) is randomly initialized, while zero-velocity initialization [12] is applied to initialize the velocity of particles (v_j). The best fitness of j particle and global best fitness are defined as $pbestfit_j$ and $gbestfit$, respectively. The fitness is a measure of how well a specific particle position satisfies the objective function.

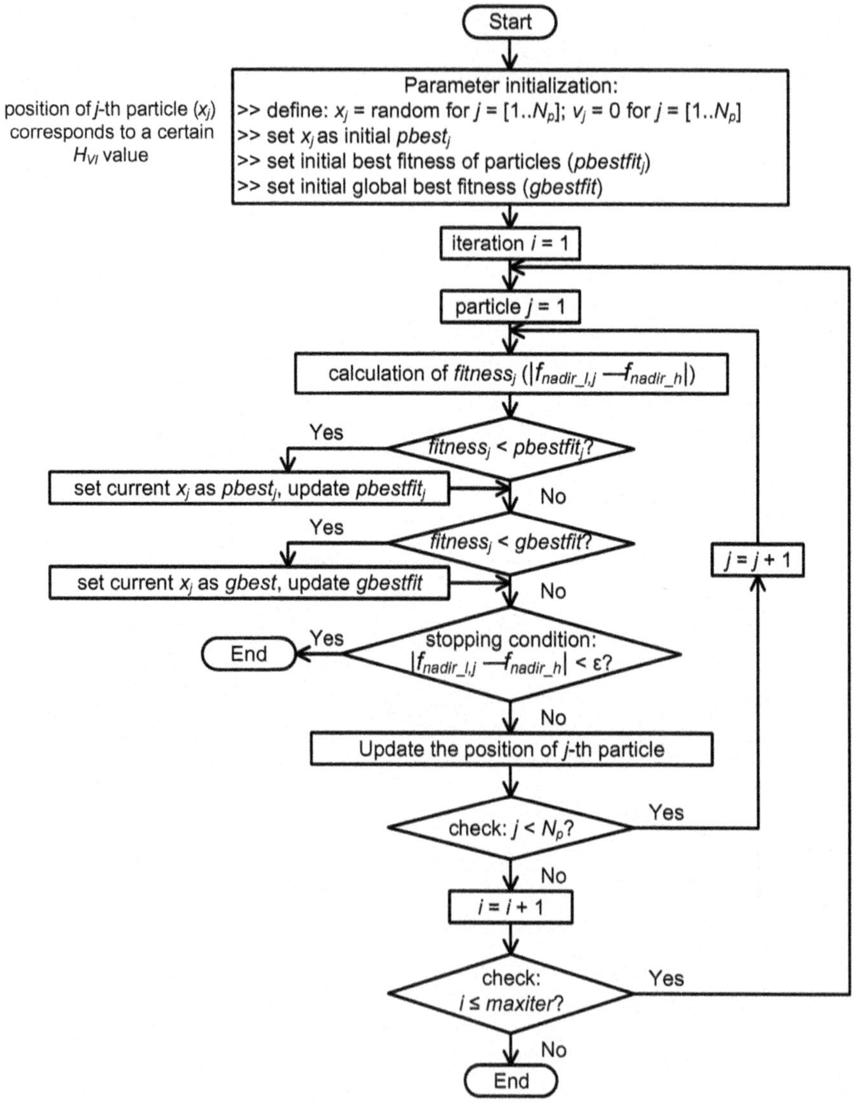

Fig. 9.2 The general flowchart of the PSO algorithm

To update the position of particles, the pair of equations given in Eqs. (9.1) and (9.2) is used.

$$v_{j,i+1} = w.v_{j,i} + c_1.r_1\left(pbest_j - x_{j,i}\right) + c_2.r_2\left(gbest - x_{j,i}\right) \tag{9.1}$$

$$x_{j,i+1} = x_{j,i} + v_{j,i+1} \tag{9.2}$$

where x_j is the position of j particle, v_j is the velocity of j particle, $pbest_j$ is the position of j particle that leads to the best fitness for the j particle, $gbest$ is position among all particles resulting in an overall best fitness, c_1 is the individual learning rate of the particles, c_2 is the global learning rate, w is inertia weight, and r_1, r_2 are random numbers between 0 and 1. The general principle to choose these values is explained in [13]. In Fig. 9.2, the use of a stopping condition is optional depending on the optimization problem to be solved.

9.3 Underfrequency Load Shedding (UFLS)

In the case of a large disturbance resulting in the loss of a large generation from the power system, the frequency of the power system would decline rapidly. The rate of the frequency decline would be affected by several factors, such as system inertia, the number of generators that could respond quickly to the disturbance (i.e. by primary frequency control from governor system), and the severity of the disturbance. If the generation-load balance could not be achieved after the disturbance, the frequency of the power system will continue declining and could lead to the system collapse. To prevent this extreme condition and maintain the frequency stability of the power system, emergency control and protection scheme might be necessary.

One of the important frequency protection schemes in the power system to restore the system frequency and prevent the system collapse is the underfrequency load shedding (UFLS) scheme. Underfrequency load shedding (UFLS) is an emergency frequency protection scheme that is commonly used in the power system to protect a power system from the frequency instability caused by rapid frequency decline subject to a large contingency in the system. In the occurrence of a large disturbance, the UFLS scheme, along with the primary frequency response from the governor would arrest the frequency decline to prevent further decline in system frequency. The prolonged operation of generation units at frequencies lower than a certain level would lead to the damage of the generation units.

The UFLS scheme is designed to curtail a certain amount of load when the system frequency falls below the specified thresholds [14]. Usually, the shedding of the loads will be divided into several stages to deal with the different generation-load mismatch subject to a disturbance. The illustration of the UFLS scheme is shown in Fig. 9.3. In Fig. 9.3, f_1, f_2, and f_3 are the frequency threshold for each stage. When the frequency threshold is exceeded for a specific time delay, the predetermined loads

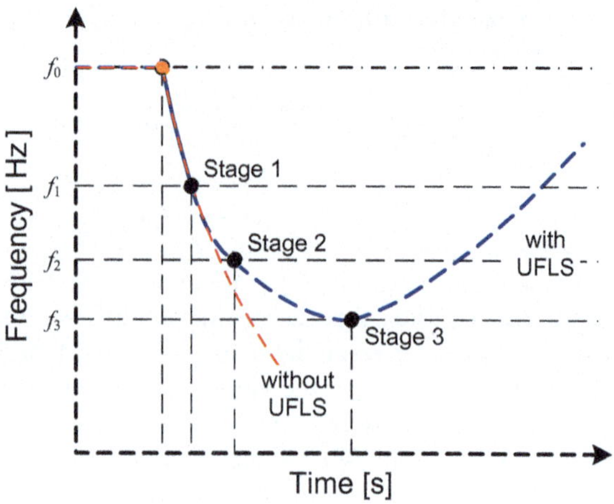

Fig. 9.3 The illustration of the underfrequency load shedding (UFLS) scheme

would be shed to prevent further frequency decline. Due to its role as an emergency protection scheme, UFLS scheme should be simple, decisive, uniformly distributed, and should not be frequently modified [15]. UFLS scheme is usually employed due to the limitations of the generation units in supplying the required active power, both in their response (e.g. the governor of generation units such as coal plants has slower response) and the availability of the spinning reserve.

In general, the UFLS scheme could be divided into static and adaptive UFLS [16]. The term 'UFLS' usually refers to the static one, in which the loads to be shed at each stage are predetermined. In an adaptive load shedding, the loads are shed depending on the magnitude of disturbance and the frequency characteristic of the system [16]. A lot of adaptive load shedding schemes use the rate-of-change-of-frequency (RoCoF) in their design. The design of the static frequency-based UFLS is described in various references, e.g. [15, 17–19], while the design of adaptive load shedding is elaborated in [14, 20–25]. The performance of different UFLS schemes is compared in [26].

A properly designed UFLS scheme could prevent the power system blackout in the case of large system frequency excursion. However, its performance might be inadequate in the modern power system with a high share of inverter-based renewable power generation. In such a system, the system dynamics in the case of large contingency could be extremely fast. The fast dynamics in the system with a high share of inverter-based renewable power generation is the result of significantly lower inertia that leads to a higher frequency deviation, lower frequency nadir, and higher RoCoF subject to the same contingency. These significant changes in the power system dynamics could lead to the insufficient response from the UFLS scheme, particularly the static UFLS scheme, to react to the contingency. Moreover, due to the variable nature of RESs, the adequacy of the UFLS scheme would be even harder to be estimated. While the modification of the UFLS scheme might solve the frequency issues caused by the high share of RESs, it is impractical to continuously modify the UFLS scheme based on the penetration level of RESs.

9.4 Design of Virtual Inertia Control Optimization Considering System Frequency Protection

In virtual inertia control, virtual inertia is emulated using a virtual inertia control system, not by 'real' inertia (i.e. the rotating mass). As a result, unlike a conventional SG, the amount of virtual inertia support from virtual inertia control could be adjusted by changing its virtual inertia constant (H_{VI}) to provide a proper inertia compensation. However, due to the inertia-less nature of the power-electronics interface and the working principle of virtual inertia control, in which virtual inertia power is injected into the system based on the frequency difference from the rated value, the virtual inertia control is not providing the inertia power to the system when there is no frequency deviation in the system. Due to these characteristics, and adding another aspect such as system nonlinearity and the effect of the governing action in the system, the proper virtual inertia constant value to retain the pertinence of the existing frequency protection scheme in low inertia condition could not be easily obtained. Therefore, optimization techniques should be employed. In this research work, the implementation of an optimization method considering the system frequency protection scheme based on particle swarm optimization (PSO) is proposed to obtain proper virtual inertia constant (H_{VI}) value of virtual inertia control.

For the proposed PSO-based method, the parameter to be optimized is the virtual inertia constant of virtual inertia control (H_{VI}). Hence, the position of particles in the PSO algorithm corresponds to a certain H_{VI} value. The objective of the proposed method is to obtain the proper H_{VI} of virtual inertia control so that the difference between the frequency nadir in low inertia condition and normal high inertia condition subject to a particular disturbance could be minimized. This objective is realized in the PSO algorithm by using the fitness function defined as $\left|f_{nadir\_l} - f_{nadir\_h}\right|$. The objective function is to minimize this fitness function, as shown in Eq. (9.3).

$$\text{objective function: minimize}\left(\left|f_{nadir\_l} - f_{nadir\_h}\right|\right) \tag{9.3}$$

where f_{nadir_l} is the frequency nadir of the system due to the same disturbance in low inertia condition and f_{nadir_h} is the frequency nadir of the system due to the same disturbance in normal high inertia condition.

Using the objective function described in Eq. (9.3), the difference between the frequency nadir in low inertia condition and normal high inertia condition subject to a particular disturbance could be minimized and hence, the level of frequency nadir achieved in normal high inertia condition can be retained at the similar level in low inertia condition. Thus, the existing underfrequency protection scheme would also be applicable in low inertia conditions without the need for modifying its previously designed setting.

The flowchart of the PSO algorithm used in this research work is illustrated in Fig. 9.4. The number of particles used (N_p) is 10 with maximum iteration number (*maxiter*) of 15. Learning rate c_1 and c_2 are both set as 2 with inertia weight (w) is 0.729.

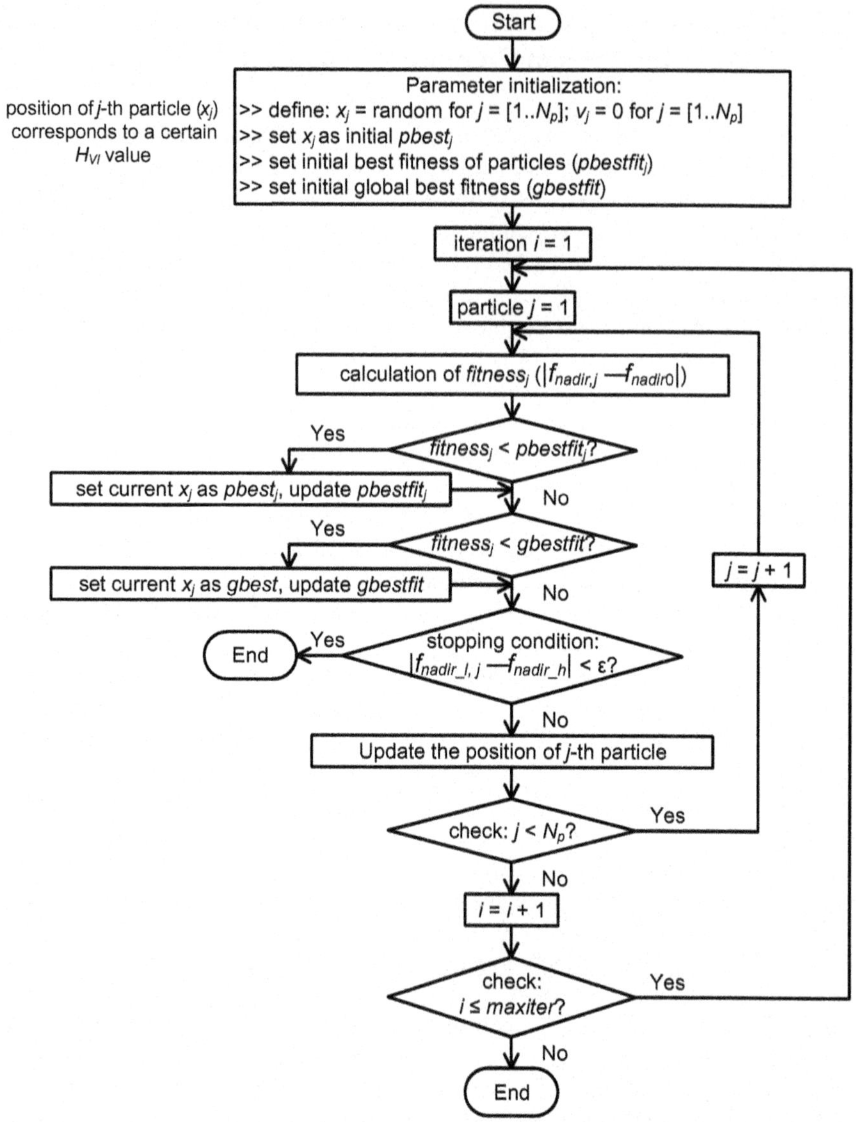

Fig. 9.4 The flowchart of the PSO-based virtual inertia control optimization method considering the system frequency protection scheme

To stop the algorithm when a desired H_{VI} value has been obtained, stopping condition described in Eq. (9.4) is applied. ε is a predetermined error between f_{nadir_l} and f_{nadir_h}, set as 0.002 Hz. Thus, the algorithm will stop when fitness $\left|f_{nadir\_l,j} - f_{nadir\_h}\right|$ is less than 0.002 Hz for a certain particle at a certain iteration. A lower ε could also be used if it is necessary. However, it will lead to a longer calculation time to obtain the desired H_{VI} value.

$$\text{stopping condition: when fitness}\left(\left|f_{nadir\_l} - f_{nadir\_h}\right|\right) < \varepsilon \tag{9.4}$$

In the flowchart, in calculating the fitness $\left|f_{nadir\_l} - f_{nadir\_h}\right|$ for a single particle j in a particular iteration, the frequency nadir for low inertia condition (f_{nadir_l}) is obtained by using time-domain simulation respect to the considered low inertia condition. The applied disturbance for determining f_{nadir_l} is the same as the considered contingency to determine the frequency nadir in the normal high inertia condition (f_{nadir_h}). In the performed simulations, the loss of generator G2 is used as the contingency for determining f_{nadir_h}. Hence, the frequency nadir in low inertia condition (f_{nadir_l}) is also determined by considering the same contingency. The selection of the contingency should consider the type of contingency used when designing the UFLS scheme. Usually, the contingency would be one of the major contingencies in the system.

In the presented method, the frequency nadir of the system is used as the only index for the objective function of the PSO. Time related-indices (e.g. frequency settling time, integral of time multiplied by squared error (ITSE) of frequency) are not utilized in the optimization process, leading to faster optimization time and reduced calculation burden. In addition, the lower and upper boundaries of 3 and 20, respectively, are defined for the virtual inertia constant value. The boundaries are used to limit the search space so that the optimal solution (i.e. the proper virtual inertia constant value) can be obtained faster. In principle, any value can be selected as the boundaries. However, in the application, the selection of the boundaries should consider maximum and minimum possible inertia in the system respect to the connected RESs so that the proper H_{VI} value could be obtained within the boundaries.

9.5 System Modeling

9.5.1 Test System

In this chapter, the system shown in Fig. 9.5 is used as the studied system. The system is based on the system model used in [27]. The operating condition of the system is described as follows. In normal conditions, the system operates in grid-connected mode. The peak demand of load L1, L2, L3, L4, and L5 is 1.85 MW, 1.7 MW, 1.75 MW, 1.9 MW, and 2.4 MW, respectively. All loads are assumed to be 80% of the peak load (i.e. 7.68 MW total). The nominal frequency of the system is 50 Hz. The inertia constant of G1 and G2 are 3 s and 1 s, respectively. The active power dispatch for G1 and G2 are 2 and 1.2 MW respectively, with G1 is equipped with a speed governor. The rated power output of the inverter of virtual inertia control is 5 MVA. The capacity of the external grid EXT is 24.13 MVA with its inertia constant (H_{EXT}) is varied depends on the considered scenario. In default high inertia condition,

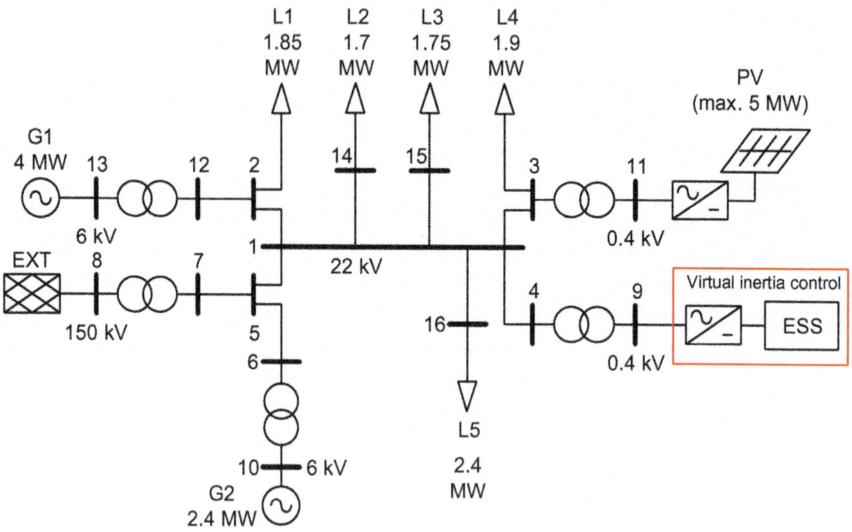

Fig. 9.5 The test system for simulation

Table 9.1 The frequency and time delay setting of the hypothetical UFLS scheme

LS stage	Frequency setting (Hz)	Time delay setting (s)	LS percentage (%)
1	49.8	0.1	10
2	49.6	0.3	10
3	49.4	0.4	10
4	49	0.5	10

the test system operates without virtual inertia control, with the inertia constant of external grid $H_{EXT} = 4$ s with zero PV penetration.

To validate the performance of the proposed virtual inertia control optimization method considering system frequency protection scheme, the proposed method will be tested in the condition when a hypothetical underfrequency load shedding (UFLS) scheme is applied in the system. The frequency and time delay settings of the hypothetical UFLS scheme are shown in Table 9.1. The UFLS scheme is assumed to be applied on load L3 and L4.

9.5.2 *Virtual Inertia Control Model*

In this chapter, the swing equation-based virtual inertia control model introduced in [28] is used. This model does not require a derivative term/block in its model and is simpler compared to another model such as the model based on the full model of a

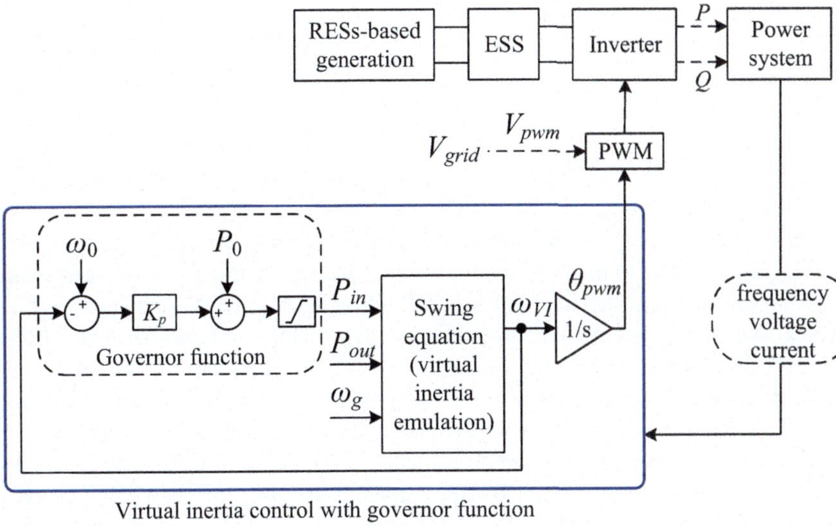

Fig. 9.6 Basic structure of swing equation-based virtual inertia control

synchronous generator (SG) (i.e. the synchronverter) [29, 30] while still effectively providing the required virtual inertia power.

In swing equation-based virtual inertia control, virtual inertia control is realized by incorporating a typical swing equation of a synchronous generator as in Eq. (9.5) into the virtual inertia control block of virtual inertia control. The swing equation-based virtual inertia control is applied in many research works, such as in [28, 31–35].

The basic structure of swing equation-based virtual inertia control is shown in Fig. 9.6. The structure is based on the structure used in [31–33, 36]. In the virtual inertia emulation block, the swing equation for virtual inertia control in per-unit defined in Eq. (9.5) is incorporated.

$$\overline{P}_{in} - \overline{P}_{out} = 2H_{VI}\frac{d\overline{\omega}_{VI}}{dt} + K_d\Delta\overline{\omega}_{VI} \tag{9.5}$$

where $\overline{P}_{in}$ is the input power of virtual inertia control [p.u.], $\overline{P}_{out}$ is the actual output power of virtual inertia control [p.u.], H_{VI} is virtual inertia constant [s], $\Delta\overline{\omega}_{VI}$ is the virtual angular rotor speed deviation [p.u.], and K_d is the virtual damping coefficient of the virtual inertia control [p.u.]. $\Delta\overline{\omega}_{VI}$ is defined as $(\overline{\omega}_{VI} - \overline{\omega}_g)$, where $\overline{\omega}_{VI}$ is virtual angular rotor speed [p.u.] and $\overline{\omega}_g$ is the angular rotor speed of the grid [p.u.].

In the virtual inertia control part, the governor function similar to the speed governor in a conventional SG is also added. The equation for the governor function in per-unit is represented as

$$\overline{P}_{in} - \overline{P}_0 = -K_p(\overline{\omega}_{VI} - \overline{\omega}_0) \tag{9.6}$$

where $\overline{P}_0$ is active power reference of virtual inertia control [p.u.], $\overline{\omega}_0$ is rated system angular frequency [p.u.], and K_p is the droop coefficient of virtual inertia control.

9.6 Simulation Results

The frequency response of the system utilizing the optimized virtual inertia control (OVIC) is compared to the case without virtual inertia control and two cases using constant-value virtual inertia control (CVIC) with virtual inertia constant of 10 s (typical inertia constant of a synchronous generator is between 2 and 10 s) and 14.25 s (based on the average of H_{VI} for three different inertia conditions in Table 9.2). The CVIC with virtual inertia constant of 10 and 14.25 s are designated as CVIC1 and CVIC2, respectively. Damping and droop coefficient K_d and K_P are both set as 0, with the active power reference (P_0) of virtual inertia control is also set at 0 MW. Using this configuration, virtual inertia control will inject the inertia power only during system dynamics and will not take part in the active power dispatch in the system.

9.6.1 *Default High Inertia Condition and the Result of Optimization*

The frequency nadir obtained from this condition will be used as the benchmark for low inertia conditions. In this condition, the test system operates without virtual inertia control, with the inertia constant of external grid $H_{EXT} = 4$ s with zero output from the PV plant. The frequency response of the system subject to the loss of generator G2 in this condition is shown in Fig. 9.7. In this condition, the frequency nadir of the system is 49.32 Hz. Accordingly, by also considering stopping condition described in Eq. (9.4), the value of f_{nadir_h} in Eq. (9.3) is set as 49.322 Hz. Thus, the obtained optimal virtual inertia constant (H_{VI}) of virtual inertia control from the PSO algorithm in low inertia condition will always yield the frequency nadir above 49.320 Hz.

By using the obtained value of f_{nadir_h}, the optimal virtual inertia constant of virtual inertia control for different system inertia could be obtained by using the proposed PSO-based optimization. The optimal virtual inertia constant of virtual inertia control with 5 MW output from PV plant ($P_{PV} = 5$ MW) subject to the loss

Table 9.2 The obtained optimal virtual inertia constant for different system inertia

H_{EXT}(s)	H_{VI} (s)	f_{nadir} (Hz)	t_{set} (s)
1	17.92	49.32	22.0
2	14.43	49.32	22.1
3	10.40	49.32	22.0

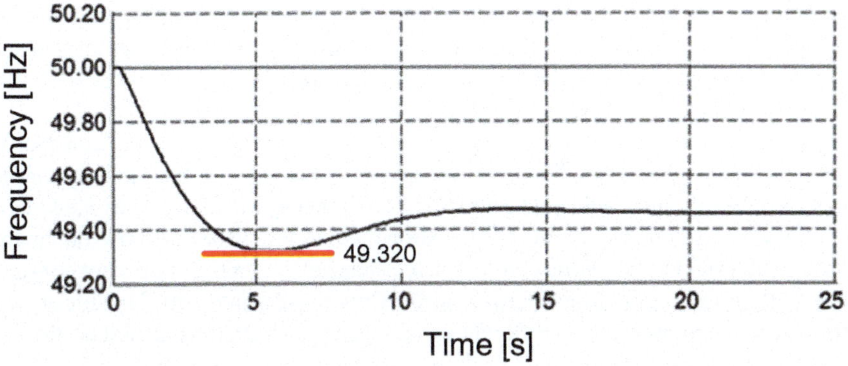

Fig. 9.7 Frequency response of the system in default high inertia condition without virtual inertia control subject to the loss of generator G2

of generator G2 as the contingency is shown in Table 9.2. f_{nadir} and t_{set} are frequency nadir and frequency settling time, respectively. t_{set} is defined as the time required for the frequency to settle within 2% of the maximum frequency deviation [5] (e.g. if maximum frequency deviation is 1 Hz and the settling frequency is 50 Hz, t_{set} is the time required for the frequency to settle within 50 ± 0.02 Hz). Table 9.2 shows that the virtual inertia constant could be properly determined in different system inertia condition by using the proposed method. In lower inertia condition, the virtual inertia constant is higher as desired.

The optimal virtual inertia constant values in Table 9.2 are the required value of virtual inertia constant so that the same level of frequency nadir as the level in high inertia condition could be achieved in low inertia condition. To show that the optimization of virtual inertia constant is necessary, the analysis of the kinetic energy of the system is performed. In this regard, since the system inertia variation is considered by varying the inertia constant of the external grid (H_{EXT}), the analysis is performed by calculating the total kinetic energy corresponds to the external grid and virtual inertia control as shown in (9.7). The result of the analysis of the kinetic energy respect to the values shown in Table 9.2 is shown in Table 9.3.

$$KE_{EXT,VSG} = H_{EXT}.S_{EXT} + H_{VI}.S_{VSG} \tag{9.7}$$

Table 9.3 The analysis of the kinetic energy

System condition index	External grid		Virtual inertia control		$KE_{EXT,VSG}$ (MVA.s)
	H_{EXT} (s)	S_{EXT} (MVA)	H_{VI} (s)	S_{VSG}(MVA)	
1	1	24.13	17.92	5	113.7
2	2	24.13	14.43	5	120.4
3	3	24.13	10.40	5	124.4

$KE_{EXT,VSG}$ is the total kinetic energy that corresponds to the external grid and the virtual inertia control. H_{EXT} and H_{VI} are the inertia constants of the external grid and virtual inertia constant of virtual inertia control, respectively. S_{EXT} and S_{VSG} are the capacity of the external grid and virtual inertia control, respectively.

Based on Table 9.3, it can be seen that the value of $KE_{EXT,VSG}$ is not constant in the considered system conditions. In other words, the value of the required virtual inertia constant value could not be directly determined by calculating the total kinetic energy in the system. Hence, the optimization method should be utilized to obtain the desired optimal virtual inertia constant value. The value of $KE_{EXT,VSG}$ is not constant due to the dynamics of virtual inertia control. Thus, the required virtual kinetic energy from virtual inertia control in different system inertia conditions could not be directly determined by using H_{VI}. S_{VSG} based on kinetic energy in the system.

9.6.2 Low Inertia Condition

Scenario 1: In this scenario, the inertia constant of the external grid (H_{EXT}) is set as 3 s with 5 MW output from PV plant ($P_{PV} = 5$ MW). The simulation result and performance comparison for each virtual inertia control configuration in this scenario are shown in Fig. 9.8 and Table 9.4. From Fig. 9.8 and Table 9.4, all cases using virtual inertia control could maintain the frequency nadir within 49.32 Hz when $H_{EXT} =$ 3 s (in the CVIC1 case, the actual frequency nadir is slightly lower than 49.320 Hz).

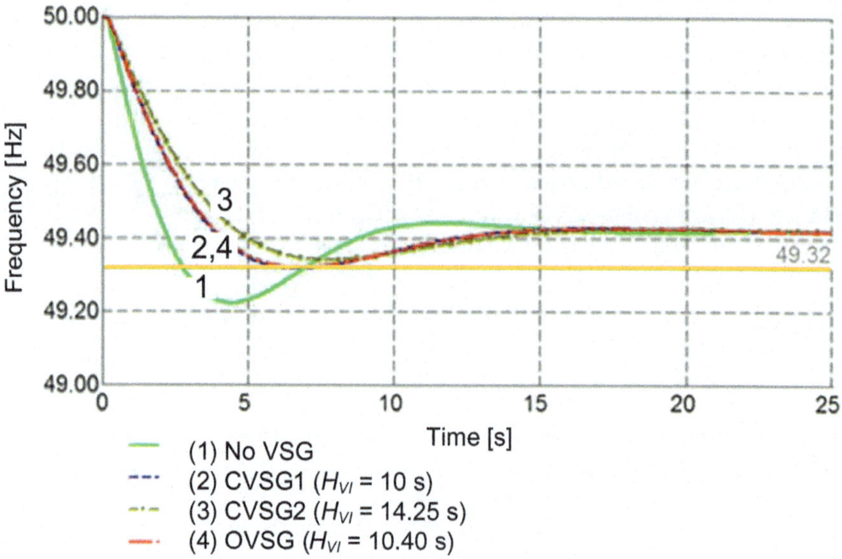

Fig. 9. Frequency response to the loss of generator G2 for $H_{EXT} = 3$ s

Table 9.4 Performance comparison for $H_{EXT} = 3$ s

Virtual inertia control configuration	H_{VI} (s)	f_{nadir} (Hz)	t_{set} (s)	Energy usage	
				MJ	MWh
No VIC	–	49.22	15.5	0	0
CVIC1	10	49.32	21.9	23.33	0.0065
CVIC2	14.25	49.34	22.6	33.10	0.0092
OVIC	10.40	49.32	22.0	24.25	0.0067

For the case without virtual inertia control, a higher frequency deviation is observed, causing system frequency to drop below 49.32 Hz.

The energy usage from the ESS is also shown in Table 9.4. Compared to CVIC2, OVIC uses about 27% less energy, while still limiting frequency nadir within 49.32 Hz as required. Thus, using the proposed method, the virtual inertia constant of virtual inertia control could be selected as low as possible to reduce the energy usage, while still fulfilling the frequency requirement of the system. For CVIC1, as described before, while the energy usage is lower than OVIC, the actual frequency nadir is lower than 49.320 Hz.

Scenario 2: In this scenario, the inertia constant of the external grid (H_{EXT}) is set as 2 s with 5 MW output from PV plant ($P_{PV} = 5$ MW). Under this condition, overall system inertia is slightly low, causing larger frequency deviation compared to the deviation in the condition when H_{EXT} is 3. The simulation result and performance comparison for each virtual inertia control configuration in this scenario are shown in Fig. 9.9 and Table 9.5. From Fig. 9.9, it can be observed that employing virtual inertia control in a low inertia system generally could improve the frequency stability of the system. However, in the case of CVIC1, system frequency could not be maintained within 49.32 Hz. On the other hand, OVIC could still maintain system frequency even though the system inertia decreases.

The energy usage from the ESS is shown in Table 9.5. In lower inertia condition, OVIC is delivering higher energy to keep the system frequency within the desired value. Without virtual inertia adjustment, virtual inertia control could not maintain the frequency as desired (see the graph for CVIC1). In the CVIC2 case, energy usage is slightly lower than OVIC while still achieving the desired frequency response. This result is obtained by chance since the virtual inertia constant for CVIC2 itself (14.25 s) is obtained by averaging optimal virtual inertia constant in three different inertia conditions. Therefore, if the average is obtained from a different set of inertia conditions, the H_{VI} for CVIC2 would also be different and might lead to insufficient performance.

Scenario 3: In this scenario, the inertia constant of the external grid (H_{EXT}) is set as 1 s with 5 MW output from PV plant ($P_{PV} = 5$ MW). Under this condition, the overall system inertia is very low, causing a large frequency deviation in the system. Based on the recent trend of increasing RESs penetration in the power system, the drop in system inertia is inevitable in the future power system. Therefore, higher virtual inertia constant is necessary to limit the frequency deviation in such a system.

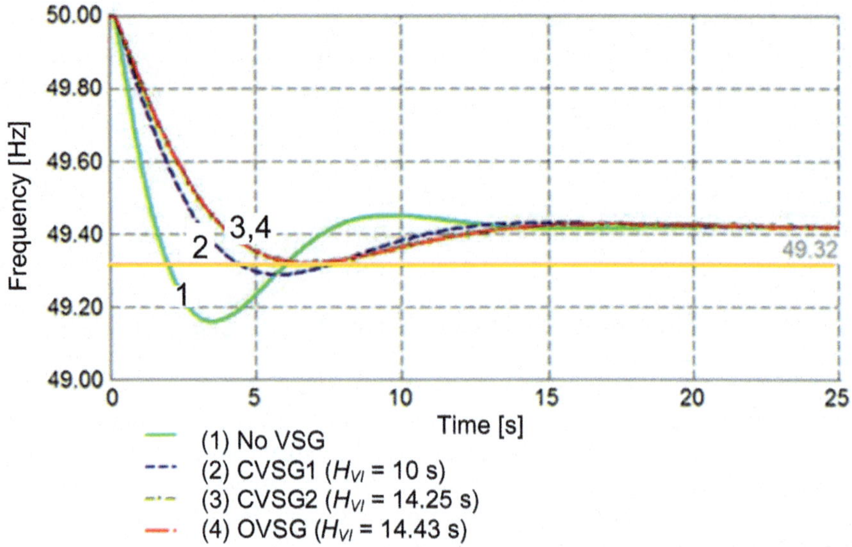

Fig. 9. Frequency response to the loss of generator G2 for $H_{EXT} = 2$ s

Table 9.5 Performance comparison for $H_{EXT} = 2$ s

VIC configuration	H_{VI} (s)	f_{nadir} (Hz)	t_{set} (s)	Energy usage	
				MJ	MWh
No VIC	–	49.16	12.9	0	0
CVIC1	10	49.29	19.9	23.34	0.0065
CVIC12	14.25	49.32	22.0	33.15	0.0092
OVIC	14.43	49.32	22.1	33.57	0.0093

The simulation result and the performance comparison for each virtual inertia control configuration in this scenario are shown in Fig. 9.10 and Table 9.6. The simulation result shows a significant frequency deviation for the no-virtual inertia control case, higher than the no-virtual inertia control case in Scenario 1 and 2 due to lower inertia. In a severely low inertia condition, a significant improvement is obtained by employing virtual inertia control in the system. However, it can be observed that although significant improvement is obtained by using CVIC1 and CVIC2, they could not maintain system frequency within the desired value. On the other hand, OVIC could still fulfill the frequency requirement in an extremely low system inertia condition, showing the effectivity of the proposed method

The energy usage from ESS is shown in Table 9.6. To achieve the performance of OVIC as shown in Fig. 9.10, higher energy is necessary to obtain the desired frequency response. In the case of CVIC1 and CVIC2, the virtual inertia support is not sufficient to maintain the system frequency within 49.32 Hz as desired.

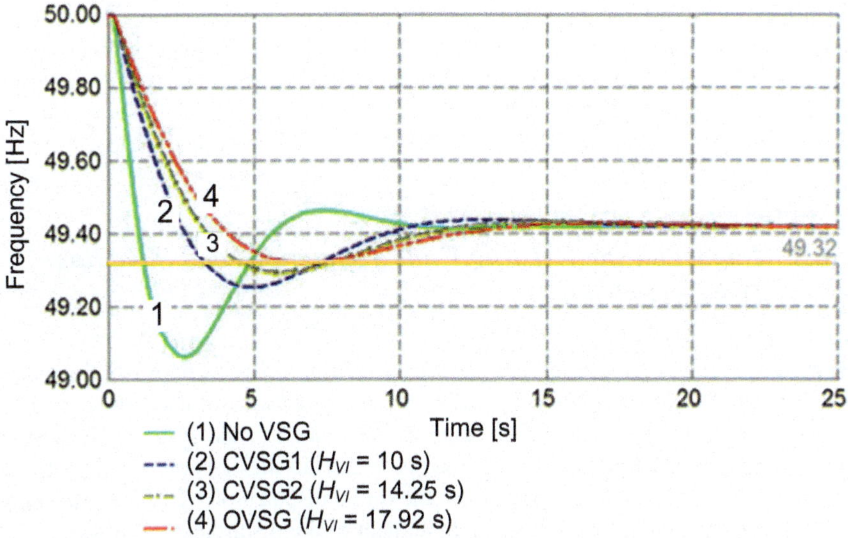

Fig. 9. Frequency response to the loss of generator G2 for $H_{EXT} = 1$ s

Table 9.6 Performance comparison for $H_{EXT} = 1$ s

Virtual inertia control configuration	H_{VI} (s)	f_{nadir} (Hz)	t_{set} (s)	Energy usage	
				MJ	MWh
No VIC	–	49.06	9.9	0	0
CVIC1	10	49.25	17.1	23.29	0.0065
CVIC2	14.25	49.29	20.2	33.17	0.0092
OVIC	17.92	49.32	22	41.64	0.0116

Scenario 4: In this scenario, the inertia constant of the external grid (H_{EXT}) is set as 1 s. The output of the PV plant is varied to show the effectiveness of OVIC in the presence of variability of RESs and different penetration level of RESs. The result is summarized in Table 9.7. It can be seen that by utilizing PSO, dynamic

Table 9.7 Optimal virtual inertia constant for different penetration level of PV power plant (with $H_{EXT} = 1$ s)

PV penetration (MW)	H_{VI} (s)	f_{nadir} (Hz)
0	11.33	49.32
1	12.9	49.32
2	14.43	49.32
3	15.87	49.32
4	16.75	49.32
5	17.92	49.32

change of virtual inertia constant subject to the change in penetration level of RESs (i.e. PV power plant in this case) could also be achieved. Higher optimal virtual inertia constant is obtained in higher penetration level of PV power plant (i.e. lower system inertia) while still maintaining system frequency as desired. It shows that the proposed method is also effective for different penetration levels of RESs.

9.6.3 Impact on the Existing Underfrequency Load Shedding (UFLS) Scheme

The performance of the proposed PSO-based virtual inertia control optimization method considering the system frequency protection scheme is also validated when the hypothetical UFLS scheme in Table 9.1 is applied in the system. The loss of generator G2 is used as the contingency. However, in this section, the dispatch of the generator G2 is set as 1.5 MW from the initial dispatch of 1.2 MW, resulting in a higher power mismatch in the system subject to the disturbance. The simulation result is shown in Fig. 9.11.

If the proposed method is not applied, the resulting frequency nadir would be lower than the frequency nadir in normal high condition subject to the same disturbance,

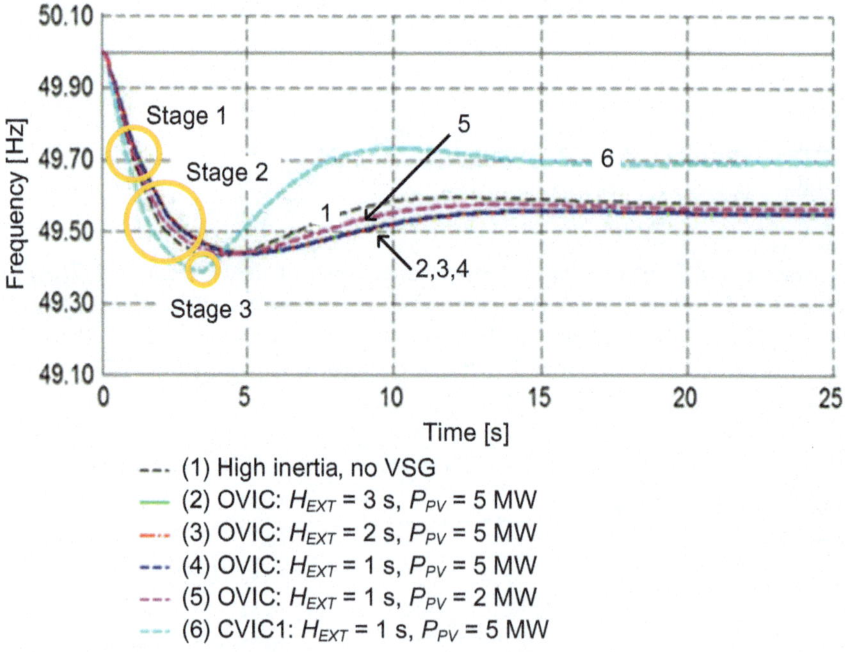

Fig. 9.11 Frequency response of the system due to the loss of generator G2 in different inertia conditions with the implementation of the hypothetical UFLS scheme

even though the same UFLS scheme is implemented, as shown in Fig. 9.11 for CVIC1. Subject to the considered disturbance, although the UFLS scheme is still sufficient to maintain the frequency stability of the system, more UFLS stage needs to be executed if CVIC1 is used instead of OVIC, leading to the interruption of more customer loads (i.e. stage-3 of UFLS scheme have to be executed for CVIC1). If the OVIC is used, the number of the executed stages in low inertia condition is the same with the number of stages for the application of the UFLS scheme in default normal high inertia condition. In a more severe system condition and disturbance, the existing UFLS scheme might fail to maintain the system frequency stability if the proposed method is not applied, even though the same UFLS scheme is adequate for the application in high inertia condition.

By applying the proposed method, a typical UFLS scheme as a default frequency protection for normal high inertia condition is still sufficient in low inertia condition. Moreover, the level of frequency nadir in low inertia condition could be retained similar to the level of frequency nadir in high inertia condition. Thus, the pertinence of the frequency protection scheme for normal high inertia conditions could also be retained in low inertia conditions without the necessity to modify the previously applied setting.

9.7 Summary

This chapter presents the implementation of particle swarm optimization (PSO) to obtain the proper value for virtual inertia constant of virtual inertia control considering the frequency protection scheme of the system. Based on the results, using the proposed method, the level of frequency nadir in normal high inertia condition could be retained at a similar level in low inertia condition. The performance of the proposed method has also been verified using a hypothetical UFLS scheme. Thus, by using the proposed method, the frequency stability of the system in the low inertia condition could be properly maintained without the necessity to reconfigure the existing setting of the frequency protection scheme.

References

1. M. Torres, L.A.C. Lopes, Virtual synchronous generator control in autonomous wind-diesel power systems, in Proc. *IEEE Electrical Power & Energy Conference (EPEC)*, 1–6 (2009)
2. S.M. Alhejaj, F.M. Gonzalez-Longatt, Investigation on grid-scale BESS providing inertial response support, in Proc. *International Conference on Power System Technology (POWERCON 2016)*, 1–6 (2016)
3. I. Serban, C.P. Ion, Microgrid control based on a grid-forming inverter operating as virtual synchronous generator with enhanced dynamic response capability. Int. J. Electr. Power Energy Syst. **89**, 94–105 (2017)

4. P. Rodriguez, E. Rakhshani, A. Mir Cantarellas, D. Remon, Analysis of derivative control based virtual inertia in multi-area high-voltage direct current interconnected power systems, IET Gener. Transm. Distrib. **10**(6), 1458–1469 (2016)
5. M. Hajiakbari Fini, M.E. Hamedani Golshan, Determining optimal virtual inertia and frequency control parameters to preserve the frequency stability in islanded microgrids with high penetration of renewables. Electr. Power Syst. Res. **154**, 13–22 (2018)
6. J. Alipoor, Y. Miura, T. Ise, Stability assessment and optimization methods for microgrid with multiple VSG units. IEEE Trans. Smart Grid **9**(2), 1462–1471 (2018)
7. X. Cao, I. Abdulhadi, A. Emhemed, C. Booth, G. Burt, Evaluation of the impact of variable system inertia on the performance of frequency based protection, in Proc. *IET International Conference on Development in Power System Protection (DPSP)*, 1–6 (2014)
8. J. Kennedy, R. Eberhart, Particle swarm optimization, in Proc. *IEEE International Conference on Neural Network*, 1942–1948 (1995)
9. F.S. Rahman, T. Kerdphol, M. Watanabe, Y. Mitani, Optimization of virtual inertia considering system frequency protection scheme. Electr. Power Syst. Res. **170**, 294–302 (2019)
10. Z.-L. Gaing, A particle swarm optimization approach for optimum design of PID controller in AVR system. IEEE Trans. Energy Convers. **19**(2), 384–391 (2004)
11. H. Bevrani, F. Habibi, P. Babahajyani, M. Watanabe, Y. Mitani, Intelligent frequency control in an AC microgrid: Online PSO-based fuzzy tuning approach. IEEE Trans. Smart Grid **3**(4), 1935–1944 (2012)
12. A. Engelbrecht, Particle swarm optimization: Velocity initialization, in Proc. *IEEE World Congress on Computational Intelligence*, 1–8 (2012)
13. I. Trelea, The particle swarm optimization algorithm: convergence analysis and parameter selection. Inf. Process. Lett. **85**, 317–325 (2003)
14. H. Bevrani, G. Ledwich, J.J. Ford, On the use of df/dt in power system emergency control, in Proc. *IEEE/PES Power Systems Conference and Exposition*, 1–6 (2009)
15. C. Concordia, L.H. Fink, G. Poullikkas, Load shedding on an isolated system. IEEE Trans. Power Syst. **10**(3), 1467–1472 (1995)
16. H. Bevrani, *Robust Power System Frequency Control*, 2nd ed. (Springer, New York, 2014)
17. J. R. Jones, W.D. Kirkland, Computer algorithm for selection of frequency relays for load shedding. IEEE Comput. Appl. Power **1**(1), 21–25 (1988)
18. H.E. Lokay, V. Burtnyk, Application of underfrequency relays for automatic load shedding. IEEE Trans. Power Appar. Syst. **PAS-87**(3), 776–783 (1968)
19. L. Sigrist, I. Egido, L. Rouco, A method for the design of UFLS schemes of small isolated power systems. IEEE Trans. Power Syst. **27**(2), 951–958 (2012)
20. L. Sigrist, A UFLS scheme for small isolated power systems using rate-of-change of frequency. IEEE Trans. Power Syst. **30**(4), 2192–2193 (2015)
21. V.V. Terzija, Adaptive underfrequency load shedding based on the magnitude of the disturbance estimation. IEEE Trans. Power Syst. **21**(3), 1260–1266 (2006)
22. H. You, V. Vittal, Z. Yang, Self-healing in power systems: an approach using islanding and rate of frequency decline-based load shedding. IEEE Trans. Power Syst. **18**(1), 174–181 (2003)
23. D.L.H. Aik, A general-order system frequency response model incorporating load shedding: analytic modeling and applications. IEEE Trans. Power Syst. **21**(2), 709–717 (2006)
24. D. Prasetijo, W.R. Lachs, D. Sutanto, A new load shedding scheme for limiting under frequency. IEEE Trans. Power Syst. **9**(3), 1371–1378 (1994)
25. P.M. Anderson, M. Mirheydar, An adaptive method for setting underfrequency load shedding relays. IEEE Trans. Power Syst. **7**(2), 647–655 (1992)
26. B. Delfino, S. Massucco, A. Morini, P. Scalera, and F. Silvestro, Implementation and comparison of different under frequency load-shedding schemes, in Proc. *IEEE Power Engineering Society Transmission and Distribution Conference*, 307–312 (2001)
27. T. Kerdphol, K. Fuji, Y. Mitani, M. Watanabe, Y. Qudaih, Optimization of a battery energy storage system using particle swarm optimization for stand-alone microgrids. Int. J. Electr. Power Energy Syst. **81**, 32–39 (2016)

28. K. Sakimoto, Y. Miura, T. Ise, Stabilization of a power system with a distributed generator by a virtual synchronous generator function, in Proc. *International Conference on Power Electronics-ECCE Asia*, 1498–1505 (2011)
29. Q.C. Zhong, G. Weiss, Synchronverters: inverters that mimic synchronous generators. IEEE Trans. Ind. Electron. **58**(4), 1259–1267 (2011)
30. Q.C. Zhong, G.C. Konstantopoulos, B. Ren, M. Krstic, Improved synchronverters with bounded frequency and voltage for smart grid integration. IEEE Trans. Smart Grid **9**(2), 786–796 (2018)
31. J. Liu, Y. Miura, H. Bevrani, T. Ise, Enhanced virtual synchronous generator control for parallel inverters in microgrids. IEEE Trans. Smart Grid **8**(5), 2268–2277 (2017)
32. J. Liu, Y. Miura, T. Ise, Comparison of dynamic characteristics between virtual synchronous generator and droop control in inverter-based distributed generators. IEEE Trans. Power Electron. **31**(5), 3600–3611 (2016)
33. F.S. Rahman, T. Kerdphol, M. Watanabe, Y. Mitani, A study on the placement of virtual synchronous generator in a two-area system, in Proc. *IEEE Innovative Smart Grid Technologies Asia (ISGT Asia)*, 782–786 (2018)
34. F.S. Rahman, T. Kerdphol, M. Watanabe, Y. Mitani, Active power allocation of virtual synchronous generator considering multiple operating scenarios, in Proc. *International Conference Electric Power Energy Conversation System (EPECS)*, 1–6 (2018)
35. F.S. Rahman, T. Kerdphol, M. Watanabe, Y. Mitani, Active power allocation of virtual synchronous generator using particle swarm optimization approach. Energy Power Eng. **9**(4B), 414–424 (2017)
36. H. Bevrani, T. Ise, Y. Miura, Virtual synchronous generators: a survey and new perspectives. Int. J. Electr. Power Energy Syst. **54**, 244–254 (2014)

Chapter 10
Technical Challenges and Further Research in Virtual Inertia Control

Abstract The integration of the significant number of virtual inertia control systems could create new challenges in the implementation of virtual inertia control systems in today and future power systems. Since the large amounts of RESs/DGs have been integrated by the electrical industry into power systems, considerable effort is required to effectively manage the installed virtual inertia control units. This chapter presents the technical challenges and further research needs regarding the virtual inertia control system. The key aspects of the challenges are how to manage changes in system topology created by various virtual inertia control systems as new control units in the network along with the resulted change in system dynamics and how to make the robust power grid by taking the advantage of the potential flexibility of the decentralized/distributed virtual inertia control units. The challenges such as the optimization of the virtual inertia control system and real-time inertia estimation are emphasized for supporting the reliable operation of virtual inertia control units for their application in low inertia power systems with high penetration of RESs. Finally, some important challenges related to the market for inertia service and the regulation of the virtual inertia control system are discussed for further research.

Keywords Control · Energy storage · Fast frequency response · Frequency control · Inertia estimation · Inertia service · Optimization · Stability · Reliability · Virtual inertia control

10.1 Introduction

From the previous chapters, the synthesis of the virtual inertia control and the implementation of various techniques to enhance the virtual inertia control have been elaborated. However, there are still many technical challenges and needs for further research in virtual inertia control to achieve a resilient inverter-dominated power system. This chapter will discuss such challenges and further research in virtual inertia control.

The challenges could be divided into two main aspects. The first aspect includes the main technical aspects such as the modeling, aggregation, and control of virtual inertia control, the complementary aspects to further to improve the flexibility of

T. Kerdphol et al., *Virtual Inertia Synthesis and Control*, Power Systems,
https://doi.org/10.1007/978-3-030-57961-6_10

virtual inertia control, and the aspects related to the energy storage system and its energy management, in which as the source of energy, has an integral role in the virtual inertia control system. The second aspect is the supporting aspects which are necessary for the integration of large-scale virtual inertia control units in the future power systems dominated by inverter-based generation units. It includes the aspects related to the economic and market structure in the power system utilizing virtual inertia control and also the standard and regulation in utilizing the virtual inertia control system. The challenges and further research scope on the virtual inertia control system have been briefly discussed in [1, 2].

10.2 Main Technical Aspects of Virtual Inertia Control

In this section, the challenges and future research outlook on the main technical aspects of virtual inertia control will be discussed.

10.2.1 Improvement in Modeling, Aggregation, and Control of Virtual Inertia Control

An appropriate mathematical model of virtual inertia control which represents the complicated dynamics in power system is necessary to study the behavior of virtual inertia control in the power system. Such a model is also important for the proper parameter tuning of virtual inertia control [2]. The small-signal model and sensitivity analysis of virtual inertia control for the application in the islanded operation is described in [3]. In [4–6], the modeling and sensitivity analysis of virtual inertia control for the application in the interconnected system is discussed. Applying eigenvalue analysis, the model described in [3–6] could be used to assess the parametric sensitivity of the virtual inertia control gain in the power system. A proper and effective modeling and aggregation methods, along with a more complete dynamic frequency model, are important for a deeper analysis of the virtual inertia control and its impact into the power system, so that large-scale integration of virtual inertia control units into a future power system with high penetration of RESs/DGs could be accommodated [1].

Also included in the important aspects for further research related to the improvement in the modeling, aggregation, and control of virtual inertia control, are the study on the behavior and coordination of virtual inertia control units respect to the already installed synchronous generator (SG) units [1, 2]. Due to its characteristics, the virtual inertia control system is able to provide fast active power injection into the grid to rapidly recover the system frequency subject to a disturbance resulting in a large generation-load imbalance. However, since the active power injected by virtual inertia control units during the system dynamics is only last for a short time, the primary frequency control in the SG units still have to compensate the active power imbalance afterward. In this regard, the fast active power injection from virtual inertia control units might slow down the response of the SG units. Hence, coordinated operation between virtual inertia control and traditional SG units are necessary [1].

10.2.2 Optimization of Virtual Inertia Control

There are several aspects related to the optimization of virtual inertia control. The first aspect is related to the sizing of the energy storage system (ESS) required to support the virtual inertia control units. The sizing of the ESS should be properly determined to obtain a proper capacity which is both economically and technically efficient. Larger than required ESS capacity would lead to a higher investment cost for the virtual inertia control. On the other hand, the ESS capacity which is too small would lead to an insufficient amount of energy to perform a required active power compensation. Hence, the optimal sizing of the ESS, along with the optimal sizing of the power-electronics inverter in the virtual inertia control units, is necessary to obtain a proper balance between cost and performance. One of the efforts regarding the sizing of ESS for providing inertial response into the power system is described in [7].

The second aspect that requires optimization is the parameters of the virtual inertia control units. To achieve optimal performance from virtual inertia control units, the optimization of virtual inertia control parameters is important. Several efforts to determine the optimal virtual inertia control parameters, particularly the virtual inertia constant/gain, have been described in [8–12]. For further research, the development of the improved parameter optimization techniques for the virtual inertia control units which are effective, efficient, and practical, is critical for the integration of large-scale virtual inertia control units in the power system, especially the one with a high share of RESs/DGs.

The third aspect regarding the virtual inertia control optimization is the optimization of its placement location. The placement of virtual inertia control units might also play an important role in the application of virtual inertia control units in the future power system with a high share of RESs/DGs. Due to its role to provide additional inertia into the power system, the large-scale integration of virtual inertia control units may interact with the already installed traditional SG units and bring significant impacts to the already present system dynamics. The placement of virtual inertia control units may have a close relationship with their impact on system dynamics. The results in [13] suggest that the resilience of a power system is dictated by the placement of inertia and the location of disturbance, rather than the total inertia in a power system. The placement of virtual inertia in the power system may also affect the inter-area oscillation mode in the power system [14]. Meanwhile, several available methods for optimal virtual inertia control placement in the power system are described in [13, 15].

Further, the placement of virtual inertia control units is directly related to the placement of energy storage units, since a virtual inertia control unit requires energy storage unit as the source of energy to emulate virtual inertia. In this regard, the placement of energy storage units should also be considered so that the energy usage from the energy storage units is efficient. Hence, the placement of virtual inertia control units should also consider the optimal placement of the energy storage units so that the virtual inertia control units are properly placed respect to its impact on the system dynamics and the energy usage. Further research is also required for the analysis of the optimal placement of virtual inertia control units considering the impact of widespread penetration of RESs/DGs, along with their intermittency. With the intermittent nature of RESs, the interaction in the power system will also be determined by the connected RESs/DGs at a particular time. Hence, the virtual inertia control units should be placed by considering the intermittency introduced by the RESs/DGs so that the overall optimal placement of virtual inertia control units could be achieved, even in the presence of intermittency.

10.2.3 System Inertia Estimation

The virtual inertia control unit is developed to provide additional inertia into a power system through a fast injection of active power from an energy storage unit by using a proper control technique to emulate virtual inertia. Unlike the conventional SG units, the emulation of virtual inertia is achieved through system control. Hence, the amount of inertia supplied by the virtual inertia control units could be flexibly allocated by changing the virtual inertia constant/gain. The benefit from this feature of virtual inertia control units could be greatly enhanced if the allocation of virtual inertia constant/gain is performed based on the actual condition of system inertia. Here, the methods for estimating system inertia would be important.

Initial effort for estimating the inertia constant of a power system is described in [16]. In [16], the transients of the frequency during various events in the system are used to estimate the inertia constant and the capacity of the spinning reserve in the power system. The method in [16] is proposed for the inertia constant estimation in the traditional power system. In the modern power system with a high share of RESs/DGs, there is a challenge in how to accurately determine system inertia subject to the intermittency introduced by the integration of RESs/DGs into the power system. With the integration of RESs/DGs, system inertia would regularly change depends on the number of RESs-based generation units that are connected to the system at a particular time. The effect of the inertia change would be more significant with higher penetration of RESs-based generation units.

Due to the constant change of system inertia and system dynamics in the power system with a high share of RESs, a real-time synchronized measurement data such as the data obtained from phasor measurement units (PMUs) would play an extremely important role. The examples of the inertia estimation method using PMUs are discussed in [17–19]. Meanwhile, the analysis regarding system inertia estimation in the power system with high penetration of RESs is discussed in [20, 21].

To further improve the accuracy of the inertia estimation methods in the system with high shares of RESs-based generation units, the real-time data regarding the available resources of RESs in the grid, along with the forecast data of the RESs such as photovoltaic and wind, would be important complementary data [2]. An accurate estimation of system inertia would be useful for various purposes, such as the allocation of virtual inertia constant/gain and the adequacy of the spinning reserve at a particular system condition, both in the power system operation and planning.

10.3 Supporting Aspects for the Integration of Virtual Inertia Control Systems

In this section, the challenges and future research outlook on the supporting aspects of the integration of virtual inertia control units will be discussed.

10.3.1 Economic Valuation for Inertia Service

One of the key issues related to the virtual inertia control is how to properly value the inertia support provided by the virtual inertia control units. In a traditional power system with traditional SG units, the inertia is inherently provided by the rotating mass of the SG units and certain loads. Thus, the amount of inertia in the SG units is not particularly considered as a valuable resource to be quantified in the traditional power system. However, this viewpoint would change in the power system with high penetration of inverter-based generation units. Since such generation units would contribute to the reduction in the overall system inertia, the role of inertia response in the power system would significantly increase and in turn, would increase the demand for the 'inertia service'. Consequently, an approach to fairly quantify and price the inertia response from the generation and energy storage units would be necessary.

The study on the market-based approaches in valuing and delivering inertia or fast frequency response is presented in [22, 23]. In [24], the economic valuation of different energy storage units is performed in terms of virtual inertia emulation. The reference also suggests that the inertia service should be traded in the future power system due to the additional costs required for the provision of virtual inertia, with the inertia (kg.m^2) as the suggested trading unit. In [25], the economic valuation tool is developed to determine the economic value of the inertia in the power system.

For further research, an appropriate model that could be used to fairly quantify the price of inertia response is important. Both the investment cost and the provisional income of trading the inertia should be considered to obtain an objective price for the 'inertia service'. In this regard, the study is also required on various related aspects, such as who has to pay for the inertia service (e.g. the end-user, the grid operator?), the penalty factor related to the inertia service, and who is eligible to provide the inertia service.

10.3.2 Standard and Regulation

Standard and regulation might also require some changes to accommodate the large-scale integration of RESs and also virtual inertia control units. With large-scale integration of RESs, the characteristics of the power system would change regularly, both in terms of system inertia and the impact of intermittency. Hence, the adjustment of the standard and regulation is necessary to ensure that a certain level of reliability could be maintained in the system operation in the presence of RESs-based generation and virtual inertia control units. To avoid misinterpretation, the standards should be comprehensive, transparent, and explicit [1]. The standards should be developed regarding the integration of virtual inertia control units. The required standards are including the standards related to the interconnection procedures of both generation and virtual inertia control units and the requirement for the interconnection.

In addition to the adjustment in the standards, the change in the regulation and policy related to the integration of virtual inertia control units would also be necessary. Some of the important technical aspects that should be considered include deregulation policies, the amount and location of VSG units, operation technology, and the size and characteristics of the grid [1]. The change in the regulation should also consider the previously discussed economic aspects of the virtual inertia.

10.4 Summary

This chapter presents the technical challenges and further research needs related to the virtual inertia control. All of the challenges presented here should be considered to efficiently integrate the virtual inertia control units in the power system, both in terms of technical performance and cost. The modification in the existing standard and regulation would also be necessary to enable large-scale integration of RESs-based generation units and virtual inertia control units without sacrificing the reliability of the power system.

References

1. H. Bevrani, T. Ise, Y. Miura, Virtual synchronous generators: a survey and new perspectives. Int. J. Electr. Power Energy Syst. **54**, 244–254 (2014)
2. U. Tamrakar, D. Shrestha, M. Maharjan, B. Bhattarai, T. Hansen, R. Tonkoski, Virtual inertia: current trends and future directions. Appl. Sci. **7**(7), 654 (2017)
3. S. D'Arco, J.A. Suul, O.B. Fosso, Small-signal modeling and parametric sensitivity of a virtual synchronous machine in islanded operation. Int. J. Electr. Power Energy Syst. **72**(1), 3–15 (2015)
4. T. Kerdphol, F. Rahman, Y. Mitani, Virtual inertia control application to enhance frequency stability of interconnected power systems with high renewable energy penetration. Energies. **11**(4), 981 (2018)
5. E. Rakhshani, P. Rodriguez, Inertia emulation in AC/DC interconnected power systems using derivative technique considering frequency measurement effects. IEEE Trans. Power Syst. **32**(5), 3338–3351 (2017)
6. E. Rakhshani, D. Remon, A.M. Cantarellas, J.M. Garcia, P. Rodriguez, Modeling and sensitivity analyses of VSP based virtual inertia controller in HVDC links of interconnected power systems. Electr. Power Syst. Res. **141**(1), 246–263 (2016)
7. V. Knap, S.K. Chaudhary, D.I. Stroe, M. Swierczynski, B.I. Craciun, R. Teodorescu, Sizing of an energy storage system for grid inertial response and primary frequency reserve. IEEE Trans. Power Syst. **31**(5), 3447–3456 (2016)
8. I. Serban, C.P. Ion, Microgrid control based on a grid-forming inverter operating as virtual synchronous generator with enhanced dynamic response capability. Int. J. Electr. Power Energy Syst. **89**(1), 94–105 (2017)
9. P. Rodriguez, E. Rakhshani, A. Mir Cantarellas, D. Remon, Analysis of derivative control based virtual inertia in multi-area high-voltage direct current interconnected power systems. IET Gener. Transm. Distrib. **10**(6), 1458–1469 (2016)
10. M. Hajiakbari Fini, M.E. Hamedani Golshan, Determining optimal virtual inertia and frequency control parameters to preserve the frequency stability in islanded microgrids with high penetration of renewables. Electr. Power Syst. Res. **154**, 13–22 (2018)
11. J. Alipoor, Y. Miura, T. Ise, Stability assessment and optimization methods for microgrid with multiple VSG units. IEEE Trans. Smart Grid. **9**(2), 1462–1471 (2018)
12. F. S. Rahman, T. Kerdphol, M. Watanabe, Y. Mitani, Optimization of virtual inertia considering system frequency protection scheme. Electr. Power Syst. Res. **170**, 294–302 (2019)
13. B.K. Poolla, S. Bolognani, F. Dorfler, Optimal placement of virtual inertia in power grids. IEEE Trans. Automat. Contr. **62**(12), 6209-6220 (2017)
14. F.S. Rahman, T. Kerdphol, M. Watanabe, Y. Mitani, A study on the placement of virtual synchronous generator in a two-area system, in Proc. *IEEE Innovative Smart Grid Technologies Asia (ISGT Asia)*, 782–786 (2018)
15. N.H. Israjuddin, L. Chao-Yuan, L. Chih-Wen, Optimal placement of energy storage with synthetic inertia control on a grid with high penetration of renewables using mean-variance mapping optimization, in Proc. *Asia-Pacific Power and Energy Engineering Conference (APPEEC)*, 1–6 (2019)
16. T. Inoue, H. Taniguchi, Y. Ikeguchi, K. Yoshida, Estimation of power system inertia constant and capacity of spinning-reserve support generators using measured frequency transients. IEEE Trans. Power Syst. **12**(1), 136–143 (1997)
17. Y. Zhang, J. Bank, Y.-H. Wan, E. Muljadi, D. Corbus, Synchrophasor measurement-based wind plant inertia estimation, in Proc IEEE Green Technologies Conference, 494–499 (2014)
18. P.M. Ashton, C.S. Saunders, G.A. Taylor, A.M. Carter, M.E. Bradley, Inertia estimation of the GB power system using synchrophasor measurements. IEEE Trans. Power Syst. **30**(2), 701–709 (2015)
19. P. Wall, P. Regulski, Z. Rusidovic, V. Terzija, Inertia estimation using PMUs in a laboratory, in Proc. *IEEE PES Innovative Smart Grid Technologies Conference Europe*, 1–6 (2015)

20. A. Fernández-Guillamón, A. Vigueras-Rodríguez, Á. Molina-García, Analysis of power system inertia estimation in high wind power plant integration scenarios, IET Renew. Power Gener. **13**(15), 2807–2816 (2019)
21. J.D. Lara-Jimenez, J.M. Ramirez, Inertial frequency response estimation in a power system with high wind energy penetration, in Proc. *IEEE Eindhoven PowerTech*, 1–6 (2015)
22. O. Agranat, I. Macgill, A. Bruce, Fast frequency markets under high penetrations of renewable energy in the Australian national electricity market, in Proc. *Conference on Asia- Pacific Solar Research* (2015)
23. J. Banks, A. Bruce, I. Macgill, Fast frequency response for high renewable energy penetrations in the future Australian NEM, in Proc. *Conference on Solar Research*, 1–12 (2017)
24. H. Thiesen, C. Jauch, A. Gloe, Design of a system substituting today's inherent inertia in the European continental synchronous area. Energies **9**(8), 1–12 (2016)
25. T. Xu, W. Jang, T. Overbye, An economic evaluation tool of inertia services for systems with integrated wind power and fast-acting storage resources," in Proc. *Annual Hawaii International Conference on System Sciences*, 1–10 (2016)

Appendix

Answers for Chap. 2

2.1 (1) 3.028 MW·s/MVA
(2) 1059.8 MW·s

2.2 The graph of frequency change over time is provided as:

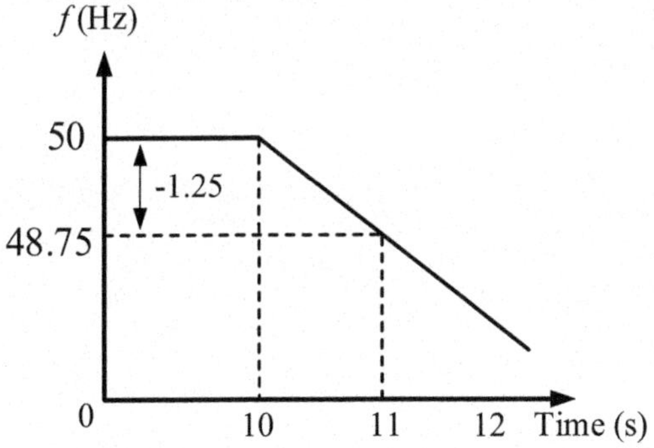

2.3 (1) 49.85 Hz (at 3.8 s)
(2) 60 Hz (at 16 s)
(3) 10 MW (at 16 s)

2.4 (1) 49.816 Hz (Area 1), 49.834 Hz (Area 2), 49.807 Hz (Area 3)
(2) At 30 s, 0 MW (Area 1), 68.4 MW (Area 2), 117 MW (Area 3)
(3) 117 MW (Area 1), 28.8 MW (Area 2), 124.2 MW (Area 3)

2.5 (1) 49.799 Hz (Area 1), 49.703 Hz (Area 2), 49.793 Hz (Area 3)
(2) At 30 s, 147.6 MW (Area 1), 0 MW (Area 2), 88.2 MW (Area 3)
(3) 124.2 MW (Area 1), 171 MW (Area 2), 50.4 MW (Area 3)

2.6 (1) 3500 MW
(2) 2275 MW

T. Kerdphol et al., *Virtual Inertia Synthesis and Control*, Power Systems,
https://doi.org/10.1007/978-3-030-57961-6

Answers for Chap. 3

3.1 2.083 MW·s (calculated from Eq. (3.2))

3.2 (1) 59.48 Hz
(2) 60 Hz (at 20 s)
(3) 10 MW
(4) 10.5 MW (at maximum), 0 MW (at steady-state)

3.3 (1) 59.65 Hz
(2) 59.8 Hz (at 20 s)
(3) 8.1 MW (at maximum), 7.7 MW (at steady-state)
(4) 8.4 MW (at maximum), 2.2 MW (at steady-state)

Answers for Chap. 4

4.1 (1) 49.879 Hz (Area 1), 49.870 (Area 2)
(2) 50 Hz (at 45 s in both areas)
(3) 0 MW (Area 1), 79 MW (Area 2)
(4) 52 MW (Area 1), 28 MW (Area 2)
(5) 55 MW

4.2 (1) 49.915 Hz (Area 1), 49.923 (Area 2)
(2) 49.91 Hz (in both areas)
(3) 30 MW (Area 1), 33 MW (Area 2)
(4) 25 MW at maximum/4 MW at steady-state (Area 1), 49 MW at maximum/7 MW at steady-state (Area 2)
(5) 72 MW

Answers for Chap. 5

5.1 The answer is explained by Sect. 5.3.
5.2 The answer is explained by Fig. 5.4 and Table 5.1.
5.3 $u(t) = 2.5 + t/7$
5.4 $u(t) = 3 + t/6.5$

5.5 (1) 59.68 Hz
(2) 59.79 Hz
(3) $K_P = 2.289$, $K_I = 9.112$
(4) 59.904 Hz
(5) $K_P = 7.501$. $K_I = 51.72$, $K_D = 0.122$
(6) 59.965 Hz

5.6 (1) 60.90 Hz
(2) 60.49 Hz
(3) $K_P = 6.308$, $K_I = 92.980$

(4) 60.077 Hz
(5) $K_P = 12.522$. $K_I = 96.186$, $K_D = 0.039$
(6) 60.048 Hz.

The manufacturer's authorised representative in the EU is Springer Nature Customer Service Centre GmbH, Europaplatz 3, 69115 Heidelberg, Germany. If you have any concerns regarding our products, please contact ProductSafety@springernature.com

Printed and bound by CPI Group (UK) Ltd, Croydon, CR0 4YY
30/09/2026
02225491-0005